Solid Oxide Fuel Cells

T0334656

Electrochemical Energy Storage and Conversion

Series Editor:
Jiujun Zhang

For more information about this series, please visit: www.crcpress.com

Solid Oxide Fuel Cells
From Fundamental Principles to Complete Systems

By
Radenka Maric and Gholamreza Mirshekari

CRC Press
Taylor & Francis Group
Boca Raton London New York

CRC Press is an imprint of the
Taylor & Francis Group, an **informa** business

First edition published 2021
by CRC Press
6000 Broken Sound Parkway NW, Suite 300, Boca Raton, FL 33487-2742

and by CRC Press
2 Park Square, Milton Park, Abingdon, Oxon, OX14 4RN

Library of Congress Cataloging-in-Publication Data
Names: Maric, Radenka, author. | Mirshekari, Gholamreza, author.
Title: Solid oxide fuel cells : from fundamental principles to complete
systems / by Radenka Maric and Gholamreza Mirshekari.
Description: First edition. | Boca Raton, FL : CRC Press/Taylor & Francis
Group, LLC, 2020. | Series: Electrochemical energy storage & conversion |
Includes bibliographical references and index. | Summary: "This book
is as a valuable resource for beginners as well as for experienced
researchers and developers of solid oxide fuel cells. It provides a
fundamental understanding of SOFCs by covering the present
state-of-the-art as well as ongoing research and future challenges to be
solved. It discusses current and future materials and provides an
overview of development activities with a more general system approach
toward fuel cell plant technology, including plant design and economics,
industrial data and advances in technology"— Provided by publisher.
Identifiers: LCCN 2020043637 (print) | LCCN 2020043638 (ebook) |
ISBN 9781466561168 (hardback) | ISBN 9780367639792 (pbk) |
ISBN 9780429100000 (ebook)
Subjects: LCSH: Solid oxide fuel cells.
Classification: LCC TK2933.S65 M37 2020 (print) | LCC TK2933.S65 (ebook) |
DDC 621.31/2429—dc23
LC record available at https://lccn.loc.gov/2020043637
LC ebook record available at https://lccn.loc.gov/2020043638

ISBN: 978-1-4665-6116-8 (hbk)
ISBN: 978-0-429-10000-0 (ebk)

Typeset in Times
by codeMantra

Contents

Preface

Sustainability is a critical global factor in the long-term viability of energy infrastructure. Today, there are more electricity generation technologies with both affordable and clean characteristics, including solar photovoltaic (PV), wind turbines, batteries, and fuel cells. Global leaders are faced with the challenge of determining how to provide energy security in an era of increased energy demand and population growth/aging, while simultaneously reducing their environmental footprint. To mitigate climate change, clean power alone is insufficient due to the complexity of decarbonizing heat and transport. Hydrogen technologies such as fuel cells have experienced cycles of overpromising, and suffered a "lost decade" after high expectations in the 2000s failed to materialize.

For me, the past several decades have been devoted to bringing innovations in fuel cell technologies closer to reality. The study of fuel cells and batteries has occupied me fully over the last 30 years and provided many challenges and the fulfillment while working for industry, federal laboratories, and as a faculty member in academia over the last ten years. The wonder of material processing and catalysis has intrigued me since my early days as a university student. I have had the good fortune to work in the laboratory of Professor Paul Hideo Shingu at the Kyoto University in Japan, who taught me that complex and useful surfaces and electrochemical devices would be a frontier field of science for decades. He also taught me that it is hard to bring to bear a disruptive technological solution if you do not have a full understanding of how to scale up the technology and the importance of economies of scale to drive cost down – in this case through manufacturing and distribution. The opportunity to work with Japan Fine Ceramics Center and Toyota Motors in Japan and the small company nGimat in Atlanta expanded my horizons in understanding innovation, technology, scale-up, and commercialization.

While there continues to be rapidly evolving knowledge and advances in fuel cell technology, the principles that defined fuel cells and batteries are based on the structure and interaction among atoms and molecules at the interfaces and remain the basis to understanding the behavior of the matter. The lengthiest chapter in this book, Chapter 2, has been dedicated to these phenomena, material development, and integration into the stack to provide power. With the Solid Oxide Fuel Cell (SOFC) technology developing at fast pace, no one can rest on their laurels, as the last ten years have seen a rapid development of the low-temperature SOFC with a core technology in proton-conducting ceramics and metal-supported cells architecture. Although the basic principle remains similar, significant emphasis has been placed on novel materials and novel processing techniques to reduce cost and improve performance. This book attempts to distill new results related to SOFC development and commercialization, presents new concepts, and, by necessity, builds on many of the concepts that are well known in SOFC.

In the last ten years, a significant number of units have been commercialized and deployed at a large scale, in particular in Japan and the US. This indicates there are strong grounds for believing that hydrogen and fuel cells can experience a cost reduction and broad adoption similar to those of solar PV and batteries, but several challenges – in particular cost and durability – must still be overcome for hydrogen and fuel cells to finally live up to their potential. The adoption of fuel cell cars, such as the Toyota Mirai ("mirai" in Japanese means future), will serve as one of the first indicators of the broader adoption and technology competitiveness. Like all others attempting to disrupt the status quo, this technology still needs to demonstrate that without government support and incentives, a competitive cost is achievable to make transport and buildings that appeal to consumers sustainable. Hydrogen and fuel cell technologies offer greater choice in the transition to a low-carbon economy, given consumer experience with fossil-fueled technologies and similarities that lead to easier adoption.

Sincere appreciation is extended to my fellow co-author, Dr. Gholamreza Mirshekari, as well as to many of my colleagues all throughout the world for the benefit of many fruitful discussions with them, visiting their laboratories and production facilities, and making some results and commercialization data available to us prior to publication. I would like to acknowledge the contribution to this book of many of my colleagues working in the field of SOFC whose results are displayed in many figures and tables in the text.

We are also sincerely grateful for the patient support of our families.

We wish all readers well, in particular during this unprecedented time of COVID-19, with a hopeful eye on the summer 2021 Olympic Games in Tokyo "The First Hydrogen Olympic Games". While these Olympics had to be postponed due to the global pandemic, they are a clear indication that Japan is setting its sights on "leaving a hydrogen society as its legacy" and demonstrating the transformational potential of fuel cells and the hydrogen economy.

Dr. Radenka Maric

Authors

Dr. Radenka Maric currently serves as the Vice President for Research, Innovation, and Entrepreneurship (VPRIE) across UConn's campuses, including the university's academic medical center, UConn Health. She is responsible for overseeing the $260 million annual research enterprise at the state's flagship public university, as well as developing and implementing a strategic vision that supports and grows groundbreaking discoveries, startup companies, creative endeavors, and innovative scholarship.

In addition to her duties as VPRIE, Dr. Maric also maintains an active research program, focusing on innovations for the fuel cell and energy industry. She is CT Clean Energy Fund Professor of Sustainable Energy in the Departments of Chemical & Biomolecular Engineering and Materials Science & Engineering. The hallmark of her research are novel manufacturing processes for battery and fuel cell development and efficient and sustainable use of precious metals in demanding reactions, such as proton exchange fuel cells, alkaline fuel cell, and the water-gas shift reforming reactions. She has been working on Solid Oxide Fuel Cell (SOFC) materials and process development since 1996, while working at the Japan Fine Ceramics Center in Nagoya, Japan. She developed novel materials and structures to improve durability and performance at the cell and system level in fuel cells and batteries.

With over 300 articles in refereed journals and conference proceedings, 2 books under preparation, 21 book chapters or invited review articles in major journals, 7 patents issued, 11 published patent disclosures, and over $30 million in competitive research funding during her career, Dr. Maric is an experienced, respected scientist with a background in academia, industry, national laboratories, and federal agencies in the US and abroad.

She is the recipient of many awards and honors from the international research community for both her preeminence in her field and her dedication to educating future generations of scientists. Some of these honors include election into the Connecticut Academy of Science and Engineering in 2012 and being named the 2015 Woman of Innovation in the Research category by the Connecticut Technology Council, a fellow of the American Association for the Advancement of Science (AAAS) in 2019, National Academy of Inventors (NAI) fellow 2019, 2020 Women in Business Award, Hartford Business Journal, and Fellow of the International Association of Advanced Materials (FIAAM, Sweden) 2020.

Dr. Maric earned her B.S. from Belgrade University and her M.S. and Ph.D. in Materials Science and Energy from Kyoto University. She holds certifications in leadership from the National Council of Canada, North Carolina Center for Leadership Development, and Lean Sensei International in Lean Manufacturing.

While a fervent fan of the University of Connecticut, Dr. Maric looks beyond UConn and engages with stakeholders around the world to have a lasting impact for her field of study and education. She is co-Chair of the CTNext Higher Education Advisory Committee with the goal of strengthening innovation and entrepreneurship within Connecticut's public and private higher education institutions while fostering collaboration and providing economic value. She serves on the Board of Directors of the Connecticut Research Council and the Board of Trustees of Solomon Schechter Day School, and she is a regular reviewer for the Department of Energy, National Science Foundation, European Commission, and Horizon 2020, just to mention a few.

Dr. Gholamreza Mirshekari received his B.Sc. in Materials Engineering (area of concentration: Industrial Metallurgy) from Shahid Chamran University in 2011, his M.Sc. in Material Engineering (area of concentration: Identification, Selection, & Manufacturing of Materials) from Isfahan University of Technology in 2014, and his Ph.D. in Engineering (area of concentration: Mechanical Engineering) from Tennessee Technological University in 2018. In 2019, he joined the University of Connecticut, Center for Clean Energy Engineering as a postdoctoral research associate. Dr. Mirshekari's research interests are in the development of electrochemical energy conversion and storage devices with a focus on design, fabrication, and characterization of catalysts, and membrane and electrode assemblies (MEAs) for fuel cells and electrolyzers. He has been involved as a key personnel in multiple academic and industrial projects, funded by the US government or national and international companies. Dr. Mirshekari has also authored various scientific papers, published in diverse prominent international journals.

1 Fundamental Aspects of Solid Oxide Fuel Cells

1.1 BACKGROUND AND PRINCIPLES OF SOFCs

Fuel cells have emerged as energy conversion devices that produce electrical power directly from electrochemical reactions by combination of gaseous fuel with an oxidant. Fuel cells have first successfully been used for space applications in the 1960s [2]. During the past decades, fuel cells have been developed and offered numerous advantages compared to conventional electrical power generation systems such as high-energy conversion efficiency, high power output, low noise, and zero environmental pollution, which have made them a promising technology for mobile and stationary power generation applications [1,2]. Today, fuel cells are widely utilized in spacecraft, automobiles, home power generation systems, etc.

Nowadays, there are different types of fuel cells classified by chemical characteristics of the electrolyte used, which in turn determines the operating temperature. Table 1.1 illustrates the technical characteristics of the main types of fuel cells that exist today. Fuel cells are categorized as alkaline fuel cell, direct methanol fuel cell, phosphoric acid fuel cell, solid acid fuel cell, proton exchange membrane fuel cell, molten carbonate fuel cell, solid oxide fuel cell (SOFC), and protonic ceramic fuel cell. The first five types have low to medium operating temperatures (50°C–210°C) with relatively lower electrical generation efficiencies (40%–55%). The other three types, however, operate at much higher temperatures (600°C–1000°C) with higher electrical generation efficiencies (45%–60%) [3]. Among the listed types of fuel cells, SOFCs are the most demanding for use as a power generation system from a materials point of view and due to their exceptional features such as:

- SOFCs offer high energy-conversion and electrical generation efficiencies (fuel input to electricity output).
- SOFCs have good fuel flexibility (e.g., natural gas and carbon-based fuels) and simplicity of design.
- Since SOFCs have a solid construction with no moving parts, they operate very quiet with minimal noise, and thereby, they can be installed indoors.
- The SOFCs' high operating temperature leads to high-quality byproducts, and exhaust heat is used for co-generation and a variety of processes.
- Since precious metals are not used in SOFCs, the price is reasonable enough for high-volume manufacturing.
- The high efficiency and operating temperature of SOFCs result in low CO_2 emission.
- SOFCs do not need to work with corrosive liquid electrolyte, making them durable with a life expectancy of 40,000–80,000 h [1,3].

TABLE 1.1

Technical Characteristics of Different Fuel Cells

Types of Fuel Cell	Electrolyte	Operating Temperature (°C)	Fuel	Oxidant	Efficiency (%)
Alkaline fuel cell (AFC)	Potassium hydroxide	50–200	Pure hydrogen, or hydrazine	O_2/Air	50–55
Direct methanol fuel cell (DMFC)	Polymer	60–200	Liquid methanol	O_2/Air	40–55
Phosphoric acid fuel cell (PAFC)	Phosphoric acid	160–210	Hydrogen from hydrocarbons and alcohol	O_2/Air	40–50
Sulfuric acid fuel cell (SAFC)	Sulfuric acid	80–90	Alcohol or impure hydrogen	O_2/Air	40–50
Proton exchange membrane fuel cell (PEMFC)	Polymer, proton exchange membrane	50–80	Less pure hydrogen from hydrocarbons or methanol	O_2/Air	40–50
Molten carbonate fuel cell (MCFC)	Molten salt (e.g. nitrate, sulfate, carbonate)	630–650	Hydrogen, carbon monoxide, natural gas, propane, marine diesel	CO_2/O_2/Air	50–60
Solid oxide fuel cell (SOFC)	Ceramic as stabilized zirconia and doped perovskite	600–1000	Hydrogen, natural gas or propane, other hydrocarbons	O_2/Air	45–60
Protonic ceramic fuel cell (PCFC)	Thin membrane of barium cerium oxide	600–700	Hydrocarbons	O_2/Air	45–60

Source: Data with minor modification from Ref. [3].

In 1838, the first fuel cell was invented by Sir William Robert Grove, a Welsh judge, inventor, and physicist, when he found a system that operates in the opposite direction of water electrolysis phenomena. Later in 1839, Grove developed his system to an electrochemical device which combines H_2 or H_2/CO fuels and an oxidant gas in the presence of an ion-conducting electrolyte and generates electricity and heat directly from the chemical energy [4,5]. Between 1853 and 1932, Friedrich Wilhelm Ostwald provided significant information about the fundamentals and theories of fuel cells, and experimentally determined the roles of fuel cell components [6]. Since then, a worldwide extensive fuel cell research has been carried out on all fuel cell types.

SOFC was first achieved by Emil Baur, a Swiss scientist, and his colleagues in the late 1930s when they conducted many experiments on SOFCs using electrolytes of clay and metal oxides such as zirconium, yttrium, cerium, lanthanum, and tungsten oxide and operating at 1,000°C [3,7,8]. In the 1940s, Davtyan, a Russian researcher, redesigned the SOFC structure to improve its mechanical strength and conductivity by adding monazite sand to a mixture of tungsten trioxide, sodium carbonate, and soda glass; however, this caused unexpected chemical reactions and less cell durability. Later on, by the late 1950s, different companies such as Central Technical Institute in the Netherlands, and Consolidation Coal Company and General Electric in the US were investing on the SOFC research due to its promising high operating temperature which would tolerate carbon monoxide, and its high stability by using a stable solid electrolyte. In 1962, the first federally funded research contract was awarded to Westinghouse to study zirconium oxide- and calcium oxide-based fuel cells. Over the recent decades, by increasing the energy demand and price and developments in materials technology, SOFCs have become more attractive for industry. A recent report indicates that more than 40 companies and many research institutes around the world are working on SOFCs to develop a new cell design and performance for the low-temperature SOFC [3]. Large-scale, utility-based SOFC power generation systems have reached commercial demonstration stages in the US, Europe, and Japan. The most aggressive in commercialization of SOFC is Japan. To establish a hydrogen-based society, Japan's Ministry of Economy, Trade and Industry (METI) has set a residential-use fuel cell target of 1.4 million units by 2020 and 5.3 million units by 2030 [10]. In 2017, METI has invested on the research and development of industrial use of SOFC systems in order to make them commercially available. Hence, Kyocera Corporation launched the industry's first 3 kW SOFC cogeneration system for institutional applications in this year [3,9,10].

Small-scale SOFCs are being developed for military, residential, industrial, and transportation applications [11,12]. Here are some examples for new relevant projects trending today: (1) establishment of mass hydrogen marine transportation supply chain derived from unused brown coal, (2) development of smart community technology by utilization of hydrogen Co-Generation System, and (3) hydrogen and natural gas co-firing gas turbine power generation facilities R&D for a low-carbon society. SOFC's history is now about 90 years old, and discoveries are still continuing in electrolyte modification, electrode and interconnecting materials development, ceramic processing to cell thickness with an extended area, operating temperature reduction down to 600°C–700°C, along with developing commercial applications [9].

1.2   DESIGN AND OPERATION OF SOFCs

SOFCs are different from other fuel cells in many aspects such as using a solid-state electrolyte and materials, which results in no restriction on cell configuration, and higher operation temperature where certain oxide electrolytes become highly oxygen-ion conducting [3]. SOFCs are mainly being configured in two different ways: tubular/rolled tube and flat-plate cells. The latter has recently been adopted by electronics companies [9].

The SOFC structure generally consists of a dense solid-state oxygen-ion-conducting electrolyte sandwiched between two porous electrodes (i.e., anodes and cathodes). Figure 1.1 shows a schematic diagram of SOFC. Hydrogen fuel is fed to the anode side in which hydrogen is combined with the oxygen, from the air, entering the cell through the cathode side. On the anode side, the hydrogen-containing fuel burns which results in a drastic reduction of the oxygen concentration on the cathode side. The oxygen ions, passing through the crystal lattice of the ceramic electrolyte [e.g., yttria-stabilized zirconia (YSZ)], react with the oxidized fuel, thereby producing electrons. The generated electrons then pass through the external circuit (from the anode to the cathode). Pure water and heat are the only byproducts of this process. The SOFC reactions are as follows:

At the cathode side:

$$\frac{1}{2}O_2 + 2e^- \rightarrow O^{2-} \tag{1.1}$$

At the anode side:

$$H_2 + O^{2-} \rightarrow H_2O + 2e^- \tag{1.2}$$

The required hydrogen can be extracted from natural gas through either external or internal reforming. The internal reforming of the fuel within the fuel cell

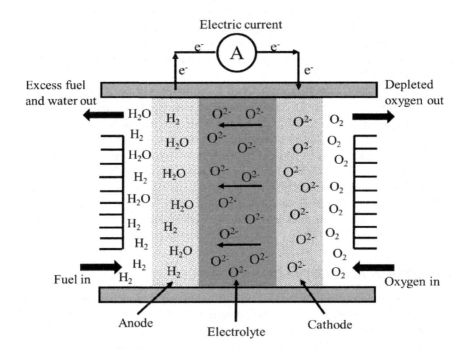

FIGURE 1.1 Schematic diagram of SOFC based on oxygen-ion-conducting electrolyte (drawn not to scale).

eliminates the necessity of external fuel reforming as it is needed for other types of fuel cells [3,5,13–15]. The fuel reforming reaction is an endothermic reaction. The required heat for the fuel reforming reaction can be provided by the overpotential loss and entropy change heat with the high-temperature fuel cells such as SOFCs. For instance, the internal methane steam reforming reaction and the water gas-shifting reaction are as follows:

Methane reforming:

$$CH_4 + H_2O \rightarrow CO + 3H_2 \tag{1.3}$$

Gas-shifting:

$$CO + H_2O \rightarrow CO_2 + H_2 \tag{1.4}$$

The CO is oxidized by oxygen ions at the anode side which produces CO_2 and electrons [15]:

$$CO + O^{2-} \rightarrow CO_2 + 2e^- \tag{1.5}$$

Figure 1.2 shows the internal reforming of the hydrocarbons in the SOFC system. As can be seen, some parts of the hydrocarbon fuel are internally reformed in an indirect reforming unit; however, the other parts are directly reformed within the fuel cell. The heat generated due to electrochemical reactions is utilized for the internal reforming process. The depleted combustible fuel containing hydrogen and CO is sent to a combustor for oxidation.

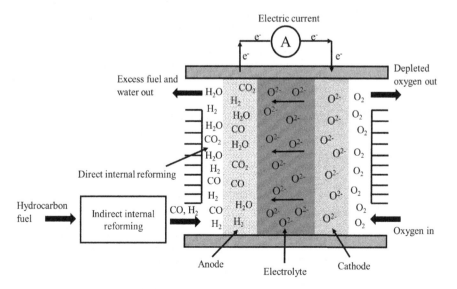

FIGURE 1.2 Schematic diagram of SOFC that features internal fuel reforming in the system (drawn not to scale).

SOFC requires to operate at high temperatures (600°C–1000°C) in order to facilitate oxygen ions migration through the electrolyte, thus achieving high ionic conductivity (~0.1 S cm⁻¹ at operating temperature) [5]. Lowering operating temperature of SOFCs not only causes reduction in ionic conductivity (i.e., increase in ohmic loss) but also decreases catalytic activity of the electrodes which has a negative impact on the cell performance. However, there are main challenges with operation of SOFCs at very high temperatures (800°C–1000°C) which are:

- Reaching the operating temperature increases fuel burning time period, thus a long start-up time.
- Exerting sealing problems and a need for expensive materials to be used as interconnects for the SOFC stack.
- Inducing thermal stresses on the SOFC materials at the electrolyte– electrode interfaces [5].

Lowering the SOFC operating temperature down to 600°C and even below can properly address the abovementioned problems. Hence, research has recently focused on lowering the SOFC operating temperature with regards to its potential barriers. In the past decade, there has been a growing interest toward proton-conducting ceramics due to their high proton conductivity with low activation energies at temperatures below 600°C [16,17]. Figure 1.3 shows the schematic of an SOFC operating with the proton-conducting electrolyte. The proton-conducting ceramic materials, termed as high-temperature proton conductors (HTPCs), exhibit proton conductivity under hydrogen and/or steam atmospheres, and the material that is commonly used as the

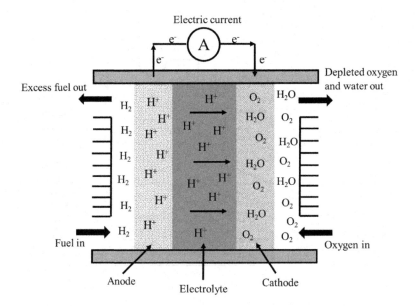

FIGURE 1.3 Schematic diagram of SOFC based on proton-conducting electrolyte (drawn not to scale).

electrolyte is based on doped $BaCeO_3$ and $BaZrO_3$ materials [18–22]. HTPC electrolytes must meet a wide range of requirements, including high ionic (protonic) conductivity, excellent thermodynamic stability, high ceramic and mechanical qualities, and acceptable thermal and chemical compatibility with other functional materials [23,24]. However, the state-of-the-art HTPC representatives still possess such disadvantages as the highly conductive cerates, $(BaCeO_3)$, and low chemical stability against C- and S-containing gas components [24]. The reaction with C-gas components causes severe degradation of the electrolyte and precludes applications in fuel cells based on hydrocarbon fuels [25–27].

The optimization of proton conductivity significantly depends on the type of dopants that are used to enhance proton transport and chemical stability of the materials. The determination of proton conductivity relies on the introduction of defects into the perovskite structure and their distribution in the crystalline lattice. To improve the key features of $BaCeO_3$, persistent efforts have shown that dopants with a low ionic charge (higher ionic size) can increase its proton conductivity because doping with an acceptor admixture (In^{3+} and Y^{3+}) compared to the lower ionic size Ce^{4+} or Zr^{4+} leads to the formation of oxygen vacancies [28]. Low-temperature solid oxide fuel cells (LT-SOFCs), from 400°C to 650°C, have seen considerable research and development and are widely viewed as the "next-generation" technology [29]. Low operating temperature is also potentially useful for reversible SOFCs because reducing the temperature can shift the H_2O-CO_2 co-electrolysis product composition to one with substantial CH_4.

In order to achieve a better insight into decreasing the SOFC operating temperature, one must explore and assess kinetics of reactions and losses occurring in the SOFC, as well as thermodynamics and electrochemistry of the reactions. The following section will discuss these aspects of SOFCs.

1.3 KINETICS OF ELECTROCHEMICAL REACTIONS AND THERMODYNAMICS OF SOFCs

As discussed in the previous section, the oxygen reduction reaction takes place at the cathode where the oxygen molecules consume electrons and are reduced to oxygen ions, as shown in Equation (1.1). This reaction can also be presented in the Kröger–Vink notation as follows:

$$\frac{1}{2}O_2 + 2e^- + V_O^{\bullet\bullet} \rightarrow O_O^{\times} \qquad (1.6)$$

where $V_O^{\bullet\bullet}$ is an oxygen vacancy site with double positive charge and $O_O^{\times}$ is a lattice oxygen, i.e., an oxygen in the lattice oxygen site with a net charge of zero. Hence, there are three basic requirements for the oxygen reduction reaction to take place at the cathode side: (1) presence of oxygen, (2) presence of electrons, and (3) diffusion of oxygen ions from the cathode to the electrolyte. To provide these requirements, the electrode materials should be highly electronic conductive, while the electrolyte must be highly ionic conductive for completion of the circuit. The porosity of the electrode also plays a key role to facilitate gas permeation along the reaction sites. Therefore, a triple-phase boundary (TPB) is defined at the adjoining sections of electrode, electrolyte, and oxidant gas. When oxygen ions pass through the electrolyte to the anode side, at the same time, the oxygen vacancies migrate from the anode side to the cathode side through the electrolyte. The fuel (H_2 or CO) oxidation reaction occurs at the anode after these migrations through the electrolyte, as described in Equations (1.2) and (1.5). During the oxidation reaction at the anode, the generated electrons are carried by a current collector to an external load using an electrical circuit connected.

The driving forces for the abovementioned electrochemical reactions can be oxygen partial pressure (PO_2) or oxygen chemical potential across the electrolyte or oxygen ion concentration gradient [5]. Therefore, an SOFC can be defined as an oxygen concentration cell where the theoretical reversible voltage (E_{th}) of the cell can be calculated using the Nernst equation as follows:

$$E_{th} = \frac{RT}{nF} \ln\left(\frac{PO_2 \text{ cathode}}{PO_2 \text{ anode}} \right) \quad (1.7)$$

where T is the temperature (K), R is the universal gas constant ($R = 8.314\,\text{J K}^{-1}\text{mol}^{-1}$), F is Faraday's constant ($F = 96,485\,\text{C mol}^{-1}$), and n is the number of electron transferred in the cell reaction. Considering hydrogen gas generally fed into the anode side and air fed into the cathode side ($PO_2 = 0.21$ atm) in laboratory experiments for SOFCs, the overall cell reaction can be represented as follows:

$$H_2(g) + \frac{1}{2}O_2(g) \rightarrow H_2O\ (g) \quad (1.8)$$

The equilibrium constant for the above reaction can be expressed as:

$$K_{eq} = \frac{PH_2O}{(PH_2)(PO_2)^{1/2}} \quad (1.9)$$

Thus, at the anode side, assuming no water condensation between the water bubbler and the anode in the gas delivery system, the Gibbs free energy change can be given by the following equation:

$$\Delta G_T = \Delta H_T - T\Delta S_T = -RT \ln\left(K_{eq}\right) \quad (1.10)$$

where ΔH_T is heat exchange, and ΔS_T is entropy exchange. The oxygen partial pressure can also be predicted as:

$$PO_2 = \left(\frac{PH_2O}{PH_2} \right)^2 \exp\left[\frac{2(T\Delta S_T - \Delta H_T)}{RT} \right] \quad (1.11)$$

In practice, multiple individual cells are combined in series as a stack with adjoining cathodes and anodes separated by interconnects to increase the cell voltage and power output. Due to some irreversible electrochemical losses, the actual cell voltage, however, is always lower than the cell theoretical voltage (Nernst value) or the open-circuit voltage $\left(-\dfrac{\Delta G_f}{nF}\right)$. Figure 1.4 demonstrates the ideal cell performance with no loss and actual cell performance operating at low and high temperatures. The initial drop in voltage is small for cell operating at high temperatures such as SOFCs. In addition, a more linear behavior is observed for the cell operating at high temperatures. The main sources of loss in a SOFC cell causing the difference between the theoretical and actual voltages (overpotential, 'η') are: (1) activation (charge transfer) loss due to the slow reaction kinetics on the surface of electrode, (2) ohmic loss due to the resistance offered to the electrons and ions flowing through the electrodes and electrolytes, respectively, and (3) concentration (mass transport) loss due to the change in the reactant concentration at the surface of electrodes. The fuel crossover loss due to fuel migration through the electrolyte is prominent in the fuel cells operating at low temperatures [30]. Thus, the total cell voltage can be summarized as:

$$\text{Total cell voltage } V = \text{Open circuit voltage } E - \text{Losses} \qquad (1.12)$$

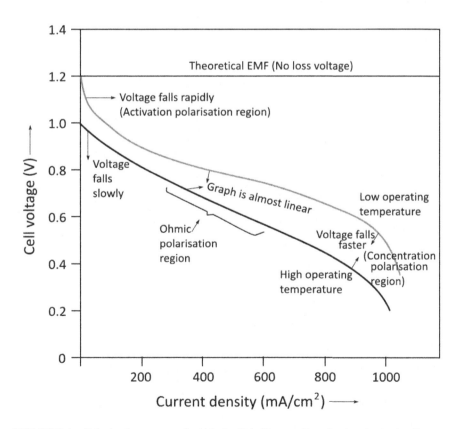

FIGURE 1.4 Polarization curves of an ideal cell (with no voltage loss) and actual cells operating and low and high temperatures. (Reprinted with permission from Ref. [30].)

1.3.1 ACTIVATION LOSS

The activation loss is caused by the slowness of the reactions occurring on the electrode surface in the presence of reacting species due to the electrochemical reaction kinetics. Thus, for the reaction to proceed, it is needed to overcome the activation energy barrier for the rate-determining step of the reaction. The required extra energy is explained by the Butler–Volmer equation:

$$i = i_0 \left\{ \exp\left(\frac{\alpha n F \eta_{act}}{RT}\right) - \exp\left(\frac{-(1-\alpha)nF\eta_{act}}{RT}\right) \right\} \qquad (1.13)$$

where i is the electrode current density, i_0 is the exchange current density, α is the charge-transfer coefficient, n is the number of electron involved in the electrode reaction, and η_{act} is the activation overpotential/loss $(E - E_{th})$. This equation describes how the electrical current density on an electrode depends on the electrode potential (activation overpotential). The activation overpotential is also termed as charge-transfer overpotential because it arises during charge transfer between the electrode/electrolyte interfaces. In the low current density regime where activation overpotential is dominant, the Butler–Volmer equation can be written as:

$$\eta_{act} \approx \left(\frac{RT}{nFi_0}\right) i = R_{ct} i \qquad (1.14)$$

where R_{ct} given by:

$$R_{ct} = \frac{RT}{nFi_0} \qquad (1.15)$$

is the intrinsic charge-transfer resistance $(\Omega \text{ cm}^2)$. This is a function of electrochemical properties of the electrode/electrolyte interfaces and also the TPB line length. Since the oxygen reduction reaction at the cathode has significantly slower kinetics compared to that of the fuel cell oxidation reaction at the anode, the activation loss is typically an issue for the cathode [5,31–38]. The voltage drop due to activation loss is also given by the Tafel equation as follows:

$$\eta_{act} = \frac{2.3RT}{\alpha n F} \log\left(\frac{i}{i_0}\right) \qquad (1.16)$$

The factor of $\dfrac{RT}{\alpha n F}$ is small for fast electrochemical reactions, while it is high for slow electrochemical reactions. It should be noted that the Tafel equation is used only for $i > i_0$.

1.3.2 OHMIC LOSS

The ohmic loss arises from the resistance to flow ions in the electrolyte or electrons in the electrodes and electron conductors, and also from the contact resistance between the electrodes and electrolytes. The main portion of ohmic loss, however, is related to the electrolyte, the magnitude of which is directly proportional to the cell current density. The cell interconnects also contribute to the ohmic loss.

The ohmic loss can be expressed as:

$$\eta_{ohm} = ir \tag{1.17}$$

where i is the current density and r is the area-specific resistance (ASR) corresponding to $1\,cm^2$ area of the cell. The ASR includes ionic, electronic, and contact resistances of the cell. The ohmic overpotential increases with an increase in the cell current density. In order to decrease the ohmic overpotential, the electrodes and the electrolytes should exhibit high electronic and ionic conductivities, respectively. Also, the ohmic overpotential can be reduced by decreasing the electrolyte thickness (electrode separation) [5,30].

1.3.3 CONCENTRATION LOSS

The concentration (mass transport) loss which is predominant in high-temperature fuel cells takes place due to the solution of reactants into the electrolyte and their diffusion to the reaction sites, dissolution of products out of the electrolyte from the reaction sites, and slow gas diffusion into the electrode pores. During the electrochemical reactions within the cell, the reactants are consumed at the surface of electrodes, leading to a concentration gradient formation and inability of maintaining the initial bulk concentration. This results in a voltage loss in the cell, known as the concentration loss. The decrease in hydrogen and oxygen concentrations fed into the anode and cathode, respectively, depends on the circulation of the gases around the electrodes. Hence, the partial pressures of hydrogen and oxygen affect the concentration. Thus, the reduction in gas partial pressure results in a cell voltage drop. The water removal can also be a prominent reason for the concentration loss [5,30].

The concentration loss can be written as:

$$\eta_{conc} = \frac{RT}{nF} \ln \frac{C_B}{C_S} \tag{1.18}$$

where C_B and C_S are bulk and surface concentrations, respectively. The concentration overpotential is also given by:

$$\eta_{conc} = -\frac{RT}{nF} \ln\left(1 - \frac{i}{i_l}\right) \tag{1.19}$$

where i_l is the limiting current density, i.e., the current density at which the rates of fuel supply and fuel consumption are equal.

Assuming negligible crossover fuel loss at high temperatures, the total effective cell voltage with regard to the losses for high-temperature fuel cells is given by [30]

$$V = E - (\text{Total losses}) = E - \left(\eta_{act} + \eta_{ohm} + \eta_{conc} \right) \tag{1.20}$$

$$V = E - \frac{2.3RT}{\alpha nF} \log\left(\frac{i}{i_0} \right) - ir + \frac{2.3RT}{nF} \log\left(1 - \frac{i}{i_l} \right) \tag{1.21}$$

Although the fuel cell losses are inevitable, it is possible to minimize them by a proper materials selection and an optimized cell components design/fabrication which will be discussed later in the next chapters.

1.3.4 THERMODYNAMICS OF IDEAL REVERSIBLE SOFC

In a fuel cell system, the fuel enters the system through an inlet and the gases leave the system through an outlet. A schematic of a reversible fuel cell system is shown in Figure 1.5. The SOFC system can be considered as a reversible system as fuel and air enter the system separately and leave the fuel non-mixed. The total system enthalpy is $n_i H_i$, entering the cell, and is $n_j H_f$, leaving the system. Heat, Q, is extracted from the cell reversibly and transferred to the environment to reach the equilibrium condition. If heat is given by the fuel cell system, the value of Q is negative, whereas it is positive if heat is transferred to the cell. Work done, W, is negative if work is done by the system, whereas it is positive if work is done on the system by the surrounding. The first law of thermodynamics is presented as:

$$\Delta H = \Delta Q + \Delta W \tag{1.22}$$

where ΔH is the enthalpy change of reaction. For a reversible reaction of $\int ds = 0$, the entropy change of reaction (ΔS) with respect to the second law of thermodynamics is as follows:

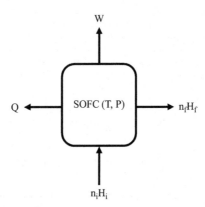

FIGURE 1.5 SOFC as a reversible system. (Adapted from Ref. [30].)

$$\Delta S = \frac{\Delta Q}{T} \tag{1.23}$$

The Gibbs free energy (ΔG), which is usually equal to the reversible work done, is given by:

$$\Delta G = \Delta H - T\Delta S \tag{1.24}$$

The efficiency of fuel cell, defined as the ratio of Gibbs free energy and enthalpy change, is expressed as:

$$\eta = \frac{\Delta G}{\Delta H} = \frac{\Delta H - T\Delta S}{\Delta H} = 1 - \frac{T\Delta S}{\Delta H} \tag{1.25}$$

The efficiency of fuel cell can also be determined by the fuel utilization factor $\left(U_f\right)$ as follows:

$$U_f = \frac{\text{Fuel}_{in} - \text{Fuel}_{out}}{\text{Fuel}_{in}} \tag{1.26}$$

In the SOFC system, if the concentration of the product (i.e., H_2O at the anode) increases, the fuel utilization factor increases, subsequently.

The reversible power (P) is defined as:

$$P = VI \tag{1.27}$$

where V and I are the voltage and current of the cell, respectively. P can also be given as the product of molar of hydrogen flow $\left(n_{H_2}\right)$ and the Gibbs free energy:

$$P = \Delta G \cdot n_{H_2} \tag{1.28}$$

The reversible cell voltage is also presented as:

$$V = \frac{-\Delta G \cdot n_{H_2}}{n_e F} \tag{1.29}$$

where n_e is the molar of electrons flow and F is Faraday's constant. In the SOFC system, when the fuel mixing happens during fuel utilization, the system remains no longer reversible. Therefore, partial pressures are taken into consideration, and Equation (1.24) can be written as:

$$\Delta G(T, p) = \Delta H(T, p) - T\Delta S(T, p) \tag{1.30}$$

For an ideal gas, Equation (1.30) can be obtained as:

$$\Delta G(T, p) = \Delta H(T) - T\Delta S(T, p) \tag{1.31}$$

$$\rightarrow \Delta S(T, p) = \frac{\Delta H(T) - \Delta G(T, p)}{T} \tag{1.32}$$

Considering the term of gas constant (R), the entropy and the Gibbs free energy are given by:

$$\Delta S(T, p) = \Delta S(T) - R \ln K \tag{1.33}$$

$$\Delta G(T, p) = \Delta G(T) + RT \ln K \tag{1.34}$$

where K is the equilibrium constant. Hence, Equation (1.29) can be written as:

$$V = \frac{-\{\Delta G(T) + RT \ln K\} \cdot n_{H_2}}{n_e F} \tag{1.35}$$

assuming a small change in enthalpy and entropy of reaction with the temperature, the reversible cell voltage increasing with an increase in system pressure and a decrease in system temperature.

1.4   HEAT TRANSFER IN SOFC

In high-temperature fuel cells, all three heat transfer modes including convection, conduction, and radiation occur. The heat transfer in the SOFC porous system (diffusion layers, or substrates, in planar SOFCs) is considerably effective due to the fast diffusion mass transfer inside the porous system at high operating temperature. In addition, chemical heat sources due to fuel reforming reactions are present. The general energy equation is given by:

$$\rho c_P \frac{\partial T}{\partial t} = -u_{x_i} \rho c_P \frac{\partial T}{\partial x_i} + \frac{\partial}{\partial x_i} \left(\lambda \frac{\partial T_s}{\partial x_i} - \sum_j D_j \frac{\partial \left(\frac{\rho_j}{M_j} \right)}{\partial x_i} H_{T, j} \right) + Q^* \tag{1.36}$$

$$H_{T, j} = \int_{T_1}^{T_2} c_{p, j} \, dT$$

where Q^* is the heat generation within the element, λ is the thermal conductivity, and ρc_P is the thermal heat capacity. Since there is significant energy transport through the porous system of SOFCs, the general energy equation for SOFCs can be modified as:

$$\frac{\partial}{\partial t}\left(\varphi\rho_f c_{P,f}T_f +(1-\varphi)\rho_s c_{P,s}T_s\right)=-u_{x_i}\rho_f c_{P,f}\frac{\partial T_f}{\partial x_i}+\frac{\partial}{\partial x_i}\left(\lambda\frac{\partial T_s}{\partial x_i}\right)$$

$$-\varphi\frac{\partial}{\partial x_i}\left[\sum_j D_j \frac{\partial\left(\dfrac{\rho_j}{M_j}\right)}{\partial x_i}H_{T_f,j}\right]$$

$$+\varphi Q_f^* +(1-\varphi)Q_s^* \tag{1.37}$$

where Q_s^* is the heat generation within the solid skeleton matrix, Q_f^* is the heat generation within the pore element, and φ is the porosity factor [39].

Most of the heat managements have been studied on planar-type SOFCs due to the design simplicity, as shown in Figure 1.6. In SOFCs, the endothermic methane reforming reaction, Equation (1.3), is normally used for hydrogen production at the anode side. Therefore, the heat management and temperature distribution within the SOFC depend on where methane reforming happens. The internal methane reforming is important since the cooling effect from this reaction considerably influences the operating condition of the SOFC cell. If we assume that methane or its reforming products are used as a fuel instead of hydrogen, the mass balance for the fuel channels is made with respect to methane, hydrogen, CO, CO_2, and water vapor and that for air channel with respect to oxygen.

The general mass balance equation can be presented as:

$$\frac{1}{M}\frac{\partial N_i}{\partial x}=v_i k_{CH_4}\frac{1}{h_f}+v_i\Delta N_{shift}+v_{i,el}\frac{j}{2Fh_f} \tag{1.38}$$

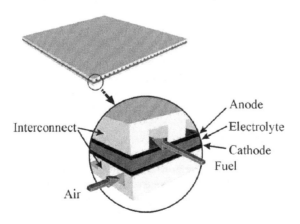

FIGURE 1.6 Schematic of a planar SOFC. (Reprinted with permission from Ref. [40].)

where k_{CH_4} is the reaction constant for methane reforming [39]. The computation of the molar conversion rate of the gas-shifting reaction, Equation (1.4), is expressed as:

$$\Delta N_{shift} = X_1 \pm \sqrt{X_1^2 - X_2} \tag{1.39}$$

$$X_1 = \frac{K_p\left(\dfrac{N_{CO}}{M_{CO}} + \dfrac{N_{H2O}}{M_{H2O}}\right) + \dfrac{N_{CO2}}{M_{CO2}} + \dfrac{N_{H2}}{M_{H2}}}{2(K_p - 1)}$$

$$X_2 = \frac{K_p\left(\dfrac{N_{CO}}{M_{CO}}\dfrac{N_{H2O}}{M_{H2O}}\right) - \dfrac{N_{CO2}}{M_{CO2}}\dfrac{N_{H2}}{M_{H2}}}{(K_p - 1)}$$

$$K_p = \exp\left(-\frac{\Delta G^0}{RT}\right)$$

The energy balances for the gas flows for both fuel and air are given by:

$$\frac{1}{M_{fuel}}\frac{\partial\left(N_{fuel}c_{P,fuel}T_f\right)}{\partial x} = \frac{h\left(T_s - T_f\right) + k_{prod}c_{P,prod}T_s - k_{educt}c_{P,educt}T_f}{h_f} \tag{1.40}$$

$$\frac{1}{M_{ox}}\frac{\partial\left(N_{ox}c_{P,ox}T_f\right)}{\partial x} = \frac{h\left(T_s - T_f\right) + k_{prod}c_{P,prod}T_s - \dfrac{j}{4F}c_{P,O_2}T_f}{h_a} \tag{1.41}$$

where h is the convection heat transfer coefficient [39]. In high-temperature fuel cells, in addition to heat transfer by convection and conduction, thermal radiation also occurs. The thermal radiation for the planar SOFC can be obtained from the Stefan–Boltzmann law as follows:

$$Q_{em} = 4\sigma_{SB}\left(\frac{T_1 + T_2}{2}\right)^3 A(T_1 - T_2) \tag{1.42}$$

where σ_{SB} is the Stefan–Boltzmann constant (5.669×10^{-8} Wm^{-2}K^{-4}) [39].

1.5 MASS TRANSFER IN SOFC

Mass transfer in SOFC, which is very effective on the electrochemical reactions and performance of the cell, refers to gas species diffusion and transport in porous electrode structure. Mass balance equation is given by:

$$\nabla\left(\rho_{eff}UY_i\right) = \nabla\left(-D_{i,eff}\nabla Y_i\right) + S_{g,i} \tag{1.43}$$

where $D_{i,eff}$ is the mass diffusion coefficient, $S_{g,i}$ is the mass source term of species i, Y_i is the mass fraction of species i, and ρ_{eff} is the effective density of multi-component gas species, expressed as:

$$\rho_{eff} = \sum_i X_i \times \rho_i \tag{1.44}$$

where X_i is the molar fraction of gas species i. The mass source term of species i, depending on the reaction mechanism, can be expressed as follows for reforming and electrochemical reactions:

$$S_{g,i} = \left(mR_r + nR_s\right) \times M_i \tag{1.45}$$

$$S_{ele,i} = \text{TPB} \times \frac{i_{act}}{nF} \times M_i \tag{1.46}$$

where R_r and R_s are the reaction rates for the methane reforming reaction and gas-shifting reaction, respectively, i_{act} is the active current density, and TPB is the triple-phase boundary area per volume for electrochemical reaction [41].

In the surface elementary reaction mechanism of methane reforming on the anode, the reaction rate of each gas species can be written as:

$$R_{g,i} = \sum_{j=1}^{K_r} v_{ij} k_j \prod_{i=1}^{K_g+K_s} [X_i]^{v_{ij}} \tag{1.47}$$

where k_j is the reaction constant, v_{ij} is the stoichiometric coefficient of products and reactants, K_r, K_g, and K_s are the numbers of total reactions, gas, and surface species, respectively, and $[X_i]$ is the gas species concentration [41].

Fick's model, a simple approach to describe the diffusion between different gas species, along with the Knudsen diffusion, to fit for the gas diffusion on the porous structure, is the most common model for mixed gas diffusion in SOFCs. Therefore, the mass diffusion coefficient can be represented as:

$$D_{i,eff} = \frac{\varepsilon}{\tau} \frac{D_{i,gm} \times D_{i,k}}{D_{i,gm} + D_{i,k}} \tag{1.48}$$

where τ is the tortuosity of porous materials, ε is the porosity, and $D_{i,gm}$ is the diffusion of multi-component gas species, given by:

$$D_{i,gm} = \frac{1 - Y_i}{\sum_{i \neq j} \dfrac{Y_i}{D_{i,j}}} \tag{1.49}$$

where $D_{i,j}$ is the binary diffusion coefficient of gas species i. This is controlled by the pore size and the free path of diffusion gas molecules [41].

REFERENCES

1. Yamamoto, O. 2000. Solid oxide fuel cells: fundamental aspects and prospects. *Electrochimica Acta* 45:2423–2435.
2. Mirshekari, G.R., and Rice, C.A. 2018. Effects of support particle size and Pt content on catalytic activity and durability of Pt/TiO_2 catalyst for oxygen reduction reaction in proton exchange membrane fuel cells environment. *Journal of Power Sources* 396:606–614.
3. Boudghene Stambouli, A., and Traversa, E. 2002. Solid oxide fuel cells (SOFCs): a review of an environmentally clean and efficient source of energy. *Renewable and Sustainable Energy Reviews* 6:433–455.
4. Grove, W.R. 1839. On voltaic series and combination of gases by platinum. *The London, Edinburgh, and Dublin Philosophical Magazine and Journal of Science* 14:127–130.
5. Mahato, N., Banerjee, A., Gupta, A., Omar, S., and Balani, K. 2015. Progress in material selection for solid oxide fuel cell technology: a review. *Progress in Materials Science* 72:141–337.
6. Spiegel, C. 2017. History of fuel cells. Fuel Cell Store. https://www.fuelcellstore.com/blog-section/history-of-fuel-cells.
7. Baur, E. 1939. Über das Problem der elektromotorischen Verbrennung der Brennstoffe. *Brennsto Chemie* 20:385–387.
8. Baur, E., and Preis, H. 1937. Über Brennstoff-Ketten mit Festleitern. *Zeitschrift fur Elektrochemie* 43:727–732.
9. Kendall, K., and Kendall, M. 2016. *High-Temperature Solid Oxide Fuel Cells for the 21st Century: Fundamentals, Design and Applications.* Elsevier Ltd, London, UK.
10. Kyocera. 2017. KYOCERA develops industry's first 3-kilowatt solid-oxide fuel cell for institutional cogeneration. https://global.kyocera.com/news-archive/2017/0702_bnfo.html.
11. Laosiripojana, N., Wiyaratn, W., Kiatkittipong, W., Soottitantawat, A., and Assabumrungrat, S. 2009. Reviews on solid oxide fuel cell technology. *Engineering Journal* 13(1):65–83.
12. Choudhury, A., Chandra, H., and Arora, A. 2013. Application of solid oxide fuel cell technology for power generation—a review. *Renewable and Sustainable Energy Reviews* 20:430–442.
13. Fergus, J.W., Hui, R., Li, X., Wilkinson, D.P., and Zhang, J. 2009. *Solid Oxide Fuel Cells: Materials Properties and Performance.* CRC Press, Taylor & Francis Group, Boca Raton, FL USA.
14. Singhal, S.C., and Kendall, K. 2003. *High Temperature Solid Oxide Fuel Cells: Fundamentals, Design and Applications.* Elsevier Ltd, Oxford, UK.
15. Zhang, X., Chan, S.H., Li, G., Ho, H.K., Li, J., and Feng, Z. 2010. A review of integration strategies for solid oxide fuel cells. *Journal of Power Sources* 195:685–702.
16. Steele, B.C., and Heinzel, A. 2001. Materials for fuel-cell technologies. *Nature* 414:345–352.
17. Fabbri, E., Pergolesi, D., and Traversa, E. 2010. Materials challenges toward proton-conducting oxide fuel cells: a critical review. *Chemical Society Reviews* 39:4355–4369.
18. Iwahara, H., Esaka, T., Uchida, H., and Maeda, N. 1981. Proton conduction in sintered oxides and its application to steam electrolysis for hydrogen production. *Solid State Ionics* 3–4:359–363.
19. Kreuer, K.D. 2003. Proton-conducting oxides. *Annual Review of Materials Research* 33:333–359.
20. Haugsrud, R., and Norby, T. 2009. Proton conduction in rare-earth ortho-niobates and ortho-tantalates. *Nature Materials* 5:193–196.
21. Fabbri, E., Magrasó, A., and Pergolesi, D. 2014. Low-temperature solid-oxide fuel cells based on proton-conducting electrolytes. *MRS Bulletin* 39:792–797.

22. Tao, S.W., and Irvine, J.T.S. 2006. A stable, easily sintered proton-conducting oxide electrolyte for moderate-temperature fuel cells and electrolyzers. *Advanced Materials* 18:1581–1584.
23. Malavasi, L., Fisher, C.A.J., and Islam, M.S. 2010. Oxide-ion and proton conducting electrolyte materials for clean energy applications: structural and mechanistic features. *Chemical Society Reviews* 39:4370–4387.
24. Bi, L., Boulfra, S., and Traversa, E. 2014. Steam electrolysis by solid oxide electrolysis cells (SOECs) with proton-conducting oxides. *Chemical Society Reviews* 43:8255–8270.
25. Ryu, K.H., and Haile, S.M. 1999. Chemical stability and proton conductivity of doped $BaCeO_3$–$BaZrO_3$ solid solutions. *Solid State Ionics* 125:355–367.
26. Kreuer, K.D. 1997. On the development of proton conducting materials for technological applications. *Solid State Ionics* 97:1–15.
27. Shafi, S.P., Bi, L., Boulfra, S., and Traversa, E. 2015. Y and Ni Co-doped $BaZrO_3$ as a proton-conducting solid oxide fuel cell electrolyte exhibiting superior power performance. *Journal of the Electrochemical Society* 162(14):F1498–F1503.
28. Lankhorst, M.H.R., Bouwmeester, H.J.M., and Verweij, H. 1997. High-temperature coulometric titration of $La_{1-x}Sr_xCoO_{3-\delta}$: evidence for the effect of electronic band structure on nonstoichiometry behavior. *Journal of Solid State Chemistry* 133(2):555–567.
29. Gao, Z., Mogni, L.V., Miller, E.C., Railsbacka, J.G., and Barnett, S.A. 2016. A perspective on low-temperature solid oxide fuel cells. *Energy & Environmental Science* 9:1602–1644.
30. Kaur, G. 2016. *Solid Oxide Fuel Cell: Interfacial Compatibility of SOFC Glass Seals.* Springer, Switzerland.
31. Virkar, A.V., Chen, J., Tanner, C.W., and Kim, J.-W. 2000. The role of electrode microstructure on activation and concentration polarizations in solid oxide fuel cells. *Solid State Ionics* 131:189–198.
32. Xia, C., Rauch, W., Chen, F., and Liu, M. 2002. $Sm_{0.5}Sr_{0.5}CoO_3$ cathodes for low-temperature SOFCs. *Solid State Ionics* 149:11–19.
33. Horita, T., Yamaji, K., Sasaki, N., Xiong, Y., Kato, T., Yokokawa, H., and Kawada, T. 2002. Imaging of oxygen transport at SOFC cathode/electrolyte interfaces by a novel technique. *Journal of Power Sources* 106:224–230.
34. Sasaki, K., Tamura, J., Hosoda, H., Lan, T.N., Yasumoto, K., and Dokiya, M. 2002. Pt-perovskite cermet cathode for reduced-temperature SOFCs. *Solid State Ionics* 148:551–555.
35. Chan, S.H., Low, C.F., and Ding, O.L. 2002. Energy and exergy analysis of simple solid oxide fuel-cell power systems. *Journal of Power Sources* 103:188–200.
36. Ishihara, T., Kudo, T., Matsuda, H., and Takita, Y. 1995. Doped $PrMnO_3$ perovskite oxide as a new cathode of solid oxide fuel cells for low temperature operation. *Journal of Electrochemical Society* 142:1519–1524.
37. Horita, T., Yamaji, K., Sakai, N., Yokokawa, H., Weber, A., and Ivers-Tiffée, E. 2001. Oxygen reduction mechanism at porous $La_{1-x}Sr_xCoO_{3-d}$ cathodes/$La_{0.8}Sr_{0.2}Ga_{0.8}Mg_{0.2}O_{2.8}$ electrolyte interface for solid oxide fuel cells. *Electrochimica Acta* 46:1837–1845.
38. Mogensen, M., and Skaarup, S. 1996. Kinetic and geometric aspects of solid oxide fuel cell electrodes. *Solid State Ionics* 86:1151–1160.
39. Vielstich, W., Lamm, A., and Gasteiger, H.A. 2003. *Handbook of Fuel Cells: Fundamentals Technology and Applications*, Volume 1: Fundamentals and Survey of Systems. John Wiley & Sons, Chichester, England.
40. Beale, S.B. 2005. Numerical models for planar solid oxide fuel cells. *WIT Transactions on State of the Art in Science and Engineering* 10:43–82.
41. Yang, C. 2012. Mass and heat processes in solid oxide fuel cell (SOFC). 2012 MVK160 Heat and Mass Transport, Lund, Sweden.

2 Materials

Electrolytes, Anodes, Cathodes, Interconnects, and Sealants

2.1 OVERVIEW OF SOLID OXIDE FUEL CELL (SOFC) MATERIALS

SOFCs have been developed worldwide for many years. A very high technical refinement has been achieved by continuous improvement in materials, cell design, and manufacturing processing for SOFCs. To commercialize SOFC systems and make them more economically viable, multiple factors are involved, including high cost of the cell materials, susceptibility of the cell materials failure because of mechanical shock, oxidation, and rapid thermal transients, unreliable cell sealing, and large production issues with ceramic parts. Therefore, due to the importance of materials selection and development on SOFC performance, many noteworthy companies and centers all around the world such as Bloom Energy (USA), Risø National Labratoary (Denmark), Ceramic Fuel Cells Limited (Australia), Tokyo Gas (Japan), Mitsui Engineering & Shipbuilding Corporation Limited (Japan), Mitsubishi Heavy Industries & Electric Power Development Company (Japan), Energy Research Center/Innovation Dutch Electroceramics (Netherlands), Forschungszentrum Jülich (Germany), Versa Power (Canada), and Space Weather Prediction Center (USA) have been working on design and processing of SOFC component materials [1–4]. The materials development for the five key components of SOFC, i.e., electrolyte, anode, cathode, interconnect, and sealant, along with the related issues are presented in Figure 2.1. The general outline of the materials requirements for cell components is given below:

- In order to prevent the mixing of reducing and oxidizing gases, electrolyte and interconnect must be fully dense, whereas the electrodes must be porous to maximize gas mobility for the reaction.
- Both anode and cathode electrodes and interconnect must have high electrical conductivity, whereas electrolytes must have high ionic conductivity. The electrodes could also be mixed conductor (electronic and ionic) in which oxygen-ion diffusivity is very high.
- During cell operation, chemical stability is necessary for the adjoining cell components to withstand oxidizing and/or reducing environments at cell operating temperatures.
- During cell long-term operation, each component must have chemical, dimensional, morphological, and phase stabilities.

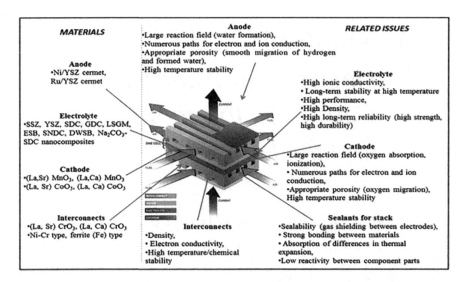

FIGURE 2.1 Materials and related issues for SOFC. (Reprinted with permission from Ref. [2].)

- Each component must have high thermal and mechanical shock resistance.
- The supporting cell components, i.e., electrolyte- and electrode-supported, must have high toughness and mechanical strength during cell operation.
- In order to avoid delamination and crack formation and to reduce internal stresses during fabrication and cell operation, all cell components must have a close thermal expansion coefficient.
- The low fabrication cost and easily available technologies for each cell component must be addressed.

A comprehensive list of different materials, having been used for each component, is shown in Figure 2.2. Table 2.1 presents an overview of the SOFCs' diversification, along with some of the main developers involved. This chapter will discuss technical achievements to date for the SOFC's five key component materials.

2.2   ELECTROLYTES

2.2.1   Oxygen-Ion-Conducting Electrolytes

The electrolyte of SOFC mainly conducts oxygen ions (O^{2-}) from cathode to anode where it reacts with fuel (hydrogen or hydrocarbons) to form H_2O and CO_2, thereby completing the overall electrochemical reaction and producing electron, passing through an external circuit. The oxygen-ion conduction occurs where the oxygen ion in a thermally activated process moves from one crystal lattice site to its neighbor site. To achieve a high ion-conducting electrolyte, its crystal structure must contain a high level of oxygen vacancy sites, and low migration energy barrier, certainly less than 1 eV. Since the oxygen ion with an ionic radius of 0.14 nm is the largest

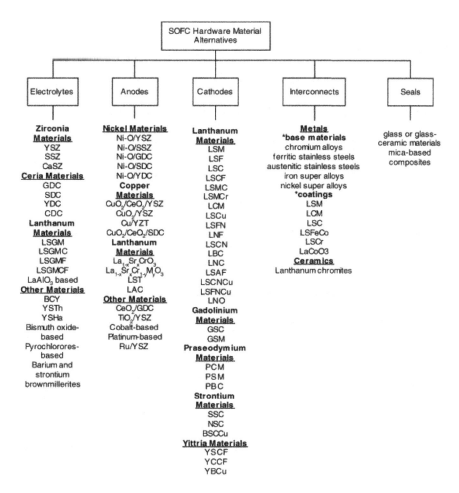

FIGURE 2.2 A list of different materials used for each of five key components in SOFC. (Reprinted with permission from Ref. [3].)

TABLE 2.1

Developers of SOFC Materials for Planar and Tubular Cell Designs and Corresponding Fabrication and Design Details

Company	Component	Material	Production Process	Thickness (µm)
Risø National Labratoary	Cathode	$(La,Sr)MnO_3 + YSZ$	Screen printing; Wet	50
	Anode	Ni/YSZ	powder spraying	200–300
	substrate	YSZ	Tape casting	10–25
	Electrolyte		Wet powder spraying	

(Continued)

TABLE 2.1 (*Continued*)
Developers of SOFC Materials for Planar and Tubular Cell Designs and Corresponding Fabrication and Design Details

Company	Component	Material	Production Process	Thickness (μm)
Ceramic Fuel Cells	Cathode	(La,Sr)MnO$_3$	Screen printing	-
Limited	Anode	Ni/YSZ	Tape casting	500–700
	substrate	Ni/YSZ	Screen printing	50
	Anode	YSZ	Reactive magnetron	10–30
	Electrolyte	3YSZ, 8YSZ	sputtering;	100
			Lamination and	
			sintering	
			Tape casting	
Tokyo Gas	Cathode	(La,Sr)MnO$_3$	Screen printing	150
	Electrolyte	3YSZ	Tape casting	50–100
	Anode	Ni/(Ce,Y)SZ	Screen printing	30
Mitsui Engineering &	Cathode	(La,Sr)(Mn,Cr)O$_3$	Painting	150
Shipbuilding	Electrolyte	8YSZ	Tape casting	300
Corporation Limited	Anode	Ni/YSZ	Painting	150
Mitsubishi Heavy	Cathode	LaCoO$_3$	Plasma spraying	150–200
Industries & Electric	Electrolyte	(La,Sr)(Mn,Cr)O$_3$	Slurry coating	100–150
Power Development	Anode	YSZ	Plasma spraying;	-
Company	Interconnect	Ni/YSZ	Slurry coating	80–100
		NiAl/Al$_2$O$_3$	Slurry coating	80–100
		(Ln,AE)TiO$_3$	Plasma spraying	-
			Slurry coating	
Energy Research	Cathode	(La,Sr)MnO$_3$+YSZ	Screen printing	-
Center/Innovation	Anode	Ni/YSZ	Tape casting	500–800
Dutch Electroceramics	substrate	Ni/YSZ	Screen printing	3–7
	Anode	YSZ	Screen printing	7–10
	Electrolyte			
Forschungszentrum	Electrolyte	YSZ	Vacuum slip casting	5–30
Jülich	Anode	YSZ	Reactive magnetron	2–10
	substrate	Ni/YSZ	sputtering	200–500
	Anode	Ni/YSZ	Tape casting	1500
		Ni/YSZ	Warm pressing	5–15
			Vacuum slip casting	
Versa Power	Cathode	(La,Sr)MnO$_3$	Screen printing	40
	Anode	Ni/YSZ	Tape casting	1000
	substrate	YSZ	Vacuum slip casting;	10
	Electrolyte		Screen printing	
Space Weather	Cathode	Doped LaMnO$_3$	Extrusion and	2200
Prediction Center	tube	YSZ	sintering	40
	Electrolyte		Electrochemical	
			vapor deposition	

Source: Data adapted and modified from Ref. [4].

component in the crystal lattice, it seems that it is difficult to attain a small barrier for oxygen-ion migration.

Thus, it would be expected that the smaller metal cations in such a metal oxide structure would be more likely to possess an appreciable movement in the lattice and thereby carry the current. However, metal cations are not capable of free movement in the lattice due to their large charge valence. In addition, in a certain metal oxide crystal structure, the number of oxygen vacancies is predominant; thus, oxygen ions move in the electric field. Some noteworthy examples of these metal oxides with partially occupied oxygen sites in the crystal structure are ZrO_2-, CeO_2-, and Bi_2O_3-based oxides with the fluorite structure, $LaGaO_3$-based perovskites, $Bi_4V_2O_{11}$- and $La_2Mo_2O_9$-based derivatives, $Ba_2In_2O_5$-derived perovskite- and brownmillerite-like phases, pyrochlores-$Gd_2Zr_2O_7$, $Gd_2Ti_2O_7$, and rare-earth-based apatite-$La_9SrSi_6O_{26.5}$ [5,6].

Among ion-conducting oxides, only a few selected oxides have mostly been developed to be used as the electrolyte for SOFCs due to the variety of essential requirements for the electrolyte component, including high oxygen-ion conductivity ($\sim$0.1 S cm^{-1} at operating temperature), low electronic conductivity, and electronic transfer number ($<10^{-3}$), chemical and thermodynamic stabilities over a wide range of temperatures and oxygen partial pressures, negligible volatilization, thermal expansion compatibility with adjoining components, sufficient mechanical properties, and negligible interaction with the electrode material during processing and service. Figure 2.3 shows the oxygen-ion conductivity of the selected oxide electrolytes.

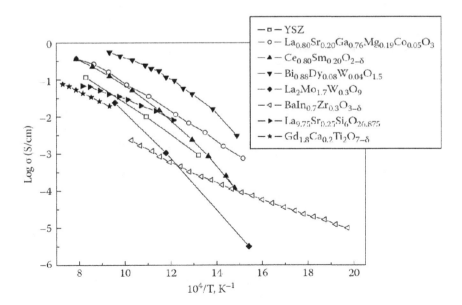

FIGURE 2.3 Oxygen ionic conductivity of various solid-state electrolytes. (Reprinted with permission from Ref. [5].)

2.2.2 Proton-Conducting Electrolytes

Proton-conducting electrolytes, carrying proton (H^+) from anode to cathode where it reacts with oxygen, thus produce water and electrons, find applications for low-temperature SOFCs since most of them decompose at 300°C [7]. It has demonstrated that the SOFC systems working with the proton-conducting electrolytes possess higher thermodynamic efficiency than the SOFC systems working with the oxygen-ion-conducting electrolytes. The proton-conducting materials derive their conductivity from the protonic defects in their crystal structure which have high mobility. Some examples of proton-conducting perovskites used as the SOFC electrolyte are $BaCeO_3$, $BaZrO_3$ and $SrCeO_3$, $SrZrO_3$, $CaZrO_3$, $BaCe_{0.8}Y_{0.2}O_3$, and $BaZr_{0.4}Ce_{0.4}In_{0.2}O_3$. The perovskite materials usually have a general formula of $AB_{1-x}M_xO_3$, where M is a trivalent rare-earth element dopant. At elevated temperatures and in the presence of water, these perovskites show proton conductivity. Thus, the level of humidity is important for a proton conductor. Since the size of protons is small, they cannot occupy the interstitial sites in the perovskites structure. Therefore, protons are mainly embedded in the electron cloud of an oxygen ion.

The Kröger–Vink notation for protonic conductors can be expressed as:

$$H_2O + O_O^\times + V_O^{\cdot\cdot} \rightarrow 2(OH)_O^{\cdot} \tag{2.1}$$

where the water molecule dissociates to form two hydroxide ions, forming two protonic defects by replacing the oxide ions in the crystal lattice [7]. Figure 2.4 shows the proton conduction for different materials.

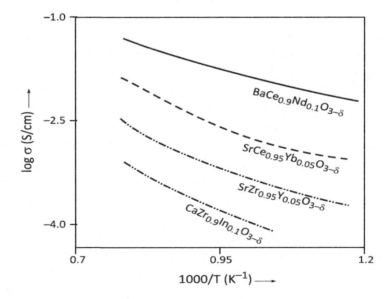

FIGURE 2.4 Conductivity of different perovskite-based proton-conducting electrolytes. (Reprinted with permission from Ref. [7].)

2.2.3 ZIRCONIA-BASED ELECTROLYTES

The zirconia-based oxides are the most common fast oxygen-ion-conducting electrolytes with the crystal structure of the fluorite type, operating either at high temperatures (800°C–1000°C) or intermediate temperatures (600°C–800°C). Pure ZrO_2 has the monoclinic structure at room temperature with relatively low electronic conductivity and not a good ionic conductivity. Upon heating, monoclinic to tetragonal and tetragonal to cubic phase transformations occur at 1170°C and 2370°C, respectively. These martensitic phase transformations are reversible on cooling, however, tetragonal to monoclinic phase transformation occurs at lower temperatures (900°C–1000°C) during cooling. In addition, a large volume change leading to the disintegration of the ceramic body occurs during cooling for the monoclinic and tetragonal phase transformations. In order to increase the concentration of oxygen vacancies within the crystal structure required for ion conduction and to stabilize the cubic structure at lower temperatures, dopants are introduced into the cation sublattice [8–17]. The three crystal structures adopted by ZrO_2 are shown in Figure 2.5.

The cubic ZrO_2 structure can be stabilized to room temperature by adding dopants such as CaO, MgO, Y_2O_3, Yb_2O_3, and Sc_2O_3. This can be described by the Kröger–Vink notation:

$$MO + O_O^x + Zr_{Zr}^x \rightarrow M_{Zr}'' + V_O^{\bullet\bullet} + ZrO_2 \tag{2.2}$$

$$R_2O_3 + O_O^x + 2Zr_{Zr}^x \rightarrow 2R_{Zr}' + V_O^{\bullet\bullet} + 2ZrO_2 \tag{2.3}$$

where M is a divalent cation, R is a trivalent cation, and $V_O^{\bullet\bullet}$ is a compensating oxygen vacancy in the crystal structure [17]. For instance, 8% mol Y_2O_3 is believed as the lowest concentration to stabilize the ZrO_2 cubic phase at room temperature. The composition range over which the cubic phase exists is narrow and temperature dependent, and is affected by the type of dopant. A complex phase assemblage consisting of two or more phases may occur if the composition is below the required amount of dopant for stabilization of the cubic structure at low temperatures. The two-phase mixture in partially stabilized ZrO_2 can be beneficial since it improves the mechanical properties of the ceramic.

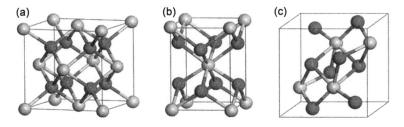

FIGURE 2.5 Crystal structures of ZrO_2: (a) cubic, (b) tetragonal, and (c) monoclinic. Dark spheres represent oxygen atoms and bright spheres are zirconium atoms. (Reprinted with permission from Ref. [16].)

Yttria-stabilized zirconia (YSZ) and scandia-stablized zirconia (ScSZ) with dopant concentrations in the range of 8–10 mol% are mainly used and developed as the effective electrolyte for the high-performance SOFC systems due to their high ionic conductivity, low cost, high mechanical and chemical stability, thermal shock resistance, and desire compatibility with adjoining components used in the SOFC system. Zirconia ceramics also show low electronic conductivity in a wide range of oxygen partial pressure (from 100 to 200 atm down to 10^{-25} to 10^{-20} atm) [5,6]. This range covers $1–10^{-18}$ atm pressure which an SOFC electrolyte has to tolerate at the anode side during cell operation at high temperatures [18]. There are various studies over a decade that tried to improve the performance of the YSZ and ScSZ electrolytes toward lowering the operating temperature of SOFC. The investigations have mainly been focused on:

- Developing the fabrication method of the electrolyte to enhance its properties and performance.
- Developing the thin-film structure electrolyte for a simple design structure of SOFC to expand the usage in transportation and portable applications.
- Modification of the electrolyte with co-dopants and bilayer structures in order to increase the amount of oxygen vacancies and the ionic conductivity.
- Improvement of mechanical and thermal properties of the electrolyte to enhance the durability of the SOFC.

2.2.3.1 Yttria-Stabilized Zirconia

YSZ is the most common oxide electrolyte material and has been the mainstay of progress to make SOFCs commercially viable. Yttria is added to stabilize the ionic conductive cubic fluorite phase and to increase the ionic conductivity. In the crystal lattice of stabilized zirconia, when Zr^{4+} cations are substituted by Y^{3+} cations, vacancy sites are formed in the oxygen sublattice due to a lower valence of Y^{3+} cations than that of Zr^{4+} cations, as presented in Figure 2.6. The vacancy production can be described by the Kröger–Vink notation:

$$Y_2O_3 \rightarrow 2Y'_{Zr} + V_O^{\bullet\bullet} + 3O_O^{\times} \qquad (2.4)$$

The reason for oxygen-ion conduction in the stabilized zirconia is the formation of the transportable oxygen vacancies, which improves defects/ion mobility inside the

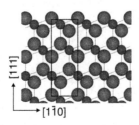

FIGURE 2.6 The side view of YSZ (111) fluorite model, large spheres (oxygen) and small spheres (Y/Zr). (Reprinted with permission from Ref. [22].)

YSZ electrolyte and reduces the energy loss by lowering the ohmic resistance in the solid electrolyte. It has been shown that the YSZ with the composition of 8 mol% Y_2O_3 (8YSZ) has the highest ionic conductivity [19–21]. Thus, this yttria content is the minimum needed to stabilize the fluorite-type cubic zirconia down to room temperature [5]. Figure 2.7 illustrates that the conductivity of YSZ increases by increasing yttria contents up to 8 mol%, whereas it then decreases for higher yttria contents due to the association of oxygen vacancies, which leads to reduction in oxygen vacancies mobility at a very high content and thus reduction in conductivity.

Although the YSZ shows desirable ionic conductivity at higher SOFC operating temperatures, the main challenge to lower the operating temperature of SOFC is maintaining the excellent conductivity property of the YSZ electrolyte since it has low conductivity at low-to-intermediate operating temperatures (<600°C). In order to enhance the conductivity of the YSZ electrolyte at lower temperatures, one solution path is to reduce the electrolyte thickness down to 10 μm, providing the minimal ohmic resistance and increasing the diffusion of oxygen ions through the YSZ electrolyte [23–25]. Additionally, introducing an alternative oxygen-ion conductor material to the YSZ electrolyte or utilizing a bilayer electrolyte design based on the YSZ electrolyte [e.g., YSZ/gadolinia-doped ceria (GDC) or YSZ/yttria-doped ceria] can improve its conductivity at low-to-intermediate temperatures and make SOFC operation more reliable at low operating temperatures. This will expand the usage of SOFC for portable and transportation applications and will also lead to the high efficiency of fuel consumption [26,27].

Since 2009, the majority of works on the YSZ electrolyte have been dedicated to fabricating a high-quality YSZ electrolyte with high performance and optimum structure to merge with other cell compartments to be used in a single cell or stack toward lowering the operating temperature. da Silva et al. [28] synthesized a nano-sized 8 mol% YSZ powders using a combustion method with urea and glycine fuels. The results showed that the dense and compact structure YSZ is produced by the glycine

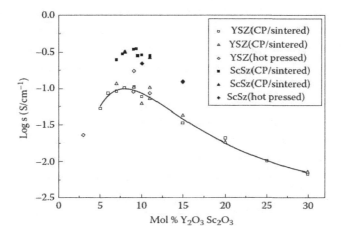

FIGURE 2.7 Conductivity of YSZ and ScSZ in air at 1000°C. (Reprinted with permission from Ref. [21].)

fuel combustion method, which is influenced by combustion flame temperature. Yin et al. [29] fabricated highly asymmetric YSZ hollow fiber membranes by modified phase inversion and a sintering technique using the internal coagulant from the mixture of N-methyl-2-pyrrolidone and water. The produced cubic fluorite structure YSZ hollow fibers possess an outer thin dense layer with a thickness of 3–5 μm and a thicker porous sub-layer with surface porosity of 28.8%, as shown in Figure 2.8. The combination of a thin dense skin layer with the porous sub-layer in the YSZ

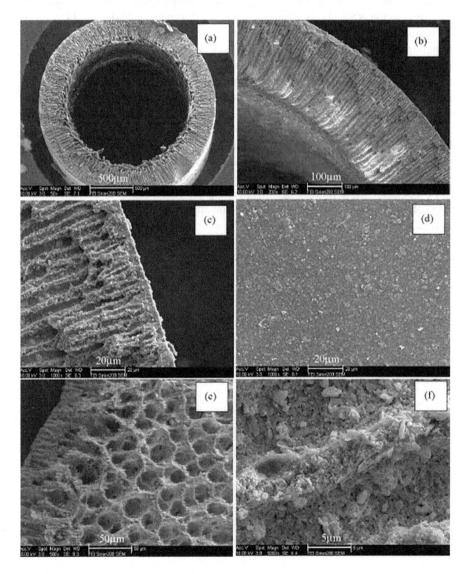

FIGURE 2.8 SEM micrographs of the YSZ hollow fiber membrane: (a) cross-sectional, (b) membrane wall, (c) outer edge, (d) outside surface, (e) inside surface, and (f) interior. (Reprinted with permission from Ref. [29].)

hollow fiber can make it suitable to be used as an electrolyte in micro-tubular SOFCs. Chen et al. [30] also synthesized a nanocrystalline 10 mol% YSZ electrolyte via a plasma spray technique. The results showed that the plasma spray technique could be a useful technique for the YSZ fabrication since the plasma-sprayed YSZ electrolyte had the ionic conductivity ~2.3 times higher than that of the conventional solid-state sintered YSZ electrolyte at 600°C in air, as presented in Figure 2.9. Most industrial

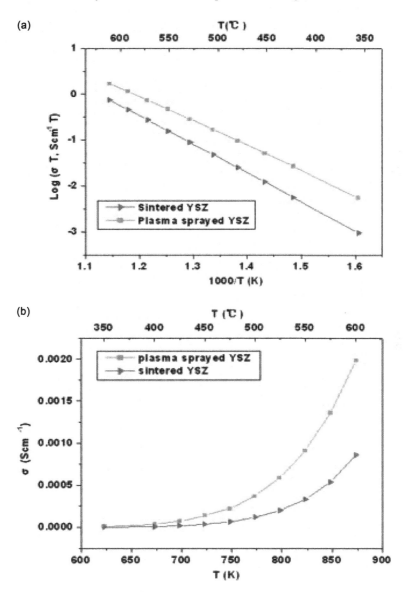

FIGURE 2.9 (a) Arrhenius plot for the total ionic conductivity and (b) ionic conductivity variation with temperature for the nanocrystalline plasma-sprayed and the conventional sintered YSZ electrolyte. (Reprinted with permission from Ref. [30].)

processes fabricate YSZ by the thick-film processing techniques such as the doctor blade method.

In order to develop a design for the SOFC stack to be applicable for portable and transportation purposes, it is also crucial to reduce the thickness of the YSZ electrolyte, which decreases ohmic resistance and improves cell performance at low-to-intermediate temperatures. There are many studies focused on the design and fabrication of the thin-layer structure of the YSZ electrolyte. Suzuki et al. [31] fabricated a dense, crack-free, single grain thick (<1 μm) YSZ used in a micro-tubular SOFC.

The results showed that the SOFC has a power density of 0.39 W cm^{-2} at operating temperature as low as 600°C. Smeacetto et al. [32] utilized the RF sputtering technique to develop a thin-film YSZ (40–600 nm) electrolyte without the use of post-deposition annealing heat treatment. The fabricated thin film was dense, continuous, and crack-free at the electrolyte/anode interface. The activation energy for ionic conduction was also found to be 1.18 ± 0.001 eV. The activation energy for conductivity is believed to result from defect-pair disassociation and oxygen vacancy migration. Table 2.2 presents the activation energy in high and low SOFC operating temperatures for different compositions of the YSZ electrolyte. The variation in conductivity with the Y_2O_3 dopant content is caused by interactions between defect pairs, the configuration of defects, and complex defect pair association [5]. Jang et al. [33] fabricated a thin-film-structured YSZ electrolyte using the atomic layer deposition (ALD) technique with higher conductivity than that of the conventional YSZ electrolytes due to the controlled ratio of yttria within the zirconia structure. In another study, the ALD-fabricated YSZ thin-film electrolyte showed a maximum power density 100% higher than that of fabricated with a sputtering method due to its more homogeneous and uniform microstructure [34].

Numerous studies have also focused on the effect of grain boundary and its homogenous distribution on the conductivity of the YSZ electrolyte and the related performance toward the low operating temperature of SOFCs. The grain boundary effect on the conductivity is important because the grain boundaries in the YSZ electrolyte can behave as a barrier for ionic transportation. A grain boundary consists of a core

---

TABLE 2.2

Activation Energy (E_a) and Conductivity (σ) Values at High and Low Temperature Range for Some YSZ Electrolytes with Different Composition

Composition	E_a (eV)		σ (1000°C) (S cm^{-1})
	400°C–500°C	850°C–1000°C	
$3Y_2O_3$	0.95	0.80	0.056
$8Y_2O_3$	1.10	0.91	0.164
$10Y_2O_3$	1.09	0.83	0.13
$12Y_2O_3$	1.20	1.04	0.068

Source: Data adapted and modified from Ref. [5].

and two adjacent space-charge layers. In the acceptor-doped zirconia electrolyte, the grain boundary core is positively charged due to the oxygen vacancy enrichment there. Therefore, the oxygen vacancies are depleted, and the acceptors are accumulated in the space-charge layer because the grain boundary core evicts oxygen vacancies while it attracts acceptor cations [35]. According to the Schottky barrier model, the space-charge potential decreases with decreasing the grain size, which results in increasing the concentration of oxygen vacancies in the space-charge layer and ionic conductivity, subsequently. The previous study showed that the fabrication of the YSZ with the grain size smaller than 10 μm can increase the conductivity by 50% than those of materials with larger grain sizes [36]. Therefore, there are many approaches to modify the microstructure, morphology, and crystallization process of the YSZ electrolyte with the various analyses. For instance, Liu and Lao [37] showed that the grain boundary conductivity of 8YSZ can be improved, especially at low-to-intermediate temperatures, by adding a small amount of ZnO, as presented in Figure 2.10. The maximum grain boundary conductivity was obtained for 5 wt% ZnO dopant at 300°C. Moreover, the 5 wt% ZnO-doped YSZ showed ~96% relative density, which was higher than that of the undoped sample with ~89% relative density.

The grain boundary conductivity is also dependent on material purity, which not only decreases by the grain boundary intrinsic blocking effect, but it can significantly decrease with impurities such as silica. Badwal and Rajendran [38] found that the grain boundary conductivity of YSZ decreases by a factor greater than 1.5, with the presence of only 0.2 wt% silica addition. Therefore, producing a high-purity YSZ can improve its grain boundary conductivity and the total ionic conductivity of the electrolyte, consequently. The grain boundary purity of the YSZ electrolyte can be controlled by the application of heat treatments, pressure, and various additives.

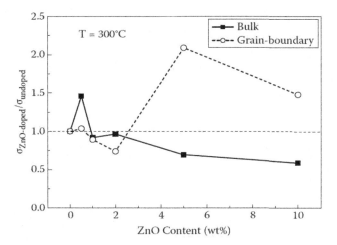

FIGURE 2.10 The grain boundary and bulk conductivities for the 8YSZ ceramics undoped and doped with ZnO dopant at 300°C as a function of ZnO content (wt%). (Reprinted with permission from Ref. [37].)

The modification of YSZ electrolyte composition with co-dopants is also an alternative solution to enhance its conductivity and performance. As discussed, the addition of 5 wt% ZnO in 8YSZ increased the grain boundary and total ionic conductivity because ZnO provides an oxygen-ion conductance channel across the grain boundary. Similar behavior has been observed for Al_2O_3-doped YSZ where minor addition of alumina not only increases the grain boundary conductivity by converting the silica impurity phase to a new phase at the grain boundary, but it also improves sintering behavior and mechanical properties. The highest grain boundary conductivity values belonged to the YSZ electrolyte doped with 4 wt% Al_2O_3. The performance of SOFC with an Al_2O_3-doped electrolyte was found better than that with a pure YSZ electrolyte [39]. In the other research, Xiao et al. [40] doped alumina to the YSZ-synthesized powder to streamline the structure of the grain boundary and reach the conductivity of 0.0344 S cm^{-1}. Adding CeO_2 in the YSZ electrolyte also reduced the defect association and migration energy, which leads to an increase in the yttria content limit for higher conductivity of the YSZ electrolyte [41]. Schmid et al. [42] also fabricated the CeO_2-doped YSZ electrolyte with higher conductivity than that of the pure YSZ due to a higher concentration of oxygen vacancies. Flegler et al. [43] modified the YSZ electrolyte with iron to enhance the ionic conductivity, densification, and stability. They found that 5 wt% Fe-doped YSZ reduced the sintering temperature and exhibited the highest oxygen-ion conductivity due to retention of the cubic phase of zirconia. Lee et al. [44] increased the ionic conductivity of the YSZ electrolyte from 0.0173 to 0.0196 S cm^{-1} by adding a trace of CuO (100 ppm) due to an increase in the amount of oxygen vacancy in the YSZ electrolyte. The cell performance was also about 0.5103 W cm^{-2} at 800°C which was 1.5 times higher compared to the cell used for the pure YSZ electrolyte. However, higher peak power density is achievable. The use of anode-based Ni-YSZ cell with a thin-film YSZ has been beneficial for maximizing the power density; for example, a Si-based free-standing thin-film SOFC with a high power density of 1.3 W cm^{-2} at 450°C has been reported [45]. Selvaraj et al. [46] fabricated the iron-/zinc-doped YSZ electrolyte with 2 mol% iron and 1 mol% zinc, which has high ionic conductivity, high hardness, and a porous-free structure. Table 2.3 presents the recent research and development on modifications of the YSZ electrolyte to improve its conductivity and single-cell SOFC performance toward lowering the operating temperature for low-to-intermediate SOFC applications.

Another issue during operation at high temperatures with the stabilized zirconia electrolytes, including the YSZ is the decrease in ionic conductivity due to aging [5,21]. This is more significant for the YSZ electrolytes with low dopant concentrations. For example, the conductivity of the as-sintered 3YSZ dropped from 0.056 to 0.049 S cm^{-1} after 5000 min annealing at 1000°C. A similar trend was observed for the 8YSZ in which the conductivity dropped from 0.164 to 0.137 S cm^{-1} [5]. Aging in the YSZ electrolyte, especially for much lower Y_2O_3 content, is attributed to the disappearance of a distorted fluorite phase (t'-phase), which has higher conductivity than the cubic phase [5,21]. Nomura et al. [47] studied the change in conductivity with the transformation of ZrO_2 using the Raman spectroscopy technique. As shown in Figure 2.11, the observed Raman spectrum [the ratio of peak intensity at

TABLE 2.3

The Conductivity and Single-Cell Performance of the YSZ Electrolyte with Various Modifications

Type of Electrolyte	Operating Temperature (°C)	Conductivity (S cm⁻¹)	Voltage (V)	Current Density (mA cm⁻²)	Power Density (mW cm⁻²)
YSZ single grain	600	0.0035	1.0	1000	390
Tri-layer YSZ	800	-	1.0	1750	935
YSZ by chemical solution	600	-	1.06	1000	452
YSZ/SDC	800	-	1.1	550	421
YSZ with an infiltrated porous anode	800	-	1.108	600	275
YSZ by sol-gel	1200	0.107	-	-	-
YSZ by inkjet printing	800	-	1.15	2500	1500
ALD YSZ	100	-	0.9	0.003	0.01
ALD YSZ	450	-	1.18	720	154.6
YSZ	750	-	1.1	1100	242
YSZ with Ni-YSZ accelerated aging	750	-	1.05	2400	1200
YSZ with Ni-YSZ tubular anode	800	-	0.8	400	500
Sputtered YSZ	800	-	1.1	2200	515
YSZ with flash sintering	850	0.056	-	-	-
YSZ with nanofiber interconnect	450	-	-	-	5.1
YSZ nanowire	375	0.01	-	-	-
CF-YSZ	550	0.477	0.85	800	187.5
YSZ with hybrid thin-film anode	850	-	1.2	1850	600
Alumina doped YSZ	800	0.00334	-	-	-
YSZ single step sintering	800	-	1.6	750	400
YSZ	550	-	1.0	27	1000
YSZ based SIMs	600	0.06	1.11	1000	628
Fe/Zn doped YSZ	300	0.0000074	-	-	-
CuO-YSZ	800	0.0196	1.2	2000	5103

Source: Data adapted and modified from Ref. [25].

$260 \, cm^{-1}$ (I_{260}) (tetragonal) and at $640 \, cm^{-1}$ (I_{640}) (cubic)] suggested that the decrease in conductivity for 8YSZ electrolyte is due to a local ordering toward the tetragonal phase formation in the cubic matrix phase by annealing at 1000°C. Similar behavior was also observed for the 8ScSZ electrolyte.

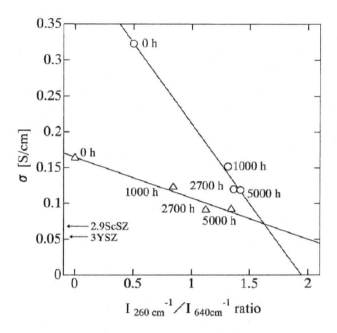

FIGURE 2.11 Change in conductivity against Raman scattering intensity ratio I_{260}/I_{640} at 260 and 640 cm^{-1}. O: 8ScSZ, Δ: 8YSZ. (Reprinted with permission from Ref. [47].)

The stabilized zirconia electrolytes, including the YSZ, have also shown good mechanical resilience, while maintaining the ionic conductivity at high operating temperatures. In order to maintain high ionic conductivity while lowering the operating temperature, one solution is to reduce the electrolyte thickness. In this case, it is crucial to fabricate a thin-film electrolyte with adequate mechanical strength to maintain cell mechanical performance. Lee et al. [48] fabricated a porous YSZ support layer on the nickel-based anode to enhance the properties of the YSZ electrolyte. The adoption of the porous layer increases the mechanical stability and performance of the electrolyte in the cell. Timurkutluk et al. [49] sealed the YSZ with glass, which increased the mechanical and thermal properties of the electrolyte. The sealed glass-ceramic composites with various YSZ additions were examined via a tensile test after thermal cycling to investigate the fracture strength. The fracture stress of the electrolyte decreased after the thermal cycling stability process. However, modification of the glass-ceramics composites with various YSZ additions improved the mechanical stability where 5 wt% YSZ content exhibited a stable mechanical performance after each thermal cycle, as shown in Figure 2.12.

Chong et al. [50] synthesized MnO$_2$-doped YSZ electrolytes, showing good mechanical stability. Figure 2.13 indicates the samples with manganese dopants have fracture toughness values of 4.1–4.6 MPa m$^{1/2}$ when sintered at temperatures above 1450°C, regardless of the dopant content. However, at the low sintering temperature, fusion and bonding of the YSZ electrolyte were promoted by the dopant dissolution. Selvaraj et al. [46] studied hardness of the iron-/zinc-doped YSZ electrolytes.

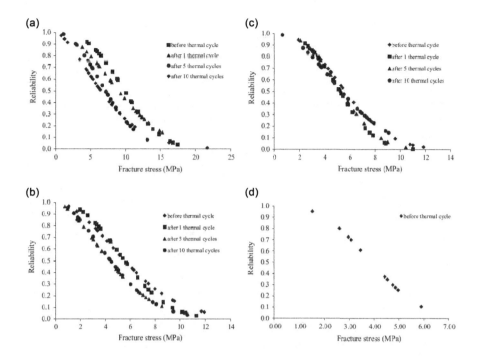

FIGURE 2.12 The effect of the YSZ addition on the fracture strength of (a) 1 wt%, (b) 3 wt%, (c) 5 wt%, and (d) 10 wt% as a function of the number of thermal cycles. (Reprinted with permission from Ref. [49].)

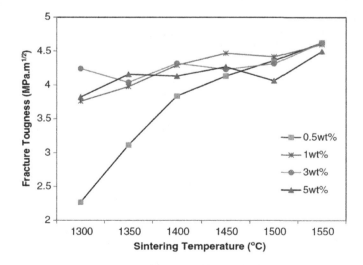

FIGURE 2.13 Fracture toughness of 8YSZ with various MnO_2 dopant concentrations sintered at 1300°C–1550°C. (Reprinted with permission from Ref. [50].)

The composite YSZ-based electrolyte showed higher hardness compared to that of the pristine YSZ electrolyte due to the good densification.

The mechanical parameter of the electrolyte can affect its ionic conductivity. Araki and Arai [51] examined the effect of mechanical stress on the ionic conductivity of the thin-film YSZ electrolyte using molecular dynamic simulation. The results indicated that the conductivity of the YSZ electrolyte can be enhanced by applying directional stresses. The introduction of external stress, which increases the conductivity, can enhance oxygen diffusion and decrease the activation energy. Hence, it is important to consider the mechanical properties of the YSZ electrolyte in order to improve its conductivity, which correlates to a higher cell power density. To reduce the ohmic resistance at low temperatures, one can fabricate a SOFC with a thin-film electrolyte (thickness <1 μm); however, the anode optimization is needed in order to improve the mechanical strength necessary for stable operation, so supporting structures are usually necessary.

2.2.3.2 Scandia-Stabilized Zirconia

The ScSZ electrolyte is considered as a promising electrolyte for intermediate temperature SOFC applications as it offers the highest ionic conductivity among any known doped stabilized zirconia electrolytes. This is shown in Figure 2.14 for the two systems (Sc_2O_3-ZrO_2 and Y_2O_3-ZrO_2) with the most interest in fuel cell applications, however, it did not have industrial application. The ionic conductivity of the Sc_2O_3-ZrO_2 is dependent on its composition, but it is difficult to obtain equilibrium in the Sc_2O_3-ZrO_2 system due to many inconsistencies in the phase diagram. In order to fully stabilize the cubic structure of zirconia, there is a need to add 8–10 mol% of Sc_2O_3. For the composition with less Sc_2O_3 content (below 6–7 mol%), monoclinic phase along with the fluorite phase exist, while for the Sc_2O_3 content above 9.5 mol%, the ordered rhombohedral phase with poor conductivity (β-phase) is formed. The compositions containing β-phase show jumps in the Arrhenius plots for the conductivity at 500°C–600°C and hysteresis effects on thermal cycling. This is shown

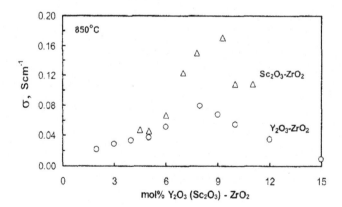

FIGURE 2.14   The electrolyte conductivity as a function of dopant concentration for Sc_2O_3-ZrO_2 and Y_2O_3-ZrO_2 systems at 850°C. (Reprinted with permission from Ref. [52].)

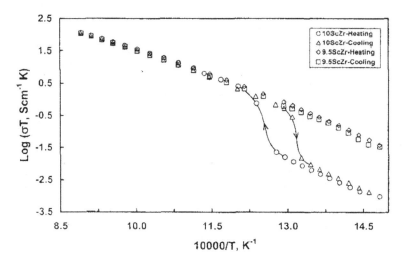

FIGURE 2.15 Arrhenius plots for two Sc_2O_3-ZrO_2 compositions showing jumps in conductivity for 10 mol% Sc_2O_3-ZrO_2 due to the presence of the β-phase and no jumps or hysteresis effect for 9.5 mol% Sc_2O_3-ZrO_2 in the absence of the β-phase. (Reprinted with permission from Ref. [52].)

in Figure 2.15 for the two Sc_2O_3-ZrO_2 systems, one with 10 mol% Sc_2O_3 containing β-phase and the other with 9.5 mol% Sc_2O_3 and almost free of the presence of β-phase. The stabilization of the Sc_2O_3-ZrO_2 cubic structure also depends on the temperature. The stability range is 9–15 mol% Sc_2O_3 content at 1200°C. At room temperature, zirconia with 10 mol% Sc_2O_3 content consists of a mixture of cubic phase and β-phase, while the β-phase transforms to the cubic phase at 1000°C [5]. Table 2.4 shows the ionic conductivity of various zirconia-based electrolytes. As seen, the ionic conductivities of Sc_2O_3-ZrO_2 electrolytes with different dopant contents are much higher than the conductivity values for the different Y_2O_3-ZrO_2 electrolytes. The conductivity maximum for the Sc_2O_3-ZrO_2 electrolyte is reached with the 9.3 mol% Sc_2O_3 at 800°C and 1000°C, while it is 8 mol% Y_2O_3 for the Y_2O_3-ZrO_2 electrolyte.

The phase structure of the Sc_2O_3-ZrO_2 electrolyte depends not only on the composition but the fabrication method. Yamamoto et al. [12] showed that for the 5 mol% Sc_2O_3, the ceramic synthesized by the spray pyrolysis technology and sintered at 1500°C for 2 h contained 60% cubic and 40% tetragonal phase, whereas it contained only 11% cubic phase when it was fabricated using the sol-gel method and sintered at 1700°C for 5 h. For the 8 mol% Sc_2O_3-ZrO_2 ceramic synthesized by sol-gel method, a single cubic phase of ZrO_2 was observed, while the 8 mol% Sc_2O_3-ZrO_2 ceramic prepared by the solid-state reaction contained monoclinic and fluorite phases [12].

The higher conductivity of the Sc_2O_3-ZrO_2 electrolytes in comparison with the Y_2O_3-ZrO_2 electrolytes is attributed to the smaller mismatch in size between Sc^{3+} and Zr^{4+} ions compared to that between Y^{3+} and Zr^{4+} ions (i.e., the ionic radius of Sc^{3+} is closer to that of the Zr^{4+}). Therefore, the steric hindrance to migration of oxygen

TABLE 2.4
Ionic Conductivity Data for the Y_2O_3-ZrO_2 and Sc_2O_3-ZrO_2 Electrolytes with Various Dopant Contents

	σ (S cm^{-1})	
Electrolyte Composition	800°C	1000°C
$(ZrO_2)_{0.98}(Y_2O_3)_{0.02}$	0.0135	0.048
$(ZrO_2)_{0.97}(Y_2O_3)_{0.03}$	0.018	0.058
$(ZrO_2)_{0.97}(Y_2O_3)_{0.03}$-10 wt% Al_2O_3	0.012	0.040
$(ZrO_2)_{0.97}(Y_2O_3)_{0.03}$-20 wt% Al_2O_3	0.008	0.028
$(ZrO_2)_{0.96}(Y_2O_3)_{0.04}$	0.023	0.076
$(ZrO_2)_{0.94}(Y_2O_3)_{0.06}$	0.036	0.116
$(ZrO_2)_{0.92}(Y_2O_3)_{0.08}$	0.052	0.178
$(ZrO_2)_{0.92}(Y_2O_3)_{0.08}$-10 wt% Al_2O_3	0.036	0.129
$(ZrO_2)_{0.92}(Y_2O_3)_{0.08}$-20 wt% Al_2O_3	0.027	0.096
$(ZrO_2)_{0.91}(Y_2O_3)_{0.09}$	0.040	0.161
$(ZrO_2)_{0.90}(Y_2O_3)_{0.10}$	0.037	0.136
$(ZrO_2)_{0.88}(Y_2O_3)_{0.12}$	0.023	0.097
$(ZrO_2)_{0.85}(Y_2O_3)_{0.15}$	0.006	0.034
$(ZrO_2)_{0.93}(Sc_2O_3)_{0.07}$	0.086	0.274
$(ZrO_2)_{0.922}(Sc_2O_3)_{0.078}$	0.120	0.310
$(ZrO_2)_{0.91}(Sc_2O_3)_{0.09}$	0.109	0.341
$(ZrO_2)_{0.91}(Sc_2O_3)_{0.09}$-10 wt% Al_2O_3	0.060	0.186
$(ZrO_2)_{0.907}(Sc_2O_3)_{0.093}$	0.120	0.354
$(ZrO_2)_{0.88}(Sc_2O_3)_{0.12}$	-	0.340

Source: Data adapted and modified from Ref. [52].

* Conductivity values are initial values in as-sintered materials obtained without aging at the stated temperatures [52].

ions/vacancies in the Sc_2O_3-ZrO_2 fluorite lattice is minimum due to very little change in the lattice volume by substitution of Zr^{4+} ions with Sc^{3+} ions [5,52]. Consequently, the activation energy for conduction in the ScSZ system is lower than that of the YSZ system in the high-temperature range, such as 850°C–1000°C. For example, the activation energy for the 7.8 mol% Sc_2O_3-ZrO_2 is 0.78 eV at 850°C–1000°C, which is lower than those listed in Table 2.2 for the Y_2O_3-ZrO_2 electrolytes in the same temperature range. However, the activation energy in the ScSz tends to increase with a decrease in the temperature, similar to that of the YSZ. At low temperatures (i.e., below 500°C), the conductivity of the ScSZ electrolyte is similar or even lower than that of the YSZ due to higher activation energy. For the 7.8 mol% Sc_2O_3-ZrO_2 electrolyte, the activation energy has been measured 1.35 eV at 400°C–500°C, which is higher than those of listed in Table 2.2 for the Y_2O_3-ZrO_2 electrolytes in the same temperature range.

Aging in the ScSZ electrolyte, which leads to a decrease in conductivity, has been an issue during operation. This is related to the disappearance of t'-phase, a distorted fluorite phase, which has higher conductivity than the cubic phase [21]. The t'-phase transforms to a tetragonal phase to increase during aging. The reduction in conductivity in ScSZ during aging is smaller compared to that for the YSZ. Nomura et al. [47] showed twice higher initial conductivity for the ScSZ than that of the YSZ. However, the conductivity of the ScSZ was the same as that of the YSZ after 5,000 h aging due to the disappearance of the t'-phase. The aging effect in the ScSZ electrolyte can be suppressed by co-doping with indium oxide or an increase in the scandia content [21]. For instance, small degradation in conductivity has been observed in 9ScSZ electrolytes, while no reduction in conductivity has been seen in the 11ScSZ, 12ScSZ, and 13ScSZ electrolytes [20,52].

The mechanical properties of the ScSZ electrolyte are similar or even better than those of the YSZ electrolyte [12,53]. It is important to produce a reliable and durable electrolyte for SOFC applications. The strength and toughness of the ScSZ electrolyte can be improved by adding oxide dispersant, e.g., alumina or niobates. In order to improve the mechanical properties and maintain conductivity, such additions must be optimized because they typically decrease the conductivity of the electrolyte.

2.2.3.3 Other Dopants and Co-Dopants for Stabilized Zirconia

In addition to Y_2O_3 and Sc_2O_3, other oxides have been used as dopants for zirconia electrolyte, such as CaO, La_2O_3, and Yb_2O_3. Figure 2.16 illustrates the conductivity of various dopants for stabilized zirconia electrolytes, where Sc and Yb doping leads to the highest conductivity. In order to achieve the maximum conductivity of the doped zirconia and to fully stabilize the cubic fluorite phase, various dopants are added. However, further addition of dopants may result in a decrease in conductivity

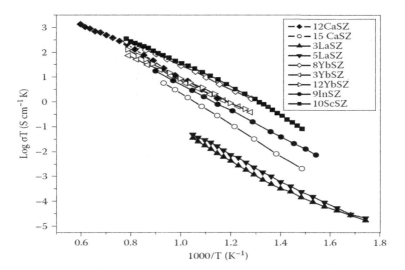

FIGURE 2.16 Conductivity of zirconia electrolytes stabilized with various dopants. (Reprinted with permission from Ref. [5].)

because of mismatch of the Zr^{4+} radius and the dopant cation radius in the zirconia lattice structure, thereby low mobility of the oxygen vacancies. Yamamoto et al. [54] studied the conductivity of Ln_2O_3-ZrO_2 electrolyte as a function of Ln^{3+} radius. Compared to Sc_2O_3-ZrO_2 electrolyte, the conductivity of Ln_2O_3-ZrO_2 electrolyte was lower due to the larger radius of Ln^{3+} cation. The conductivity of YbSZ is similar to that of the ScSZ due to an almost similar radius of Yb^{3+} cation compared to Sc^{3+} cation. Bivalent metal oxides, such as MgO and ZnO, have also been used to stabilize zirconia, but they are less effective than the other dopants.

Co-doping, involving adding a third cation to modify the YSZ or ScSZ electrolyte, has been used to improve the chemical and thermal stability, electrochemical properties, and to reduce the cost of stabilizer. Some examples of co-doping are Mn_2O_3-Sc_2O_3-ZrO_2, Bi_2O_3-Sc_2O_3-ZrO_2, and Ca_2O_3-Y_2O_3-ZrO_2 ternary systems. The addition of Bi_2O_3 to the ScSZ system can inhibit the transformation of the cubic phase to the rhombohedral phase and decrease the sintering temperature [55]. The conductivity of 2 mol% Bi_2O_3-Sc_2O_3-ZrO_2 was reported as 0.16 S cm^{-1} at 600°C [56]. Liu and Lao [37] found that the addition of ZnO to the YSZ system can promote the densification, increase the conductivity, and also decrease the sintering temperature. The densification of 5 wt% ZnO in 8YSZ was reported to be 96%. The increase in total conductivity of ZnO-Y_2O_3-ZrO_2 was attributed to an increase in grain boundary conductivity due to an oxygen-ion conductance across the grain boundary provided by the ZnO co-dopant. Similar behavior has been observed for Al_2O_3-doped zirconia [57]. In another study, Lei and Zhu [58] showed that adding 2 mol% Mn_2O_3 to 11ScSZ can inhibit cubic-rhombohedral phase transformation and reach nearly full density at a sintering temperature of 850°C. The conductivity of 2Mn_2O_3–11ScSZ was measured ~0.1 S cm^{-1} at 800°C. The conductivity of some zirconia-based ternary systems is shown in Figure 2.17.

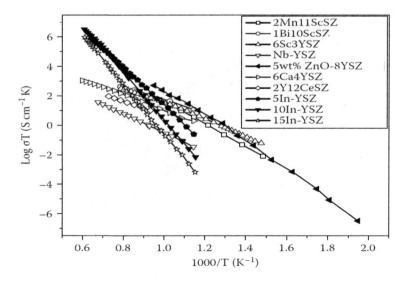

FIGURE 2.17 Conductivity of co-doped zirconia-based ternary systems. (Reprinted with permission from Ref. [5].)

2.2.4 CERIA-BASED ELECTROLYTES

Similar to zirconia, ceria (CeO_2) has the fluorite-type crystal structure over wide ranges of temperature and oxygen partial pressure and is a common electrolyte material for SOFC applications. The structure can be viewed as an face-centered cubic (FCC) array of Ce^{4+} ions with the O^{2-} ions residing in the tetrahedral holes. The ceria unit cell consists of four Ce^{4+} ions and eight O^{2-} ions, as shown in Figure 2.18. In the cubic fluorite structure of ceria, each Ce^{4+} is surrounded by eight equivalent O^{2-} ions, and each O^{2-} ion is surrounded by a tetrahedron of four equivalent Ce^{4+} ions [59]. Doped CeO_2 is a promising electrolyte material with high ionic conductivity as well as a high surface exchange coefficient. In particular, the ionic conductivities of GDC and samaria-doped ceria (SDC) are two orders higher than that of YSZ at near 500°C, and SDC also shows high surface exchange owing to the large size of the dopant ions (Sm^{3+}, 109.8 pm) compared to the other dopant ions (Gd^{3+}, 107.8 pm; Y^{3+}, 104 pm) [21,60].

Ceria has both electronic and ionic conductivities by nature. Pure ceria has higher electronic conductivity than ionic conductivity. Hence, ceria is doped to improve its oxygen-ion conductivity. Doping of ceria with a number of oxides such as CaO, Sm_2O_3, Gd_2O_3, Yb_2O_3, Y_2O_3, and La_2O_3 increases the concentration of oxygen vacancies in the crystal structure, leading to high ionic conductivity. Analogous to that of the doped zirconia electrolyte, the creation of oxygen vacancies in the crystal structure of ceria as a result of doping can be presented by the Kröger–Vink notation as follows:

$$M_2O_3 + CeO_2 \rightarrow 2M'_{Ce} + V_O^{\bullet\bullet} + 3O_O^{\times} \tag{2.5}$$

and in a reducing environment, Ce^{4+} is reduced to Ce^{3+}, which leads to electronic contribution in conductivity, expressed as:

$$O_O^{\times} \rightarrow V_O^{\bullet\bullet} + \frac{1}{2}O_2(g) + 2e' \tag{2.6}$$

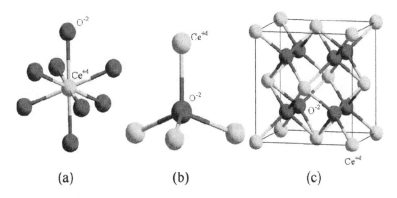

| (a) | (b) | (c) |

FIGURE 2.18 (a) Ce ion with eight oxygen ions, (b) oxygen ion with four Ce ions, and (c) fluorite structure of CeO_2. (Reprinted with permission from Ref. [59].)

Like zirconia, the highest conductivity reaches for dopant ions with the lowest size mismatch, which for cerium is gadolinium and samarium. It has been noted that the ionic radius of the dopant ions is correlated with conductivity and activation energy of the doped ceria. The minimum activation occurs for those dopants with radius most closely matching that of Ce^{4+}. Figure 2.19 illustrates the conductivities of doped ceria with some common trivalent and bivalent dopants. As observed, GDC and SDC show the highest conductivity values among those doped ceria at 800°C. Compared to pure ceria, the stability regime is remarkably higher for doped ceria [5,21,52]. Equation (2.7) shows the classical model for oxygen-ion conductors such as doped ceria and doped zirconia, which is a function of both dopant concentration and temperature:

$$\sigma = \frac{A}{T}[V_O^{\bullet\bullet}](1-[V_O^{\bullet\bullet}])\exp\left(-\frac{E}{RT}\right) \tag{2.7}$$

where E is the activation energy for conduction, R is the gas constant, T is the absolute temperature, and A is the pre-exponential factor. For trivalent dopants, the model suggests that the pre-exponential factor should be independent of temperature at low temperatures. At a given concentration, Equation (2.7) is usually presented as:

$$\sigma T = \sigma_0 \exp\left(-\frac{E}{RT}\right) \tag{2.8}$$

Ceria-based electrolytes have a higher ionic conductivity than zirconia-based electrolytes at low operating temperatures and lower polarization resistance.

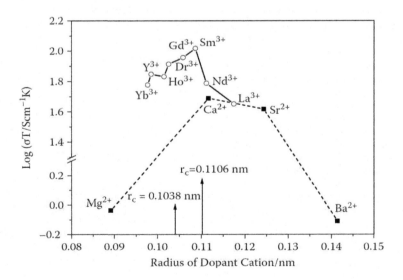

FIGURE 2.19 Ionic conductivity of doped ceria at 800°C against the radius of dopant cation, r_c shown in the horizontal axis is for the critical radius of bivalent or trivalent cation, respectively. (Reprinted with permission from Ref. [5].)

In addition, ceria-based electrolytes possess promising chemical inertness and thermal expansion match with high performing electrodes. The primary drawback of the ceria-based electrolyte is that at high temperatures (above 700°C) and low oxygen partial pressure, some of the Ce^{4+} ions are reduced to Ce^{3+} ions, which results in mixed electronic/ionic conductivity, as expressed in Equation (2.6), thereby lowering the open-circuit voltage (OCV) and internal short-circuiting [21,59]. The most widely used ceria-based electrolyte, GDC ($Ce_{1-x}Gd_xO_2$), has shown a comparable conductivity with those of the YSZ and ScSZ electrolytes, as shown in Figure 2.20. The conductivities of the $Ce_{1-x}Gd_xO_{2-x/2}$ electrolytes with various Gd contents are higher than those of the YSZ and ScSZ electrolytes. The difference is more significant at temperatures below 600°C. Similar to zirconia, the conductivity of ceria-based electrolytes increases with an increase in dopant concentration up to a maximum value (e.g., 0.2–0.25 Gd) and then bounces back. As shown in Figure 2.20, the range of conductivities for $Ce_{0.9}Gd_{0.1}O_{1.95}$ and $Ce_{0.8}Gd_{0.2}O_{1.9}$ are similar. However, at low oxygen partial pressure, $Ce_{0.9}Gd_{0.1}O_{1.95}$ has better stability than $Ce_{0.8}Gd_{0.2}O_{1.9}$. In addition to $Ce_{1-x}Gd_xO_2$, SDC ($Ce_{1-x}Sm_xO_2$) has been widely used as the electrolyte for SOFC applications below 600°C. Both materials exhibit a mixed conductivity, electronic and ionic conductivity at the temperature above 560°C. The conductivity of $Ce_{1-x}Sm_xO_2$ is similar to that of the $Ce_{1-x}Gd_xO_2$ but in a lower range. Like zirconia, co-doping can be used for improving the properties of the ceria-based electrolyte. For example, adding praseodymium and samarium can increase the conductivity of $Ce_{1-x}Gd_xO_2$.

Like zirconia, grain boundary effect is also important in ceria-based electrolytes. In both $Ce_{1-x}Sm_xO_2$ and $Ce_{1-x}Gd_xO_2$, a reduction in grain size has shown a detrimental effect on conductivity and performance. Moreover, materials with small

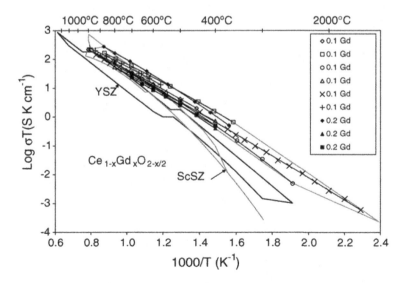

FIGURE 2.20 Conductivity of $Ce_{1-x}Gd_xO_{2-x/2}$ electrolytes with various Gd contents in air. (Reprinted with permission from Ref. [21].)

grains are more susceptible to a reduction in low oxygen partial pressures. Therefore, controlling the grain size of ceria-based electrolytes is crucial during the fabrication process. The chemical stability of ceria with cathode materials is superior to that of the zirconia, although its stability in low oxygen partial pressure is lower than zirconia. Because of this promising stability with cathode materials, ceria-based interlayers have been applied between the cathode and the YSZ electrolyte in order to prevent reaction [21].

2.2.4.1 Gadolinia-Doped Ceria

Among doped ceria electrolytes, GDC electrolyte has been known as a pioneer electrolyte for intermediate- and low-temperature SOFCs due to its Gd^{3+} closer radius to r_c which improves the electrolyte conductivity. Although the conductivity of GDC depends on the gadolinia content, it is difficult to optimize the gadolinia content to maximize the conductivity for all situations because the concentration of vacancies is a function of temperature. For example, Hohnke [61] reported the highest conductivity and the lowest activation energy for 6 mol% gadolinia at temperatures between 400°C and 800°C, while Steele [62] found that 25 mol% GDC has the highest conductivity, and Zha et al. [63] and Zhang et al. [64] observed the conductivity peak for 15 mol% gadolinia. Moreover, Babu et al. [65] showed that the conductivity of GDC is ionic at low temperatures, while it changes to mixed ionic and electronic conduction at higher temperatures due to the reduction of ceria. The impurity of the material and fabrication process have also had a great impact on the GDC conductivity as the impure ceramic electrolytes have shown high resistances at the grain boundaries, thereby lowering the total conductivity. Zhang et al. [66] identified the grain boundary conductivity, grain interior conductivity, and total conductivity for the $Ce_{1-x}Gd_xO_{2-\delta}$ ceramics ($0.05 \leq x \leq 0.3$) with and without different levels of the SiO_2 content (30–3000 ppm) at 350°C. The results demonstrated that in the ceramics with lower Gd contents, the blocking effect of grain boundaries dominates the total conductivity, even with a small amount of impurity (30 ppm SiO_2). However, in high-purity ceramics, the grain boundary behavior is usually attributed to the intrinsic blocking effect, known as the space-charge effect. Figure 2.21 shows the grain boundary conductivity at 350°C and the grain interior conductivity at 350°C and 500°C, as a function of Gd content. As can be seen, the trend of the grain interior conductivity is similar for different SiO_2 contents, although a slight decrease in grain interior conductivity is observed with the increasing SiO_2 content. The grain boundary conductivity also increases with the increasing Gd content at 350°C. Kharton et al. [67] indicated that grain boundary contribution in total conductivity of the GDC electrolyte decreases with an increase in grain size for the grains larger than 2–3 μm. However, the bulk conductivity of the electrolyte is essentially independent of the grain size.

Gd^{3+} (0.1053 nm) has a lower ionic radius than Sm^{3+} (0.1079 nm), while Sm^{3+} radius has been proposed as an ideal size for an improved ionic conductivity for the doped ceria electrolyte [65]. Hence, co-doping GDC with slightly larger ions has been found to be an effective method to improve the conductivity by reducing the size mismatch. For example, Babu et al. [65] showed that the conductivity is improved in Nd-GDC

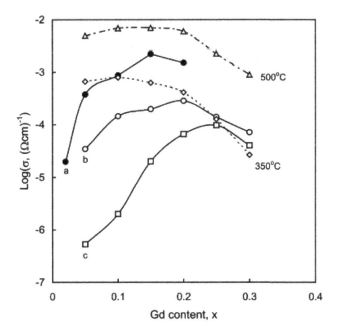

FIGURE 2.21 Grain boundary conductivity at 350°C (solid line) and grain interior conductivity at 350°C and 500°C (dashed line) as a function of Gd content: (a) with 30 ppm SiO$_2$, (b) with 200 ppm SiO$_2$, and (c) with 3000 ppm SiO$_2$. (Reprinted with permission from Ref. [66].)

with co-doping Nd^{3+} (0.1109 nm), while Eu-GDC demonstrated similar conductivity compared to that of the GDC, as shown in Figure 2.22.

2.2.4.2 Samaria-Doped Ceria

Although some of the previous research showed that the GDC electrolytes have the highest conductivity among the ceria-based electrolytes, other studies found that SDC electrolytes have the highest conductivity [65,68]. Like the GDC, the conductivity of the SDC electrolyte depends on its composition and the Sm dopant content. In Sm$_x$Ce$_{1-x}$O$_{2-x/2}$, the conductivity increases with an increase in the Sm content, reaches the maximum value at $x = 0.15$–0.2, and then decreases with a further increase in the Sm content [63,69–71]. This maximum peak behavior for conductivity with the Sm content is consistent in all SDC studies. However, the reason is interpreted in different ways. Some researchers believe that this trend in conductivity of SDC is related to the oxygen vacancies and their mobility. Doping samarium into the ceria crystal structure first increases the concentration of oxygen vacancy; thus, an increase in the ionic conductivity. However, at higher doping levels, the oxygen vacancy mobility decreases due to the interactions between the doped samarium (Sm$'_{Ce}$) and oxygen vacancy (V$_O^{\bullet\bullet}$), which leads to a reduction in ionic conductivity. For example, Jung et al. [69] studied the composition dependence of the conductivity

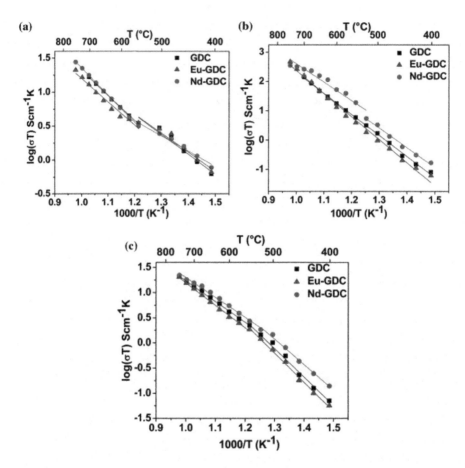

FIGURE 2.22 Arrhenius plot for (a) grain interior conductivity, (b) grain boundary conductivity and (c) total conductivity. (Reprinted with permission from Ref. [65].)

and activation energy at 600°C for $Sm_xCe_{1-x}O_{2-x/2}$ with various Sm contents, as shown in Figure 2.23, in which the maximum peak behavior in conductivity and the minimum peak behavior in activation energy are related to the oxygen vacancies mobility and defect interaction. As observed, the conductivity shows the maximum value at the concentration where the activation energy is at the lowest value.

The other reason proposed for the maximum peak behavior of conductivity at $x = 0.15–0.2$ is the grain boundary effect in SDC with a change in composition. With the increasing Sm content, the grain boundary conductivity increases substantially due to a decrease in the grain boundary effect, while the bulk conductivity decreases. These two opposite effects lead to the maximum conductivity value for $Sm_{0.2}Ce_{0.8}O_{1.9}$ [70]. Zhan et al. [70] believed that the conductivity of $Sm_{0.2}Ce_{0.8}O_{1.9}$ at low temperatures is controlled by the grain boundary, while at high temperatures, it is controlled by both grain boundary and grain interior.

It is believed that Sm and Gd co-doping of ceria electrolytes can suppress the ordering of oxygen vacancies in the lattice structure, thus improving the ionic conductivity

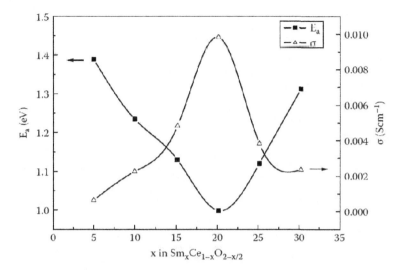

FIGURE 2.23 Composition dependence of the conductivity and activation energy at 600°C for $Sm_xCe_{1-x}O_{2-x/2}$ with various Sm contents. (Reprinted with permission from Ref. [5].)

and lowering the activation energy. For instance, Wang et al. [72] synthesized a set of co-doped ceria electrolytes with a nominal composition of $Ce_{0.85}Gd_{0.15-y}Sm_yO_{1.925}$, wherein $0 \leq y \leq a$. Figure 2.24 shows the conductivities of the samples in air at different temperatures. It can be seen that the co-doped electrolytes have apparently higher conductivity values than the single doped electrolytes. At 973 K, the conductivity of the co-doped samples reached the maximum of 0.0475 S cm^{-1}.

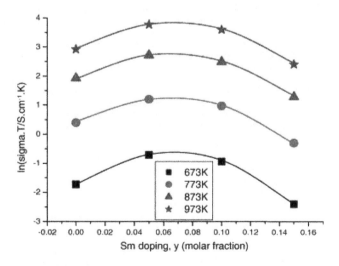

FIGURE 2.24 Effect of Sm doping (y) on the conductivity of pellet samples with nominal composition of $Ce_{0.85}Gd_{0.15-y}Sm_yO_{1.925}$ in air and at different temperatures. (Reprinted with permission from Ref. [72].)

2.2.5   Bi₂O₃-Based Electrolytes

In order to improve the ionic conductivity of the electrolyte for intermediate-temperature SOFCs, researchers have sought to find candidate materials with higher ionic conductivity and lower electronic conductivity compared to other conventional materials. δ-Bi_2O_3 (bismuth oxide) electrolyte with FCC crystal structure has shown the highest ionic conductivity than that of the zirconia-based and ceria-based electrolytes. However, the δ-Bi_2O_3 is reported to be stable only within the temperature range of 730°C–830°C. All forms of Bi_2O_3 are listed in Table 2.5. The δ-Bi_2O_3 has poor thermodynamic stability under a reducing atmosphere (oxygen partial pressure below 10^{-13} atm at <730°C) where it is reduced from Bi^{3+} to Bi^0 (metallic bismuth) [73–75]. Therefore, to stabilize the δ-Bi_2O_3 at lower temperatures, it has been doped with aliovalent metal oxides (with a smaller ionic radius than Bi^{3+} ion). Among several compositions of the $(Bi_2O_3)_{1-x}(M_2O_3)_x$ system (where M represents Sm, Gd, Nd, Y, Er, etc.), $(Bi_2O_3)_{0.75}(Y_2O_3)_{0.25}$ and $(Bi_2O_3)_{0.75}(Er_2O_3)_{0.25}$ have shown the best performance and attracted much attention. The co-doped $Dy_{0.08}W_{0.04}Bi_{0.88}O_{1.56}$ (DWSB) also showed a conductivity of 0.098 S cm⁻¹ at 500°C [76]. Another stable type of the fluorite-related Bi_2O_3-based electrolyte at low temperatures is bismuth metal (dopant) vanadium oxide, also known as BIMEVOX, with the typical composition of $Bi_2M_xV_{1-x}O_{5.5-3x/2-\delta}$. The BIMEVOX electrolytes are doped with transition metals and vanadium and are highly effective as oxygen separating membranes with promising ionic conductivity. For instance, $Bi_2VO_{5.5}$ electrolyte showed some ionic conductivity but has similar redox problems as that of the Bi_2O_3 [75]. The BIMEVOX electrolytes are also doped with Cu, Ni, Co, and Mg to not only stabilize the highly conductive γ-phase at lower temperatures but also achieve an ionic transport number close to 1.0 [17]. For example, $Bi_2V_{0.9}Cu_{0.1}O_{5.35}$ has shown the ionic conductivity of 0.1 S cm⁻¹ at 600°C [76]. There is a phase transition in the BIMEVOX materials, which led many studies to focus on stabilization of the fast ion-conducting, high-temperature γ-phase. Since BIMEVOX electrolytes have good conductivity, interest in these materials is still active. However, their intrinsic problems such as low mechanical strength, high chemical reactivity, and ability to easily reduce are yet to be addressed.

The Bi_2O_3-based materials have also been used as the electron blocking layer due to their low electronic conductivity and high ionic conductivity. The bilayer form of the electrolyte with the stabilized bismuth oxide, such as erbia-/yttria-stabilized

---

TABLE 2.5
Temperature Regions of the Stable and Metastable Forms of Bi₂O₃

Phase	α	δ	β	γ
Phase stability range (°C)	<730	730–830	330–650	500–640
Temperature (°C)	25	773	642	25
Structure	Monoclinic	FCC	Tetragonal	BCC

Source: Data adapted and modified from Ref. [75].

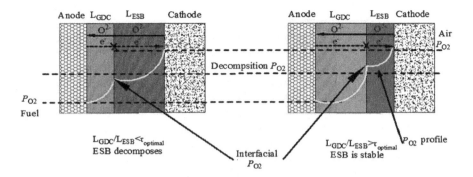

FIGURE 2.25 Schematic diagram of ESB decomposition and electronic current transport in the GDC/ESB bilayer electrolytes. (Reprinted with permission from Ref. [74].)

Bi_2O_3 (ESB/YSB), located at the cathode side has been proposed to improve the stability of the stabilized Bi_2O_3 and to block electronic current leakage of the ceria-based electrolyte, as shown in Figure 2.25. It has been proved that the bilayer electrolyte configuration increases the OCV of the cell and improves the cell performance for low-temperature SOFCs. For example, the bilayer electrolyte, $Sm_{0.075}Nd_{0.075}Ce_{0.85}O_{2-\delta}$/DWSB, showed an increase in cell OCV and power density up to 1.0 V and 3.5 W cm^{-2}, respectively [77].

2.2.6 PEROVSKITE-STRUCTURED ELECTROLYTES

2.2.6.1 Perovskite-Structured Oxygen-Ion Conductors

Over the past decades, great attention has been paid to Bi_2O_3-, ZrO_2-, and CeO_2-based ceramic electrolytes for SOFC applications. However, the perovskite-structured electrolytes have also received excellent attention due to their structural tolerance to accept various sizes of A and B cations and a large concentration of aliovalent cation dopants in sublattice sites. Furthermore, the perovskite-structured electrolytes have a great geometrical and chemical flexibility, making them a promising candidate for SOFC electrolytes. The perovskite structure with ABO_3 stoichiometry, as shown in Figure 2.26, typically consists of the A-site large cation, such as rare earth, which is 12-coordinated by the anions in the lattice and the smaller B-site cation, such as transition metal, which is 6-coordinated by the anions, forming BO_6 octahedra. The two cation sites upon which to substitute lower-valence cations leads to a wide range of possible oxygen-ion-conducting materials.

As the first perovskite-structured electrolyte, La(Ca)AlO$_3$ showed adecent performance in SOFC with the ionic conductivity of ~0.005 S cm^{-1} at 800°C. In 1994, doped lanthanum gallate (LaGaO$_3$) electrolyte was discovered by Feng and Goodenough in Texas, USA, and Ishihara and his group in Oita, Japan, and since then many researchers have focused on this type of electrolyte [75,78]. The (Sr, Mg)-doped LaGaO$_3$ electrolyte with the general composition of $La_{1-x}Sr_xGa_{1-y}Mg_yO_{3-\delta}$ (LSGM) has been widely investigated for intermediate-temperature SOFCs [79–84]. However, the extensive use of LSGM electrolytes has so far been hampered due to their high

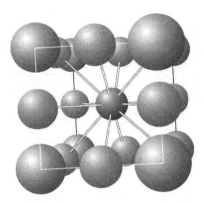

FIGURE 2.26 The perovskite structure, ABO_3, where largest spheres represent the B-site cations, centered sphere represents the A-site cation, and medium spheres represent oxygen ions. (Reprinted with permission from Ref. [78].)

reactivity with Ni on the anode, which results in the formation of the ion-insulating $LaNiO_3$ phase, the volatility of gallium at high temperatures, and the difficulty of achieving a single-phase material. The LSGM electrolyte exhibits promising ionic conductivity over a wide range of oxygen partial pressure ($10^{-20} < PO_2 < 1$). The $La_{0.9}Sr_{0.1}Ga_{0.8}Mg_{0.2}O_{3-\delta}$ electrolyte has shown higher ionic conductivities over a wide range temperature (even at a relatively low temperature) than those of the YSZ electrolyte and similar to those of the GDC at high temperatures [5,17], as shown in Figure 2.27. Although the overall conductivity of LSGM is somehow lower than that of the GDC, it is superior to GDC under low oxygen partial pressure environments since it has no reducible ion such as Ce^{4+}. It seems that the highest oxygen-ion conductivity among (Sr, Mg)-doped $LaGaO_3$ electrolytes has been obtained for the compositions of $La_{0.8}Sr_{0.2}Ga_{0.85}Mg_{0.15}O_{3-\delta}$ and $La_{0.8}Sr_{0.2}Ga_{0.8}Mg_{0.2}O_{3-\delta}$ [5,21].

In the LSGM electrolyte, doping of La sites with Sr and Ga sites with Mg occurs in order to create oxygen vacancies and improve the oxygen-ion conductivity of the electrolyte. Oxygen vacancies are created according to the following reactions:

$$2SrO + 2La_{La}^x + O_O^\times \rightarrow 2Sr_{La}' + V_O^{\bullet\bullet} + La_2O_3 \tag{2.9}$$

$$2MgO + 2Ga_{Ga}^x + O_O^\times \rightarrow 2Mg_{Ga}' + V_O^{\bullet\bullet} + Ga_2O_3 \tag{2.10}$$

As seen in Equation (2.9), the concentration of oxygen vacancies increases with an increase in Sr dopant content, resulting in higher oxygen-ion conductivity. However, the solid solubility of Sr into La sites within the $LaGaO_3$ lattice is poor. Also, for Sr contents higher than 10 mol%, secondary phases of $SrGaO_3$ or La_4SrO_7 are formed. Therefore, increasing the number of oxygen vacancies by introducing Sr dopant into La sites is limited to a low level. Hence, in order to create more oxygen vacancies and a further increase in ionic conductivity, the amount of Mg dopant increases into Ga sites, as shown in Equation (2.10), which can attain a maximum content

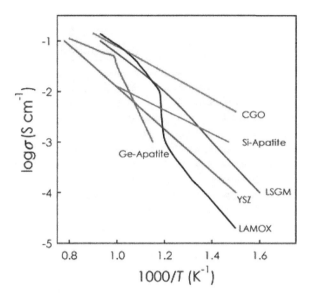

FIGURE 2.27 Total conductivities of several well-known oxygen-ion conductors as a function of inverse temperature: YSZ: $(ZrO_2)_{0.92}(Y_2O_3)_{0.08}$; CGO: $Ce_{0.8}Gd_{0.2}O_{1.9}$; LSGM: $La_{0.9}Sr_{0.1}Ga_{0.8}Mg_{0.2}O_{2.85}$; LAMOX: $La_2Mo_2O_9$; Si-apatite: $La_{10}(SiO_4)_6O_3$; and Ge-apatite: $La_{10}(GeO_4)_6O_3$. (Reprinted with permission from Ref. [17].)

of 20 mol% Mg. Due to a larger ionic radius of Mg than that of Ga, which leads to the enlarged crystal lattice of $LaGaO_3$, the solubility of Sr into $LaGaO_3$ lattice can increase up to 20 mol% in presence of Mg dopant [5,85].

Doping B sites of the LSGM lattice structure with transition metals such as iron and cobalt (i.e., $La_{1-x}Sr_x(Ga_{0.8}Mg_{0.2})_{1-y}(Co$ or $Fe)_yO_{3-\delta}$) has shown a capability of increasing the ionic conductivity. Although both dopants are effective in improving the LSGM conductivity, cobalt is more effective than iron, specifically at low temperatures [5]. Other perovskite-structured materials such as $NdGa_{0.9}Mg_{0.1}O_{2.95}$ and $NdGa_{0.9}Mg_{0.1}O_{2.95}$ also exhibit oxygen-ion conductivities comparable to that of the YSZ electrolyte [86,87]. The wide range of composition and solid solubility for the perovskite systems warrant continued research in this field.

2.2.6.2 Perovskite-Structured Proton Conductors

To operate SOFC systems at intermediate temperatures (500°C–700°C) with a metal support, significant attention has been given to proton-conducting SOFC stack with an integrated natural gas fuel processor. This operating temperature is believed to be economically optimal, considering the installed cost, the effective electric efficiency, and the projected life/degradation rate. In general, the metal-supported stack concept has been demonstrated using traditional high-temperature electrolyte materials, however, it is more ideally suited to intermediate-temperature SOFCs. Metal-supported SOFC stacks offer the potential for more automated/lower-cost manufacturing approaches (i.e., approaches consistent with the FOA system cost objective of $1500 kW^{-1}) and more rapid start-up times, while the proton-conducting electrolyte offers the potential for higher conductivity at lower temperatures than is typical of oxygen-ion-conducting solid oxide electrolyte materials. Therefore, proton-conducting electrolytes offer tremendous potential for the realization of lower temperature SOFC stacks [88].

After more than 20 years of research on proton-conducting oxide materials to meet the requirements for SOFC applications such as high dimensional and chemical stability, prevention of mixing reducing and oxidizing gases, high proton conductivity, high electronic resistivity, high thermal and chemical shock resistivity, and low cost, the perovskite-structured cerates, and zirconates (e.g., $SrCeO_3$, $BaCeO_3$, $CaZrO_3$, and $SrZrO_3$) have become well-established proton-conducting electrolytes for SOFC applications. In general, the cerate-based perovskite-type oxides show higher proton conductivity, but zirconium-based oxides are more stable, i.e., chemical and mechanical strength, than the cerate-based oxides. In the perovskite structure of these materials, as shown in Figure 2.26, protonic defects are formed by water adsorption and dissociation at the surface, needing the presence of oxygen vacancies. The ionic conductivity of these perovskite-type oxides depends on the atmospheric conditions where they should be exposed to the hydrogen-containing or water vapor atmosphere. In such an environment, during the water dissociation process, the gas-phase water dissociates to a proton, and a hydroxide ion where the oxygen vacancy is then filled with the hydroxide ion and the proton forms a covalent bond with lattice oxygen [17]. This reaction can be written by the Kröger–Vink notation:

$$H_2O_{(g)} + V_O^{\cdot\cdot} + O_O^{\times} \rightarrow 2OH_O^{\cdot} \qquad (2.11)$$

Two most widely investigated proton-conducting oxides are strontium cerates ($SrCeO_3$) and barium cerates ($BaCeO_3$), discovered by Iwahara and his group in Japan in the early 1800s [81,89]. However, pure crates and zirconates exhibit low proton conductivity unless doped with subvalent cations. The general compositions for doped $BaCeO_3$ and $SrCeO_3$ oxides can be written as $BaCe_{1-x}M_xO_{3-\delta}$ and $SrCe_{1-x}M_xO_{3-\delta}$, respectively. For example, in the $BaCeO_3$ system, Ce^{4+} ions are replaced by M^{3+} cations (typically Y^{3+}), forming oxygen vacancies:

$$M_2O_3 + 2Ce_{Ce}^{\times} + O_O^{\times} \rightarrow 2M_{Ce}' + V_O^{\cdot\cdot} + 2CeO_2 \qquad (2.12)$$

The energy of water incorporation in such systems is exothermic for both doped and pure cerates and zirconates. However, this is more exothermic in doped systems than the undoped systems. Figure 2.28 shows the total conductivities of several well-known proton conductors at different temperatures.

Among barium cerate electrolytes, yttrium-doped barium cerate (BCO), especially 20 mol% yttrium-doped barium cerate ($BaCe_{0.8}Y_{0.2}O_{3-\delta}$), has shown the best performance so far [17]. Despite many advantages of barium cerates electrolytes, the main problems in the use of these materials are their poor chemical stability and the formation of electronic species under reducing environments. Particularly, in hydrocarbon-fueled SOFCs below 800°C, barium cerate electrolytes react with CO_2 and decompose to cerium oxide and barium carbonate. The barium zirconate

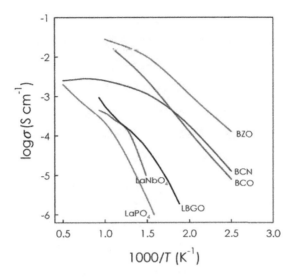

FIGURE 2.28 Total conductivities of several well-known proton conductors as a function of inverse temperature: BCO: $BaCe_{0.9}Y_{0.1}O_3$; BZO: $BaZr_{0.8}Y_{0.2}O_3$; LaPO$_4$: $La_{0.9}Sr_{0.1}PO_4$; BCN: $BaCa_{1.18}Nb_{1.82}O_{8.73}$; LaNbO$_4$: $La_{0.99}Ca_{0.01}NbO_4$; and LBGO: $La_{0.8}Ba_{1.2}GaO_{3.9}$. (Reprinted with permission from Ref. [17].)

electrolytes are more chemically stable than the barium cerate electrolytes [90,91]. However, perovskite-structured cerates and zirconates are both soluble. Therefore, developing cerates/zirconates solid electrolytes with good chemical stability and adequate proton conductivity under SOFC operating conditions are being investigated. Co-doping or double doping with several dopants such as Ti, Sm, In, Sn, Gd, Y, Ta, Nb, Ca, and Pr may be able to improve the stability of doped barium creates and zirconates without affecting their conductivities [89].

Yttrium-doped barium zirconate (BZO), especially 20 mol% yttrium-doped barium zirconate ($BaZr_{0.8}Y_{0.2}O_{3-\delta}$), has shown promising proton conductivity (see Figure 2.28) with low activation energy in intermediate temperature range and good chemical stability under CO_2-containing atmosphere among barium zirconate electrolytes. However, the proton conductivity of the BZO electrolytes can widely vary with the fabrication procedure and sintering temperature. The BZO electrolytes can possess low densities and poor conductivities, despite the use of high sintering temperatures (>1600°C) or sintering additives. Also, the Y_2O_3 secondary phase may segregate to the grain boundaries. For example, Yamazaki et al. [92] showed that through suitable control of the calcination step in the fabrication process of the nanocrystalline $BaZr_{0.8}Y_{0.2}O_{3-\delta}$ (sol-gel), the total conductivity of 0.01 S cm^{-2} at 450°C could be attained. In order to study the capability of the BZO electrolytes for low-temperature SOFC applications, the area-specific resistances (ASRs) of 15 μm thick electrolyte layers of the 8 mol% YSZ (8YSZ) and the $BaZr_{0.8}Y_{0.2}O_{3-\delta}$ were investigated. The results revealed that the ASR of the $BaZr_{0.8}Y_{0.2}O_{3-\delta}$ layer at 500°C is comparable to that of the 8YSZ at the typical oxygen-ion-conducting SOFC operating temperature of 800°C (~0.1 ohm cm^2) [93].

Another strategy for achieving high proton-conducting perovskites is to fabricate cation-off stoichiometric perovskites where the non-stoichiometric structure causes charge imbalance, which is compensated by protons. $Ba_3Ce_{1.18}Nb_{1.82}O_{8.73}$ is the most well-known material in this group. $(Ba_{1-x}La_x)_2In_2O_{5+x}$ systems have also shown significant proton conductivity due to the presence of intrinsic oxygen vacancies in the structure. The proton conductivity can increase with the La content up to 1.12×10^{-5} S cm^{-1} for $x = 0.1$ at 400°C [94]. Investigations on these systems have recently begun and offer many opportunities for developing new electrolytes with remarkable proton conductivity.

2.2.7 NEW OXYGEN-ION AND PROTON CONDUCTORS

As discussed, although the conventional oxygen-ion and proton-conducting electrolytes have shown good conductivity and some of them are already practical, several drawbacks such as partial electronic conductivity, high operating temperatures, and poor chemical and mechanical stabilities hamper their widespread SOFC applications. Therefore, researchers have been motivated to develop new electrolyte materials with improved properties that can operate in the temperature range of 450°C–650°C. However, there are still many avenues to explore further improvement in the properties. Some of these new material systems will be discussed in the following.

2.2.7.1 Silicate- and Germanate-Based Apatites

Apatite-type oxides have been considered as alternative materials for SOFC electrolytes. Silicate-based apatites with the chemical composition of $Ln_{9.33}(SiO_4)_6O_2$ (where Ln is typically a large lanthanide ion, e.g., La^{3+} or Nd^{3+}) were first discovered as fast oxygen-ion conductive materials in the mid-1990s [95,96]. The apatite structure of $Ln_{9.33}(SiO_4)_6O_2$ is shown in Figure 2.29. Since then, many studies have sought to investigate the structures and properties of these apatite oxides, specifically Si- and Ge-based apatite systems. For the apatite oxides with the typically hexagonal crystal structure, the general composition can be written as $M_{10}(XO_4)_6O_{2\pm y}$, where M is alkaline- or rare-earth cation, y is the amount of oxygen non-stoichiometry, and X is a p-block element, e.g., Si, Ge, and P [17]. So far, among the apatite-type oxides, Si- and Ge-based lanthanum apatite electrolytes have shown the highest oxygen-ion conductivity.

In the Si- and Ge-based lanthanum apatite systems, the interstitial oxygen ions, leading to interstitial-type transport mechanisms, are the key reasons for high ionic conductivity. This is in contrast with the oxygen-ion-conducting mechanism in the conventional fluorite- and perovskite-type oxygen-ion conductors where oxygen vacancies are the key defects for ionic conductivity, as previously discussed. The conductivities for the Si- and Ge-based apatite electrolytes are comparable to other common SOFC electrolytes (Figure 2.27). The maximum amount of interstitial oxygen which can be accommodated in the crystal structure is higher for the silicate than that of the germinate [97]. It has been established that as extra oxygen is introduced into the structure of rare-earth apatite silicates, the conductivity increases. For instance, the conductivity of $La_{9.33+x}(SiO_4)_6O_{2+3x/2}$ with hexagonal symmetry increased

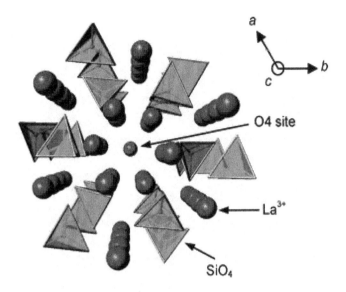

FIGURE 2.29 Apatite structure of $La_{9.33}(SiO_4)_6O_2$ viewed down the c axis showing SiO_4 tetrahedra, rows of lanthanum ions and the La/O4-containing channels. (Reprinted with permission from Ref. [17].)

from 1.1×10^{-4} S cm^{-1} for $x = 0$ to 1.3×10^{-3} S cm^{-1} for $x = 0.34$ at 500°C [98]. On the contrary, for the rare-earth apatite germinates, $La_{9.33 + x}(GeO_4)_6O_{2+3x/2}$ with triclinic symmetry, excess oxygen leads to formation of stronger oxygen bonds in lower symmetry, thereby a decrease in conductivity [99]. To stabilize the hexagonal structure while introducing a high oxygen content to enhance the conductivity, doping yttrium to form Y-doped Ge-based lanthanum apatite has been instrumental [100].

Doping Si- and Ge-based lanthanum apatite electrolytes, especially on the Si/Ge sites, is another effective way to improve the ionic conductivity. The structure of apatite shows great flexibility to accommodate a wide range of cation dopants on the both Si/Ge and La sites. This is even significantly wider than the fastest oxygen-ion conductors. Lower valent ions, especially smaller cations such as Mg^{2+}, have shown to increase the conductivity [97].

2.2.7.2 $La_2Mo_2O_9$ (LAMOX)

In 2000, the $La_2Mo_2O_9$ (LAMOX) compound was first reported as a new fast oxygen-ion conductor. In the LAMOX system, the key phase transition temperature is around 580°C, where the non-conductive α phase with a monoclinic crystal structure transforms into the highly conductive β phase with a cubic crystal structure. Pure LAMOX electrolyte has also possessed an ionic conductivity of 6×10^{-2} S cm^{-1} at 800°C. The structure of the cubic β phase of LAMOX is presented in Figure 2.30. As observed, the [La-Mo-O1] tetrahedral units are surrounded by O2 and O3 sites. The [La-Mo-O1] framework also contains O2 and O3 sites located within the channels parallel to the unit cell axes. The conductivities for the pure LAMOX electrolyte at different temperatures are shown in Figure 2.27 [101,102].

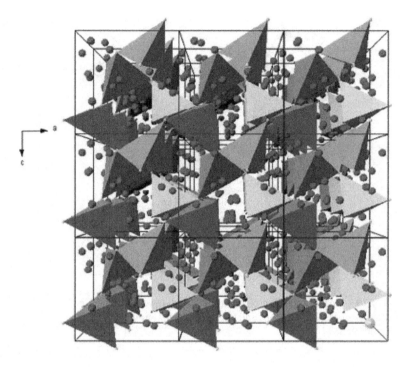

FIGURE 2.30 β-La$_2$Mo$_2$O$_9$ as a rigid framework of [La-Mo-O1] with O2 and O3 oxide ions sites embedded within its channels. (Reprinted with permission from Ref. [17].)

Although pure LAMOX in the cubic structure form has shown a high oxygen-ion conductivity, it still has some challenging drawbacks to be practical for SOFC electrolyte applications. The main problems are its low chemical stability under the reducing environment and the structural phase transition at 580°C, which results in significant volume strains. Hence, in order to stabilize the cubic β phase at room temperature without a remarkable change in conductivity, pure LAMOX has been doped with rare-earth (e.g., Nd, Gd, and Y), alkaline-earth (e.g., Ca, Sr, and Ba), and alkaline (e.g., Rb and K) cations on the La sites [102–105]. However, a systematic study on the β phase stability using various dopants for the LAMOX electrolyte is still required.

2.2.7.3 Gallium-Based Oxides (Tetrahedrally Coordinated)

Recently, lanthanum gallates with the composition of La$_{1-x}$Ba$_{1+x}$GaO$_{4-x/2}$, containing GaO$_4$ tetrahedral units, have displayed high oxygen-ion conductivity and even more significant proton conductivity (especially below 700°C) [106,107] which have made them a good candidate for SOFC electrolyte applications. Figure 2.31 shows the crystal structure of the stoichiometric compound LaBaGaO$_4$ with GaO$_4$ distorted tetrahedral. The oxygen vacancy sites are formed in the lattice structure by variation of La/Ba ratio (changing x in the general composition). The oxygen-ion migration in this compound mainly occurs between the tetrahedral units, facilitating long-range diffusion. The proton migration is also equally favorable between GaO$_4$ units within

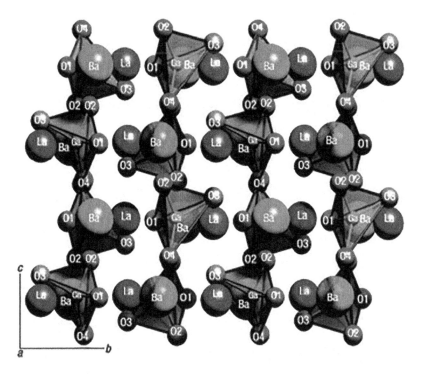

FIGURE 2.31 Crystal structure of LaBaGaO$_4$ with GaO$_4$ distorted tetrahedral and ordered layers of Ba and La. (Reprinted with permission from Ref. [17].)

a layer and between layers. The rate-determining step for long-range proton diffusion is intra-tetrahedral migration [17,106].

Li et al. [107] investigated the formation of oxygen vacancies and oxygen excess with La/Ba ratio variation in La$_{1-x}$Ba$_{1+x}$GaO$_{4-x/2}$ electrolyte. The results showed that an increase in the La content above 1, thus, introducing oxygen excess, causes impurities formation, and a decrease in conductivity. In contrast, lowering the La content ($0 \le x \le 2$) results in oxygen vacancy formation and an increase in conductivity. Figure 2.32 shows that although the conductivities of the undoped phase LaBaGaO4 ($x = 0$) are comparatively low, La$_{1-x}$Ba$_{1+x}$GaO$_{4-x/2}$ possesses significantly higher conductivities for $x > 0$. Typical proton conductivities reported for the composition of La$_{0.8}$Ba$_{1.2}$GaO$_{3.9}$ are among the highest conductivities measured at intermediate temperatures.

La$_{1+x}$Sr$_{1-x}$Ga$_3$O$_{7+x/2}$-based systems with melilite structure have been introduced as alternative oxygen-ion-conducting electrolyte materials due to their high ionic conductivity. Structural investigations have revealed that the interstitial oxygen sites in the crystal structure are responsible for ionic conductivity of the La$_{1+x}$Sr$_{1-x}$Ga$_3$O$_{7+x/2}$. The previous study showed that the conductivity of the parent compound, LaSrGa$_3$O$_7$ ($x = 0$), is relatively low. However, it increases with an increase in La/Sr ratio due to introducing oxygen excess to maintain charge balance [108]. Figure 2.33 shows the variation of conductivity and activation energy with La/Sr ratio (varying x in the general composition). The melilite structure of LaSrGa$_3$O$_7$ is also shown in

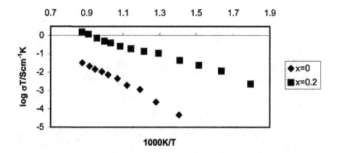

FIGURE 2.32   Conductivity of $La_{1-x}Ba_{1+x}GaO_{4-x/2}$ in air. (Reprinted with permission from Ref. [107].)

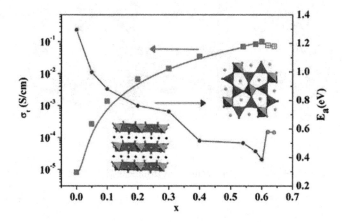

FIGURE 2.33   Variation of conductivity and activation energy with La/Sr ratio (x) in $La_{1+x}Sr_{1-x}Ga_3O_{7+x/2}$-based systems at 800°C. (Reprinted with permission from Ref. [108].)

Figure 2.34. The tetragonal symmetry of $LaSrGa_3O_7$ is composed of La and Sr positioned between GaO_4 units, which are linked to form distorted pentagonal rings [17,108].

2.2.7.4   Niobates and Tantalates

Haugsrud and Norby [109] studied several rare-earth-doped ortho-niobates and ortho-tantalates with the general composition of $RE_{1-x}A_xMO_4$ (where RE = La, Nd, Gd, Y, Er; M = Ta or Nb; A = Ba, Sr, Ca; x = 0.01–0.05) as alternative stable proton-conducting materials. However, this class of materials shows mixed protonic, native ionic, and electronic conduction depending on conditions. The proton conductivity is dominant in wet conditions below 800°C. The $RE_{1-x}A_xMO_4$ materials have a monoclinic fergusonite-type structure at low temperatures and a tetragonal scheelite-type structure at high temperatures which both show proton conduction. These two crystal structures are presented in Figure 2.35. The phase transition temperature depends on the chemical composition of $RE_{1-x}A_xMO_4$ [109].

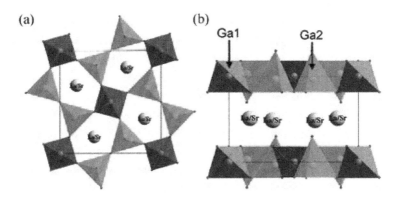

FIGURE 2.34 Melilite structure of LaSrGa₃O₇ showing two types of GaO₄ tetrahedra: (a) c axis view, showing Ga tetrahedra connected through bridging oxygen sites to form distorted pentagonal rings; (b) b axis view, showing the layered structure with interstitial oxygen sites in the center of the Ga layers with La/Sr cations. (Reprinted with permission from Ref. [17].)

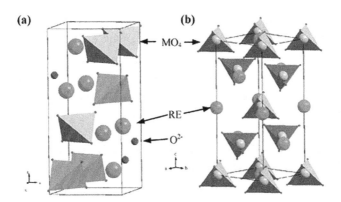

FIGURE 2.35 (a) Monoclinic and (b) tetragonal structures of ortho-niobates and ortho-tantalates REMO₄. (Reprinted with permission from Ref. [17].)

Among the chemical compositions of this class of materials investigated so far, $La_{0.99}Ca_{0.01}NbO_{4-\delta}$ has the highest conductivity of approximately 10^{-3} S cm^{-1} at 800°C–1000°C, as shown in Figure 2.36. This is due to the larger lattice volume of La-containing niobates compared to other RE-containing phases, leading to higher proton mobility. As seen, the proton conductivity also decreases with a decrease in the radius of rare-earth cations.

Although the conductivities of this class of materials are still lower than those of reported for perovskite-structured cerates, they exhibit the highest proton conductivity among the Ba- and Sr-free oxides (Ba and Sr as the main components). Therefore, $RE_{1-x}A_xMO_4$ compounds can be considered as an interesting candidate for thin-film electrolytes used in SOFCs that are capable of operating in CO_2-containing environments. However, it is still needed to consider different dopants to improve the proton conductivity of these materials [109].

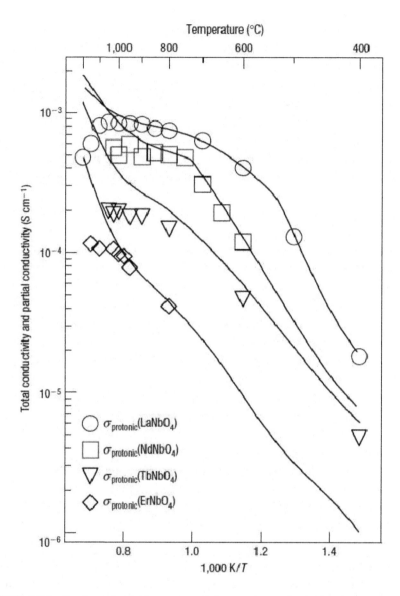

FIGURE 2.36 Total conductivities and partial protonic conductivities for selected 1% Ca-doped $RENbO_4$ niobates in wet H_2 atmosphere at different temperatures. (Reprinted with permission from Ref. [109].)

2.3 ANODES

The primary function of the anode in a SOFC system is to promote the electrochemical fuel oxidation reaction. In addition, the anode may promote internal reforming or partial oxidation of the hydrocarbon fuels such as methane. The electrochemical or chemical reactions typically occur at preferential surfaces or triple-phase boundaries (TPBs), which are defined as the collection sites where the electron-conducting anode metal phase, the electrolyte, and the gas phase all come together. The capability of anode electrode to promote these electrochemical reactions is determined by intrinsic anode surface catalytic activities and multiple key factors such as microstructure, morphology, and transport properties of the electrode materials. The design of an efficient anode with optimized mass and charge-transfer properties from the surface through the bulk electrode has been challenging for decades. For an anode-supported SOFC, the anode mechanical strength is also important since it acts as a cell structural support.

The general requirements for the SOFC anode material during cell operation can be listed as: excellent catalytic activity for fuel oxidation reactions, high electronic conductivity to transport electrons from the reaction sites to the interconnect, sufficient porosity to allow the fuel and byproducts to be delivered and removed from the reaction sites, high chemical and thermal stability, compatible thermal expansion coefficient with the adjacent cell components, good mechanical strength, sufficient ionic conductivity, tolerance to carbon deposition, sulfur poisoning and re-oxidation, low cost, and ease of fabrication.

In the early stage of SOFC development, various materials have been studied to be used as the anode material such as iron, platinum, cobalt, and nickel. Among those metals, Ni has shown the best performance due to its excellent catalytic activity for the hydrogen oxidation reaction and hydrocarbon fuel reforming, high chemical stability in SOFC operation environment, and low cost. The reason for non-acceptance of other metals may mainly be due to the effect of toxicity, corrosion, availability, and cost. However, pure Ni is not a good candidate for the SOFC anode because it suffers from significant thermal expansion mismatch and poor binding with the YSZ electrolyte and considerable coarsening during cell operation. Pure Ni metal has a high thermal expansion coefficient (16.9×10^{-6} K^{-1}), compared to that of the YSZ electrolyte (10.5×10^{-6} K^{-1}). In 1970, Spacil [110] developed an Ni-YSZ cermet material which interestingly meets most of the requirements for the SOFC anode. In a porous Ni-YSZ cermet anode, the Ni metal provides the electronic conductivity and catalytic activity for fuel oxidation reaction, while the YSZ ceramic phase controls the thermal expansion coefficient mismatch between the anode and the electrolyte (i.e., the thermal expansion coefficient of the Ni-YSZ cermet is reduced to 12.7×10^{-6} K^{-1} for 40 vol% Ni + 60 vol% YSZ compound) and prevents Ni phase coarsening and aggregation during cell operation. It also extends active regions for the anode oxidation reactions by offering oxygen-ion conduction. The mechanical strength of the Ni-YSZ cermet has been reported to be less than 100 MPa [75]. Even though many alternative materials have been proposed for the SOFC anode application, the porous Ni-YSZ cermet anodes are still extensively used in SOFCs. Table 2.6 presents some of the most important anode materials that are currently used. With the development of the low-temperature electrolytes, other anodes, viewed as the "next-generation" technology, are currently undergoing commercialization for low- and intermediate-temperature SOFCs. For example, Goodenough and co-workers have investigated using mixed-conducting oxides as the active components in SOFC anodes. One such

TABLE 2.6

Some Selected Materials Developed as the Anode for SOFC Application

Materials	Conductivity (S cm^{-1})	Advantage/Disadvantage
Ni-YSZ	250	High operating temperature
Ni-SDC	573	Coke formation
Ni-GDC	1070	Coke formation and electronic performance degradation
$La_{0.8}Sr_{0.2}Fe_{0.8}Cr_{0.2}O_3$	0.5	Low conductivity
$La_{0.8}Sr_{0.2}Cr_{0.95}Ru_{0.05}O_3$	0.6	Expensive
$(La_{0.7}Sr_{0.3})_{1-x}Ce_xCr_{1-x}Ni_xO_3$	5.03	Carbon deposition
$Sr_{0.88}Y_{0.08}TiO_3$	64	High operating temperature
$LaSrTiO_2$	360	Poor compatibility
$CrTi_2O_5$	177	Expensive
$Cu-CeO_2$	5200	Improved electronic conductivity
$Cu-GDCCrTi_2O_5$	8500	Good thermal expansion and electronic performance

Source: Data adapted and modified from Ref. [19].

material that they have identified that shows a high performance and durability are double perovskites of composition $Sr_2MgMoO_{6-\delta}$ (SMMO) [111]. The ability of these materials to support a high concentration of oxygen-ion vacancies makes them good oxygen-ion conductors, and the mixed valence of the cations makes them good n-type electronic conductors. In the last few years, the SOFCs operating at lower temperatures have seen considerable research and development, and anodes were a combination of Ni and perovskites [112]

2.3.1 NI-YSZ CERMET ANODE

Among metal-fluorite cermets, the cermet of Ni and YSZ has attracted much attention during past decades and has mostly been used as the anode material for the SOFC application [113–117]. This material is not only inexpensive but also catalytic active for the anode hydrogen oxidation reaction at the SOFC operating temperature. In addition, this material has shown excellent long-term stability (up to several thousand hours) under SOFC operating conditions [113]. To be precise, the following characteristics of the Ni-YSZ cermet make this material attractive:

- Ni is an excellent catalyst for the hydrogen oxidation reaction with high electronic conductivity ($\sim 12.7 \times 10^{-4}$ S cm^{-1} at 1000°C);
- YSZ constitutes a framework for Ni dispersion, inhibits Ni coarsening during the cell operation, and offers a significant contribution for ion conduction to the overall conductivity, thus, effectively broadens the TPB region;
- Chemically stable in reducing environments at high temperatures;

- Thermal expansion coefficient of Ni-YSZ anode can be matched with other adjacent SOFC components by varying the composition;
- Intrinsic charge transfer resistance due to low catalytic activity at the Ni/YSZ boundary;
- Ni and YSZ are essentially non-reactive with each other over a wide range of temperature [23].

The operation of the Ni-YSZ cermet anode can be defined by the species adsorption, reaction, and species transport process occurring at TPBs, while considering the effect of the microstructure, especially the density of TPB lines within the anode. The hydrogen oxidation reaction at or near the TPB can be written as:

$$H_2(g) + O_O^x(YSZ) = H_2O(g) + 2e^-(Ni) + V_O^{\bullet\bullet}(YSZ) \tag{2.13}$$

which can be divided into several steps. In some cases, it has been assumed that only charge transfer happens at the TPB with no species transport, while in the most widely accepted case, a "spillover" mechanism has been assumed where hydrogen, oxygen, or hydroxyl is transported across the Ni/YSZ surface [76]. Figure 2.37 shows the hydrogen spillover mechanism at the TPB of the Ni-YSZ anode.

Zhu et al. [118] have described the proposed hydrogen spillover mechanism which includes five elementary reactions at the Ni-YSZ TPB:

1. H_2 adsorption/desorption on the Ni surface

$$H_2(g) + 2(Ni) = 2H(Ni) \tag{2.14}$$

2. Charge-transfer reactions at the TPB

$$H(Ni) + O^{2-}(YSZ) = (Ni) + OH^-(YSZ) + e^-(Ni) \tag{2.15}$$

$$H(Ni) + OH^-(YSZ) = (Ni) + H_2O(YSZ) + e^-(Ni) \tag{2.16}$$

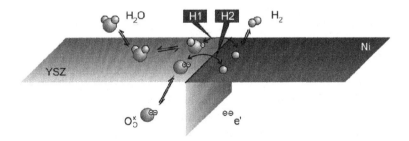

FIGURE 2.37 Hydrogen oxidation reaction at the TPB of the Ni-YSZ anode via the proposed hydrogen "spillover" mechanism. (Reprinted with permission from Ref. [76].)

3. H_2O desorption/adsorption on the YSZ surface

$$H_2O(YSZ) = H_2O(g) + (YSZ) \qquad (2.17)$$

4. Oxygen ions transfer between the surface and the bulk YSZ

$$O_O^x(YSZ) + (YSZ) = O^{2-}(YSZ) + V_O^{\bullet\bullet}(YSZ) \qquad (2.18)$$

where H(Ni) is an adsorbed atomic hydrogen on the Ni surface, (Ni) is a vacant Ni surface site, e^-(Ni) is an electron within Ni, $O_O^x(YSZ)$ is lattice oxygen, and $V_O^{\bullet\bullet}(YSZ)$ is an oxygen vacancy. $O^{2-}(YSZ)$, $H_2O(YSZ)$, and $OH^-(YSZ)$ are adsorbed species on the YSZ surface, and (YSZ) is a vacant YSZ surface site. The above mechanism has also been evaluated by other researchers based on the data achieved from the modeled Ni-YSZ electrodes [119,120].

The electrochemical performance and the electrical and mechanical properties of the Ni-YSZ anode strongly depend on its raw (starting) material microstructural characteristics, composition, and porosity. The typical processing method for the fabrication of Ni-YSZ cermet anodes with regard to the conventional ceramic powder mixing process includes the following steps: (1) homogeneous mixing of NiO and YSZ powders with appropriate sizes and compositions. The particle size of NiO powders for Ni-YSZ synthesis is usually ~1 µm, while that for YSZ powders varies from 0.2 to 0.3 µm. The Ni:YSZ volume ratio typically varies between 35:65 and 55:45. Also, a pore former is added to achieve sufficient porosity in the anode/anode support layer and to tailor the shrinkage (for the co-firing); (2) firing of anodes at high temperatures; and (3) reduction of NiO-YSZ into Ni-YSZ during SOFC operation or external reduction in the hydrogen/nitrogen mixture atmosphere at SOFC operating temperature (600°C–1000°C) which results in about ~41% volume reduction and pore formation. The amount of porosity created during NiO to Ni reduction process depends on the composition of the cermet. The porosity increases with increasing NiO content. Moreover, a negligible bulk dimension change occurs by the reduction of NiO-YSZ compound to the Ni-YSZ cermet anode when vol% of YSZ is larger than that of the NiO. The reduction in solid volume following to the NiO to Ni conversion appears as an increase in the volume fraction of porosity [5,23].

2.3.1.1   Influence of Materials Characteristics and Fabrication Conditions on Electronic Conductivity of Ni-YSZ Cermet Anode

As mentioned above, the electronic conductivity of the Ni-YSZ cermet anode depends on its composition (Ni:YSZ volume ratio), microstructural characteristics of raw materials such as particle size and mixing and distribution of NiO and YSZ powders, and sintering/reducing temperature and atmosphere. The electronic conductivity of a porous Ni-YSZ cermet anode usually varies from 0.1 to 1000 S cm^{-1}. This is due to a higher conductivity of Ni (more than 5 orders of magnitude) than that of the YSZ under SOFC operating conditions. Figure 2.38 shows the electronic conductivities of two types of Ni-YSZ cermets as a function of Ni volume content synthesized using NiO powder with 0.25 µm particle size and 3.5 m^{2}g^{-1} specific surface area, and

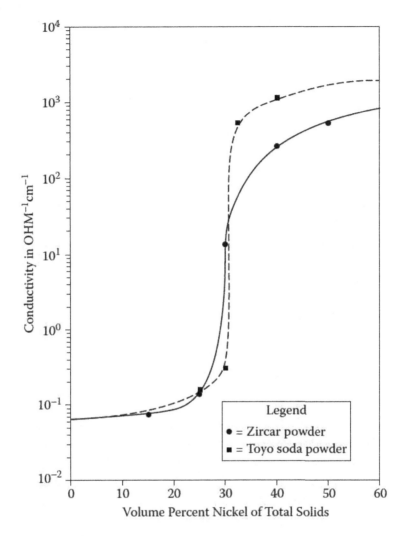

FIGURE 2.38 Electronic conductivities of Ni-YSZ cermets vs. Ni content at 1000°C. Two types of YSZ were used to synthesize the Ni-YSZ cermets: Toyo Soda powder (specific surface area: $23\,m^2g^{-1}$; particle size: 0.3 μm), and Zircar powder (specific surface area: $47\,m^2g^{-1}$; particle size: 0.1 μm). (Reprinted with permission from Ref. [5].)

two types of YSZ powders with 0.1 and 0.3 μm particle sizes, respectively. As can be observed, the electronic conductivity of Ni-YSZ cermet anode versus Ni volume content follows an S-shaped trend where the conductivity increases from 0.1 to 100–1000 S cm^{-1} with an increase in the Ni content from ~25 to ~35 vol% [5].

In general, the electronic conductivity of the Ni-YSZ cermet anode increases with a decrease in the NiO particle size. Figure 2.39 illustrates the change in the electronic conductivity of the Ni-YSZ cermets with regard to the NiO particle size.

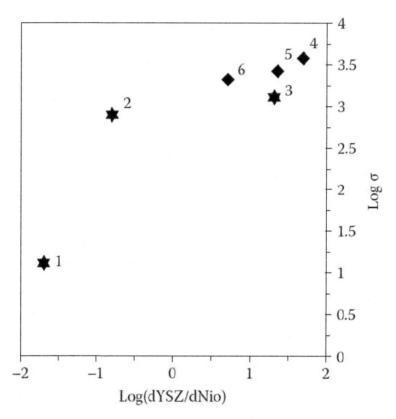

FIGURE 2.39 Electronic conductivity of six Ni-YSZ cermets vs. the YSZ to NiO particle size ratio (d_{YSZ}/d_{NiO}). The corresponding YSZ and NiO particle sizes are: 1 ($d_{YSZ} = 0.3\,\mu m$, $d_{NiO} = 16\,\mu m$), 2 ($d_{YSZ} = 0.3\,\mu m$, $d_{NiO} = 1.8\,\mu m$), 3 ($d_{YSZ} = 37\,\mu m$, $d_{NiO} = 1.8\,\mu m$), 4 ($d_{YSZ} = 50\,\mu m$, $d_{NiO} = 1\,\mu m$), 5 ($d_{YSZ} = 23\,\mu m$, $d_{NiO} = 1\,\mu m$), and 6 ($d_{YSZ} = 5\,\mu m$, $d_{NiO} = 1\,\mu m$). Samples 1–3 have a Ni:YSZ volume ratio of 55:45 and samples 4–6 have a Ni:YSZ volume ratio of 60:40. (Reprinted with permission from Ref. [5].)

Comparing samples 1 and 2, the cermet electronic conductivity increases from ~10 to ~1000 S cm^{-1} as the NiO particle size decreases from 16 to 1.8 μm for the same Ni:YSZ volume ratio and the same YSZ powder with a particle size of 0.3 μm. Similarly, the electronic conductivity of the Ni-YSZ cermet changes with the YSZ particle size. Figure 2.40 shows that the electronic conductivity of the cermet increases with an increase in the YSZ particle size when other parameters (NiO particle size and Ni:YSZ volume fraction) remain constant due to its impact on the NiO clustering [5].

In general, small particles tend to agglomerate around the large particles. Therefore, if the YSZ particles are larger than the NiO particles, it is expected to achieve high electronic conductivity. On the contrary, if the NiO particles are larger than the YSZ particles, the electronic conductivity would be lower due to the

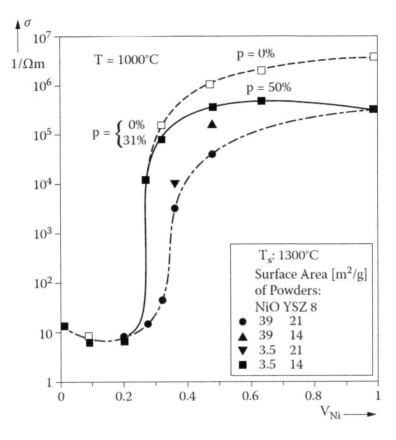

FIGURE 2.40 Electronic conductivity of the various synthesized Ni-YSZ cermets using different precursor materials as a function of Ni volume fraction. (Reprinted with permission from Ref. [5].)

agglomeration of small YSZ particles around the larger Ni particle, which electrically isolates the Ni particle.

It should be noted that the fabrication of the Ni-YSZ cermets is determined by the size of agglomerates formed during the synthesis process, not by the size of primary particles. As observed in Figure 2.40, the electronic conductivity of the Ni-YSZ cermets increases with a reduction in specific surface area of primary NiO powder from 39 to $3.5\,m^2g^{-1}$ with NiO primary particle sizes of 0.02 and 0.25 µm, respectively. This might seem in contrast with the earlier statement that the electronic conductivity of the Ni-YSZ cermets increases with a decrease in the NiO particle size (increase in specific surface area). This happens because the specific surface area measured by the Brunauer–Emmett–Teller (BET) technique may not reflect the degree of agglomeration and ultrafine particles on the order of 0.02 µm typically form agglomerates during the synthesis process. These agglomerates are hard to be broken during the normal mixing process. Therefore, it is critical to characterize the average particle sizes and the particle size distributions of raw materials rather than the BET surface area of the mixed particles.

In pursuit of achieving higher electronic conductivity for the Ni-YSZ cermets, it is important to not use very coarse YSZ powders. Using very large YSZ particles can lead to the poor mechanical strength of the cermet and deteriorate the effectiveness of YSZ on preventing Ni particles from coarsening. For example, Yu et al. [121] showed that the mechanical strength of the Ni-YSZ cermet anode with coarse YSZ particles (~8 μm) was about 20%–30% of the cermet made with fine YSZ particles (~0.8 μm) and with similar NiO particle size (~0.8 μm) and almost similar porosity, as presented in Figure 2.41.

The electronic conductivity of the Ni-YSZ cermet anode also varies with porosity. The bulk conductivity of a fully dense cermet, σ_b, can be calculated using the Bruggeman equation:

$$\sigma_e = \sigma_b \left(1 - f_p\right)^{3/2} \tag{2.19}$$

where σ_e is the effective conductivity that changes with porosity with a given composition (Ni:YSZ ratio), and f_p is the porosity of a porous cermet anode [122]. According to the Equation (2.19), the effective conductivity increases with a decrease in the

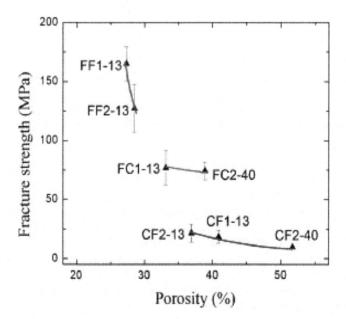

FIGURE 2.41 Fracture strength of Ni-YSZ cermets as a function of porosity. Standard deviation is superimposed on each average value. The corresponding YSZ and NiO particle sizes are: FF1-13 ($d_{YSZ} = 0.8$ μm, $d_{NiO} = 0.8$ μm), FF2-13 ($d_{YSZ} = 0.8$ μm, $d_{NiO} = 0.8$ μm), FC1-13 ($d_{YSZ} = 6$ μm, $d_{NiO} = 0.8$ μm), FC2-40 ($d_{YSZ} = 6$ μm, $d_{NiO} = 0.8$ μm), CF1-13 ($d_{YSZ} = 0.8$ μm, $d_{NiO} = 8$ μm), CF2-13 ($d_{YSZ} = 0.8$ μm, $d_{NiO} = 8$ μm), and CF2-40 ($d_{YSZ} = 0.8$ μm, $d_{NiO} = 8$ μm). (Reprinted with permission from Ref. [121].)

electrode porosity. This is also shown in Figure 2.42 for the Ni-YSZ cermets with a Ni:YSZ volume ratio of 40:60.

In addition to composition, the particle size of raw materials and porosity, the electronic conductivity of the Ni-YSZ cermet anode can strongly be affected by the synthesis parameters such as sintering and reduction conditions. It has been shown that the electronic conductivity of the Ni-YSZ anodes increases with sintering temperature in the range of 1200°C–1350°C and sintering time. Change in sintering temperature and time will affect the porosity of the anode, resulting in electronic conductivity variation. The porosity of the anode decreases with increasing sintering temperature or sintering time, leading to higher effective electronic conductivity (see Figures 2.42–2.44).

The anode reduction condition (e.g., temperature and atmosphere) was also found to be an effective factor for the electronic conductivity of the Ni-YSZ cermets. The anode reduction condition somehow determines whether the fabricated anode is conductive or insulating. Grahl-Madsen et al. [123] showed that the high-temperature electronic conductivity of the Ni-YSZ anodes reduced at 1000°C was ~2–4 times higher than those reduced at 800°C. In addition, the room-temperature electronic conductivity of the cermets decreased from 6000 to 1000 S cm^{-1} with a reduction in sintering temperature from 1000°C to 650°C. This may attribute to a change in morphology with reduction temperature. The anode reduced at 800°C revealed many crevices between the YSZ and the Ni phase with YSZ network surrounding islands of Ni particles, whereas the anode reduced at 1000°C showed a uniform interface with close contact between the YSZ and the Ni phase with a continuous Ni phase

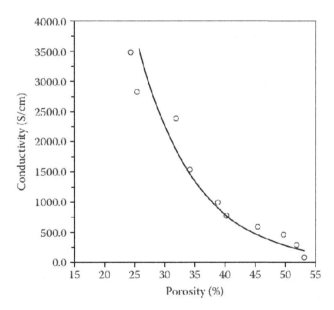

FIGURE 2.42 Electronic conductivities of the Ni-YSZ cermet anodes with a Ni:YSZ volume ratio of 40:60 as a function of electrode porosity. (Reprinted with permission from Ref. [5].)

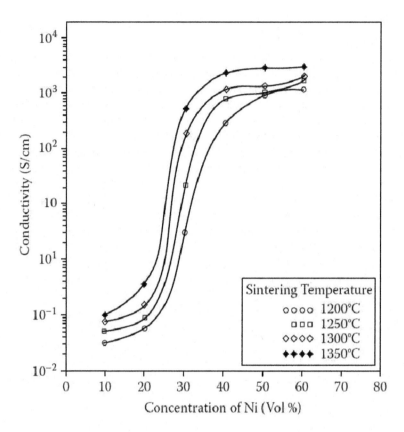

FIGURE 2.43 Electronic conductivities measured at 1000°C for the Ni-YSZ cermet anodes sintered for 2h at 1200°C, 1250°C, 1300°C, and 1350°C, respectively, as a function of Ni vol%. (Reprinted with permission from Ref. [5].)

(see Figure 2.45). Furthermore, it should be mentioned that the anode reduction at low temperatures (e.g., 200°C–400°C) not only lowers the electronic conductivity but also decreases the mechanical strength [123]. Wang et al. [124] discussed that the lower mechanical strength of the Ni-YSZ anode reduced at lower temperatures is attributed to the microcracks in the YSZ framework formed as a result of NiO conversion to Ni during the subsequent heating process since the Ni has a higher thermal expansion coefficient than that of the YSZ.

In the other study, it has been found that if hydrogen is introduced into the anode chamber at a high temperature (e.g., 750°C) after heating up the NiO-YSZ composite in air or nitrogen atmosphere to that temperature, the resulting Ni-YSZ cermet will have extremely high electronic conductivity. However, if hydrogen is introduced from room temperature when the composite is being heated to the high temperature, the reduction will occur at ~300°C–400°C, and the resulting Ni-YSZ cermet will be almost insulating at room temperature, even though NiO has fully been reduced into pure Ni [5].

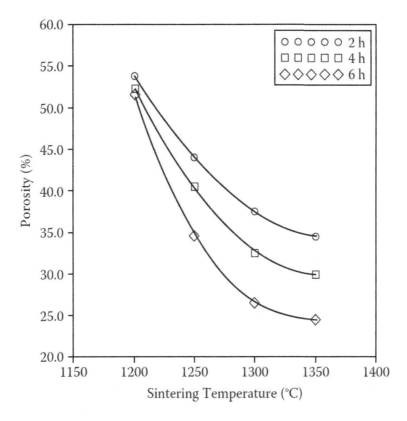

FIGURE 2.44 Porosity vs. sintering temperatures for the Ni-YSZ cermets sintered for 2, 4, and 6 h, respectively. (Reprinted with permission from Ref. [5].)

FIGURE 2.45 Microstructure of the Ni-YSZ cermets tested for 100 h and reduce at (a) 1000°C, and (b) 800°C. Ni particles are gray, while YSZ is lighter. (Reprinted with permission from Ref. [123].)

2.3.1.2   Influence of Materials Characteristics and Fabrication Conditions on Electrochemical Performance of Ni-YSZ Cermet Anode

Similar to electronic conductivity, the composition of the anode (Ni:YSZ volume ratio), microstructural characteristics of raw materials (particle size/distribution and porosity), and fabrication conditions (sintering temperature and atmosphere) also affect the anode polarization or catalytic activity toward the electrochemical oxidation of the fuel. Regarding the anode composition, the lowest anode polarization resistance at a constant current density is usually obtained for the Ni-YSZ cermet with a Ni:YSZ volume ratio of ~40:60 [125,126], as shown in Figure 2.46. In addition, in the range of 40–90 vol% of Ni, polarization resistance increases with an increase in Ni content. However, the cell power output linearly increases with an increase in Ni content, as presented in Figure 2.47. Although the maximum power output of 1.10 W was obtained for a cell using the anode with a Ni:YSZ volume ratio of 31:69, the power output increased up to 1.97 W in a cell using the anode with a Ni:YSZ volume ratio of 87:13 [126]. Moreover, a long-term cell performance (8000 h) was achieved for a double-layer anode structure in an electrolyte-supported cell with a volume ratio of 40:60 (Ni:YSZ). Hence, in order to obtain a high-performance cell, it is very important to optimize Ni:YSZ volume ratio.

Previous studies also showed that the particle size of raw/starting materials (NiO and YSZ) strongly influences the electrochemical cell performance where decreasing the particle size results in higher electrochemical cell performance [127,128]. For instance, Huebner et al. [127] found that sintering the NiO-YSZ cermet at higher temperatures results in a more rigid YSZ support, which prevents Ni particles coarsening and improves electrochemical performance. Jiang and Badwal [128] observed that the anode polarization resistance and ohmic resistance decrease with a decrease in NiO

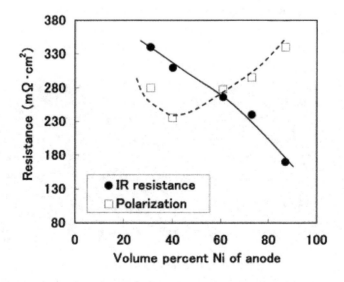

FIGURE 2.46   IR resistance and activation polarization of single cells with different Ni-YSZ anodes as a function of Ni vol%. (Reprinted with permission from Ref. [126].)

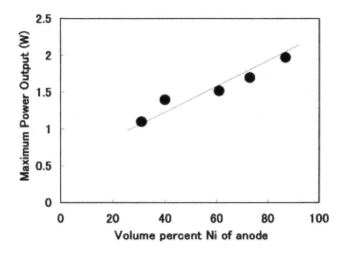

FIGURE 2.47 Maximum power output of single cells with different Ni-YSZ anodes as a function of Ni vol%. (Reprinted with permission from Ref. [126].)

particle size for the Ni-3YSZ cermet. They also found that the anode overpotential decreases with a decrease in the YSZ particle size in the range of 0.1–1.5 μm.

The effect of sintering temperature on the anode electrochemical performance is fairly clear as most of the previous studies showed that the anode performance is improved by increasing the anode sintering temperature due to a decrease in interfacial and bulk resistances [125,127,128]. Sintering the NiO-YSZ cermet at higher temperatures is essential to form a good bonding between the Ni and the YSZ particles, leading to the formation of a YSZ with a rigid structure that supports the Ni particles yet limits Ni coarsening.

For the anode-supported cells, it has been shown that low porosity in the anode significantly limits the mass transfer process and results in a large anode overpotential. Therefore, the addition of pore former is essential, particularly when ultrafine raw materials are used. For example, Primdahl et al. [129] showed that the anode interfacial resistance for the sample with 40 vol% added pore former is almost half of the sample without pore former. Moreover, Zhao and Virkar [130] found that the maximum cell power density increases from 0.72 to 1.55 W cm^{-2} when the porosity of Ni-YSZ anode increases from 32 to 57 vol%. Hence, in order to ensure high SOFC electrochemical performance, the porosity of anode-supported cells is usually kept fairly high.

2.3.1.3 Degradation Mechanisms in Ni-YSZ Cermet Anode

Degradation in the SOFC system is typically defined based on the performance loss of a single component or the interaction of different components. For the Ni-YSZ cermet anode, three different types of degradation mechanisms have been proposed: (1) material transport mechanism, (2) thermomechanical mechanism, and (3) passivation and deactivation mechanism [113]. The material transport mechanism is the most prominent type of anode degradation, including different microstructural changes due to Ni particle agglomeration and impurity particles. Thermomechanical degradation of the anode occurs because of the residual stresses between the interfaces at a small scale or stack parameters at a large scale. Passivation and deactivation mechanisms cause performance loss as a result of the impurities which are present in carbonaceous fuels. The most important impurities are sulfur and carbon. The related degradation mechanisms are also known as "sulfur poisoning" and "coking", respectively. During coking, carbon particles formed by the Boudouard reaction are settled on the Ni particles and prevent the electrochemical reactions on the Ni surfaces. Similarly, for the sulfur poisoning, the sulfur impurity (H_2S) in fuel reacts with O_2, producing S and H_2O. The produced S then settles on the Ni particles and blocks the activation sites. At SOFC operating temperatures, the Ni_3S_2 volatile compound may form, which leads to a decrease in the anode Ni content. All these degradation mechanisms of the Ni-YSZ anode result in Ni grains growth, change in Ni particles surface morphology, and Ni deactivation, leading to a reduction of the TPB density, thereby ultimately exhibiting a high overall resistance and poor performance [113]. The degradation mechanisms of the Ni-YSZ anode will be discussed in the following.

Ni coarsening is one of the most common and dominant types of degradation mechanisms in the Ni-YSZ cermet anodes. This phenomenon is related to the Ni particle agglomeration, change in Ni particles surface morphology and Ni-Ni contact loss. The Ni coarsening is either attributed to Ni transport in the gas phase or diffusion/sintering of particles, grain boundaries and vacancies, depending on the anode operating conditions (e.g., temperature and humidity). The driving force for diffusion of particles, grain boundaries, and vacancies during sintering of Ni particles at high temperatures is particle related properties such as size and surface energy. In this degradation mechanism, smaller particles with higher surface energy have more intention to agglomerate to reduce the energy. As already discussed, the increase in Ni particle size leads to a corresponding reduction of the TPBs and the polarization resistance in the anode. It also decreases the electronic conductivity due to Ni-Ni contact loss [113]. For example, Iwata [131] showed a voltage degradation rate of 14 μV h^{-1} at a current density of 0.3 A cm^{-2} after a long-term operation for 1015 h at 1000°C due to an increase in Ni particle size from 0.1 to 10 μm, thus, a decrease in specific surface area. Figure 2.48 illustrates the Ni particle size growth and the related degradation in cell performance. It has been shown that the introduction of sintering inhibitors such as CeO_2 and GDC can effectively hinder the Ni particle growth, thereby maintaining sufficient TPB density and high electrochemical performance for the long-term operation [132]. CeO_2 and GDC sintering inhibitors provided relatively low TPB density degradation rates of 0.000413 and 0.000373 $\mu m^{-2}h^{-1}$, respectively, at 1000°C.

Sulfur poisoning is another important degradation mechanism for the Ni-YSZ cermet anodes. Sulfur is an important constituent of several commercially available SOFC fuels, which can either be present as an impurity or an additive in the form of H_2S [133]. The H_2S compound interacts with the Ni-YSZ anode and blocks the fuel supply to the system, thus, the polarization losses become dominant. Previous studies showed that even a few ppm of H_2S could negatively affect the Ni-YSZ performance where the cell voltage sharply decreases in the first few hours of operation followed

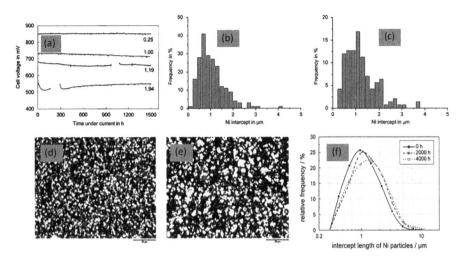

FIGURE 2.48 (a) Cell voltage degradations at different current densities, and corresponding Ni particle size distribution (b) before and (c) after long-term test at 850°C for 1500 h. SEM images of Ni-YSZ anodes (d) before and (e) after long-term test at 1000°C for 4000 h (white = Ni particles), and (f) corresponding Ni particle size distribution. (Reprinted with permission from Ref. [113].)

by a gradual deterioration of the cell [134,135]. The H_2S poisoning is also associated with hampering oxygen ions migration toward the TPBs and fuel traveling toward the anode active sites, resulting in higher polarization resistance.

Furthermore, the unrecoverable Ni_3S_2 compound formed from Ni and S interactions causes permanent damage to the cell [136]. Sasaki et al. [137] have summarized the mechanisms of sulfur poisoning, as shown in Figure 2.49. These different mechanisms associated with different types of degradations, which are typically defined by the H_2S content in the fuel, cell operating temperature, cell current density [113,133]. For relatively low sulfur concentration (ppm level), the reversible sulfur adsorption/ desorption process is considered as the predominant mechanism (Figure 2.49b). The initial decrease in cell voltage due to sulfur poisoning is considerably higher in CO-rich fuels (H_2-poor fuels), which may be resulted from more preferred adsorption of sulfur on Ni with the decreasing H_2 concentration, as shown in Figure 2.49e and f. As can be seen in Figure 2.49g, an irreversible degradation associated with the oxidation of Ni can occur at a higher sulfur concentration and/or a lower operating temperature. Comparing Figure 2.49c and d, sulfur poisoning to internal reforming reactions is also serious where much larger cell voltage drop occurs using CH_4-rich fuels. As already mentioned, the formation of Ni_3S_2 compound (with a melting point of 787°C) is also possible for H_2-poor fuels (Figure 2.49h).

So far, several techniques have been utilized to overcome the negative impacts of sulfur on the Ni-YSZ anode performance such as operation at high temperature and high current density, surface regeneration of sulfur-poisoned Ni surfaces with appropriate partial pressures of O_2 and H_2O, and cooling down the Ni-YSZ anode slowly (~2–5°C min⁻¹) from the operating temperature to room temperature in a

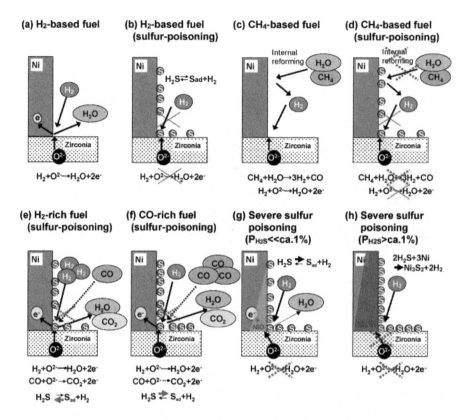

FIGURE 2.49 Possible sulfur mechanisms for the Ni-YSZ anode within the SOFC system. (a) H_2-based fuel; (b) H_2-based fuel (sulfur-poisoning); (c) CH_4-based fuel; (d) CH_4-based fuel (sulfur-poisoning); (e) H_2-rich fuel (sulfur-poisoning); (f) CO-rich fuel (sulfur-poisoning); (g) severe sulfur poisoning ($P_{H2S} \ll$ ca.1%); (h) severe sulfur poisoning ($P_{H2S} >$ ca.1%). (Reprinted with permission from Ref. [137].)

H_2-rich fuel [138–140]. Modifying composition of the Ni-YSZ cermet is another way to improve its sulfur poisoning resistivity. For example, Choi et al. [141] studied the addition of niobium oxide (Nb_2O_5) to a Ni-YSZ anode to assess sulfur tolerance. The results showed an enhanced cell conductivity, power density and stability under a H_2S environment for the Nb_2O_5-modified Ni-YSZ anode.

Coking is another responsible phenomenon for the degradation of Ni-YSZ anode and cell performance. Carbon formation or coking (in different forms of carbon ranging from graphite to hydrocarbons) on the Ni-YSZ anodes can occur in different ways. One is the dissolution-precipitation mechanism in which the dissolved C atoms into the Ni grains will diffuse through and then deposit at the outer surface. The deposition results in catalyst de-activation and cell performance degradation because of covering active anode sites. The other degradation mechanism associated with coking is the crack formation and propagation in the Ni-YSZ anode structure due to carbon dissolution into the Ni grains, thus, a significant anode dimensions change and stress induction [142–144]. Figure 2.50 shows the coking of a Ni-YSZ cermet

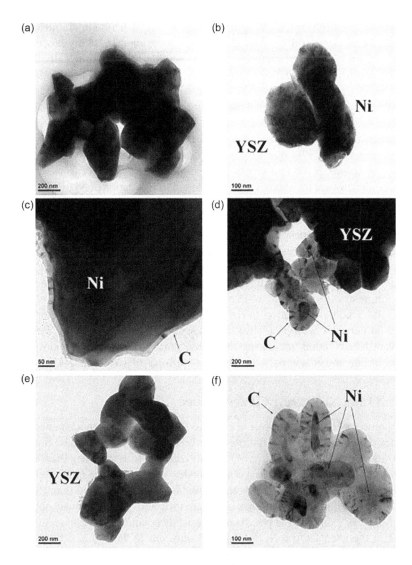

FIGURE 2.50 TEM images of the Ni-YSZ anode (a and b) operated under hydrogen atmosphere; (c) operated under a syngas atmosphere for 15 min; (d–f) operated under syngas atmosphere for 120 min. (Reprinted with permission from Ref. [145].)

anode operated under hydrogen and syngas (a mixture of H_2, CH_4, CO, CO_2, and H_2O) environments for 15 and 120 min in a commercial SOFC testing system [145]. As can be seen, the cell operated under the hydrogen atmosphere has the Ni-YSZ anode in which the Ni and YSZ are tightly bonded. However, a 10 nm thick carbon layer has been formed surrounding the nickel particles after 15 min of operation under the syngas environment. The carbon layer becomes even thicker, and the Ni particles become smaller after exposure to syngas for 120 min.

In general, controlling the cell operating conditions such as temperature and ratios of C–H–O can successfully hinder the carbon formation and coking process. Other techniques can be listed as increasing the oxygen-ion flux from the cathode toward anode by providing polarization current, alloying anode materials with other metals such as Fe or Cu, coating the anode surface with protective metals such as Au, and microstructural and compositional modifications of the anode [113].

Redox cycling is also another major issue for the anode and cell performance degradation through stress development, decrease in electrochemical performance, and the density of TPBs due to potential increase in polarization resistance and mechanical failure. If the fuel supply is interrupted due to some reasons, the oxygen ions migration toward the anode will only result in the oxidation of Ni to NiO. The NiO can also form if the partial pressure of H_2O or fuel utilization increases on the anode side. The Ni to NiO conversion is associated with an increase in the anode volume for about 70%. This phenomenon is usually known as redox cycling. Figure 2.51 illustrates the related processes during the redox cycling of a Ni-YSZ anode.

In order to reduce the damage as a result of redox cycling, several strategies have been developed, including cooling rates of greater than $3°C$ min^{-1} to prevent NiO formation, restricting the flow of extra air into the anode side, the introduction of Ni into the YSZ by infiltration to reduce the internal stresses at low temperatures, coating Ni particles with protective metals such as Y, Zr, and Ce and their oxides to prevent Ni oxidation, increasing the anode porosity, and using anodes having a mixture of fine and coarse particles with the optimized Ni:YSZ ratio [113].

2.3.2 OTHER NI-FLUORITE CERMET ANODES

Studies on cermet anodes are still in progress, and many other anode materials have been developed based on the fluorite-structured electrolyte materials. Since many low- to intermediate-temperature SOFCs utilized electrolytes other than YSZ, it is common to use other Ni-fluorite cermet anodes such as Ni-ScSZ, Ni-doped ceria, or Ni-LSGM which have higher potential for better performance than that of the Ni-YSZ anodes due to their higher conductivities and possibly lower polarization resistances. For example, the polarization resistance values for a unit length of TPB line (R^{LS}) reported for the Ni-ScSZ (5400 ohm cm at $750°C$) and the Ni-LSGM (55,000 ohm cm at $650°C$) are similar to those of the Ni-YSZ [76]. Among the Ni-doped ceria anodes, Ni-GDC and Ni-SDC anodes have shown good performance for low-temperature SOFC applications [82,83,147–154]. It has also been shown that a number of $Ni-LnO_x$ cermets as the anodes of intermediate-temperature SOFCs, where the ceramic is a rare-earth oxide (i.e., Ln = Ho, Er, Dy, and Yb), provided a low anode polarization resistance in a humidified hydrogen environment with the maximum cell power density of $600 mW$ cm^{-2} at $600°C$ [155]. This electrochemical performance is comparable to those of the commonly Ni-doped ceria cermets. The performance is somehow surprising for these oxides, whose oxygen-ion conductivity is negligible.

Regarding the Ni-SDC cermets, they have been investigated as the SOFC anode not only due to their high power densities but also ceria is a good sulfur sorbent at high temperatures given that it reacts with H_2S. Sulfur is still expected to adsorb

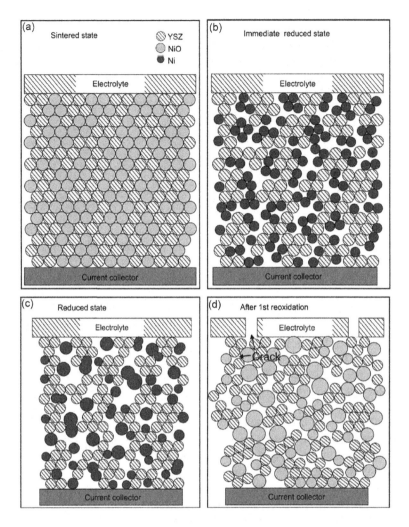

FIGURE 2.51 Model illustrating the distortion of Ni-YSZ anode microstructure during the redox cycling. (a) sintered state; (b) immediate reduced state; (c) reduced state; (d) after 1st reoxidation. (Reprinted with permission from Ref. [146].)

at the nickel surface and along with the TPB for those cells, blocking the active sites for fuel oxidation and the poisoning characterization remain similar to Ni-YSZ [139]. Chen et al. [156] found that the open porosity, electronic conductivity, thermal expansion, and mechanical strength of the cermets are susceptible to the Ni content. The Ni-SDC cermets containing 50–60 vol% Ni content showed the best performance with superior electronic conductivities of over 1000 S cm^{-1} at the intermediate operating temperature range of 600°C–800°C, open porosity of more than 30%, a moderate thermal coefficient of 12.6–13.5 × 10^{-6} at the temperature range of 100°C–800°C, and excellent bending strength of about 100 MPa. In the other study, Maric et al. [150] showed that the anodic polarization was strongly influenced by the

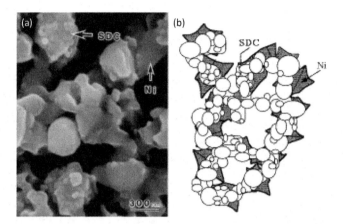

FIGURE 2.52 (a) SEM micrograph and (b) sketch of the microstructure of the Ni-SDC anode sintered at 1300°C. (Reprinted with permission from Ref. [150].)

anode microstructure. The lowest anode polarization (30 mV) was obtained for the Ni-SDC cermet sintered at 1300°C for a current density of 200 mA cm^{-2} due to the cermet microstructure in which Ni particles formed a skeleton with well-connected SDC particles finely distributed over the Ni particles, as shown in Figure 2.52.

The composition of the Ni-SDC cermet anodes has also been modified to improve their electrochemical performance and stability. For example, Zhao et al. [157] modified the composition of Ni-SDC cermet with MnO-Co ($Mn_{1.5}Co_{1.5}O_4$). The results showed 55% increase in maximum power density (1756 mW cm^{-2} at 700°C) for the modified anode compared with the original Ni-SDC anode. The higher maximum power density of the modified anode was attributed to its higher porosity and lower polarization resistance. However, the MnO-Co-modified Ni-SDC cermets revealed lower stability due to higher carbon deposition. Up until recently, this material is a great alternative anode for the intermediate-temperature SOFCs [158]. In the other study, Qu et al. [159] introduced some basic oxide additives such as CaO, BaO, MgO, SrO, La_2O_3 into the conventional Ni-SDC cermet anode. Among those materials, the CaO-modified Ni-SDC anode exhibited higher catalytic performance, higher cell power density (1009 mW cm^{-2}), and superior coking tolerance under methane steam reforming conditions at intermediate temperatures. Hence, the CaO-modified Ni-SDC anode seems to be a promising material for intermediate-temperature SOFCs due to its great cell performance and low cost of CaO.

Considering the Ni-GDC cermets, they have also been utilized as the SOFC anode with some success. On nano-scale Ni-GDC anode, it showed a polarization resistance of 0.14 ohm cm^2 in a humidified hydrogen atmosphere at 600°C [160]. Also, in conjunction with GDC electrolyte and $Ba_{0.5}Sr_{0.5}Co_{0.8}Fe_{0.2}O_{3-\delta}$ (BSCF) cathode, the cell maximum power density reached up to ~1300 mW cm^{-2} at 600°C [161]. Symmetric infiltrated Ni-GDC anodes on the YSZ electrolyte showed a polarization resistance of 0.34 ohm cm^2 in 50% H_2/50% N_2 atmosphere at 580°C, which then decreased down to 0.14 ohm cm^2 at 750°C [162]. The Ni phase has also been alloyed with other elements such as Cu and Fe to improve anode performance. For example, Sin et al. [163] tested a Ni-Cu-GDC anode on a GDC-supported electrolyte. The cell showed a good performance for 1300 h of operation in the dry methane atmosphere. During the operation, the compositional change carbon deposition effect was negligible. Similar behavior was observed for a Fe-Ni-GDC anode tested for 50 h. In addition, the infiltrated Ni-Pd-GDC and Ni-Pt-GDC anodes yielded lower polarization resistances of 0.31 and 0.11 ohm cm^2, respectively, compared to that of the infiltrated Ni-GDC in the humidified H$_2$ atmosphere at 600°C [164]. However, the main problem with GDC containing anodes still remains in that introducing gadolinium into the ceria decreases the conductivity in either metallic or bi-metallic compounds [165].

2.3.3 Alternative Anode Materials

Perovskite-structured materials have widely been investigated as the alternative SOFC anode over the past 20 years. Most of the studies on the perovskite-structured anode materials have focused on the development of the properties (i.e., conductivity and catalytic activity) of the doped strontium titanates such as lanthanum-doped strontium titanate (LST; $Sr_{1-x}La_xTiO_3$) and yttrium-doped strontium titanate (YST; $Sr_{1-x}Y_xTiO_3$), and $La_{0.75}Sr_{0.25}Cr_{0.5}Mn_{0.5}O_{3-\delta}$ (LSCM).

It was found that the local point defects in the perovskite-structured anodes can improve their performance. For LST and YST anode materials, the introduction of A-site deficiencies was discovered to be efficient at achieving higher electronic conductivity than the introduction of B-site deficiencies [165]. Li et al. [166] found that the A-site deficient LST possesses higher electronic conductivity (145 S cm^{-1} at 800°C) than that of the similar deficient YST anode. It was also found that impregnation of 20% GDC and copper in the LST anode can remarkably improve the cell power density up to 500 mW cm^{-2} at 750°C in a hydrogen atmosphere [167]. The use of <5%–10% of nickel in a LST-GDC anode has been found to significantly increase the cell maximum power density [168]. Introduction of Ga and Mn into the LST also forms a promising SOFC anode material with impressive cell performance in a wet hydrogen atmosphere at 950°C. This anode material also showed a good performance for methane oxidation at high temperatures, with OCVs in excess of 1.2 V [169]. Despite an improvement of the microstructure of the LST or YST and ceria in a 1:1 ratio, generally, the performance of these materials has been poor with a cell power density of <200 mW cm^{-2} at 900°C [165].

Although nickel has been found to have the highest catalytic activity for the hydrogen oxidation reaction due to the ideal strength of the oxygen-metal bond, other metals such as ruthenium and palladium have been found to be used as an effective replacement for Ni in the anode cermet due to their sufficient catalytic activity and stability against Sulfur poisoning or coking. For example, Lu et al. [170] showed that the use of Pd can double the cell performance of a LST-$La_{0.4}Ce_{0.6}O_2$ composite, reaching the power density of over 1000 mW cm^{-2} at 850°C with no degradation in H$_2$S environment. Kurokawa et al. [171] also found Ru as an effective replacement for Ni in the anode cermet due to a low polarization resistance of 0.8 ohm cm^2 at 800°C, which was further reduced down to 0.5 ohm cm^2 with the addition of ceria. As can be seen from the above studies, although only a small amount of these alternative metal catalysts is usually sufficient to improve the anode performance, the cost of these metals hampers their extensive use in the SOFC anode applications.

Similar compositional manipulations used for the doped strontium titanates anodes can be applied to the LSCM anode. However, since the compound structure is already doped on both A-site and B-site, the effect of manipulations cannot be easily predicted. Introduction of A-site deficiencies to the LSCM structure increases the ionic conductivity ($\sim3 \times 10^{-4}$ S cm^{-1} at 950°C), thereby a decrease in electronic conductivity [165,172]. Partial substitution of lanthanum with cerium was found to improve the catalytic activity of the anode and reduce the polarization resistance down to 0.2 ohm cm^2 at 800°C [173].

Replacement of Mn with other metals such as Co, Fe, Ru, and Ni in the LSCM composition was also found to increase the conductivity and improve the catalytic activity of the compound, thus, the cell performance due to formation of metal nanoparticles upon reduction. For example, the cell performance for both LSCRu and LSCNi anodes was <300 mW cm^{-2} at 800°C. A combination of LSCRu with GDC led to steady cell performance (400–500 mW cm^{-2} at 800°C) with no noticeable Ru nanoparticle agglomeration during 1000 h of cell operation [174–177].

Rutile-structured compounds have also attracted attention as SOFC anode materials due to their short M-M distances, promoting electronic conductivity [165]. For instance, Lashtabeg et al. [178] fabricated niobium- and chromium-doped rutile TiO$_2$ with good electronic conductivity, however, the reduction kinetics of compounds were slow where the equilibrium condition not obtained after 50 h in 5% H$_2$/Ar atmosphere at 900°C. The chromium niobate with the same structure showed faster reduction kinetics but possessed lower conductivity (~3.7 S cm^{-1}) at 900°C [179]. At high temperatures, chromium niobates start to degrade, preventing attempts to increase the conductivity with further formation of oxygen vacancies [165].

Some metal vanadates such as Ce$_{0.9}$Sr$_{0.1}$VO$_x$ ($x = 3, 4$) have been found to have a good sulfur resistance at high temperatures, making them a potential anode material for SOFC applications [180–182]. Sr- and Ca-doped CeVO$_4$ showed interesting redox stability below 600°C in 5% H$_2$/Ar atmosphere but low conductivity of <1 S cm^{-1}, which is not sufficient to be used as SOFC anode in terms of conductivity. However, Ce$_{0.85}$Sr$_{0.15}$VO$_3$ compound exhibited a higher conductivity of 2.5–6.0 S cm^{-1} between 25°C and 700°C in 5% H$_2$/Ar atmosphere, which is suitable for SOFC anode applications [183,184].

The development of alternative anode materials has recently attracted considerable interest. Although several materials exhibited good catalytic activity, sufficient conductivity, sulfur poisoning resistance, and carbon deposition tolerance, critical issues associated with each candidate material are yet to be overcome.

2.4 CATHODES

So far, the cathode is the most studied components in low- and intermediate-temperature SOFCs due to the slow kinetics of oxygen reduction reaction (ORR) occurring on the cathode side:

$$O_2 + 2V_O^{\bullet\bullet} + 4e^- = 2O_O^\times \tag{2.20}$$

where, in the Kröger–Vink notation, $V_O^{\bullet\bullet}$ is an oxygen vacancy and $O_O^\times$ is an oxygen ion on a regular oxygen site in the electrolyte (e.g., YSZ) lattice. As shown in Equation (2.20), the ORR requires the presence of oxygen and electrons as well as the possibility of oxygen ions transportation from the reaction site into the electrolyte bulk [5].

The electrochemical reduction of oxygen usually needs very high activation energy, which implies even slower kinetics of the ORR at lower SOFC operating temperatures, thus, a significant drop in SOFC performance. Hence, there has been a lot of research carried out on new cathode materials to improve the cell performance and efficiency in the low to intermediate temperature range. There is a general agreement that the cathode materials should possess both electronic and ionic conductivity properties in the intermediate temperature range to maximize cell performance. Since the ORR requires the highest activation energy, a large electronic loss is observed at the cathode with decreasing the operating temperature of SOFC. For the ORR to take place, there is a need to have a TPB, including gas (oxygen), electrode, and electrolyte. The TPB is limited to a thin region in the electrolyte surface for a pure electronic conductor cathode, determined by the morphology of the cathode. However, for a mixed ionic-electronic conductor cathode, the TPB interface (reaction zone) is extended to all contact points between the gas and solid phase (i.e., electrode surface away from the electrode–electrolyte interface) due to oxygen-ion conductivity of the cathode which allows the diffusion of oxygen ions through the entire cathode layer. This leads to an extended ORR zone and enhanced performance [185,186].

In addition, the cathode materials must be stable under oxidizing conditions at fabrication and operating temperatures. Importantly, the porosity and active surface area of the cathode should remain stable over time. The chemical stability and thermal expansion coefficient compatibility with regard to neighboring components (i.e., electrolyte and current collector) are also necessary for a long-term cell operation and to withstand the delamination issue during the heating and cooling steps. Hence, a global effort has been made to optimize the SOFC components and find a successful strategy for reducing the SOFC operating temperature [186].

To be precise, the cathode materials should possess the following functionalities for efficient operation of SOFC:

- High electronic conductivity (preferably over 100 S cm^{-1} in an oxidizing atmosphere);
- High oxygen-ion conductivity;
- High electrocatalytic activity of the ORR;
- Thermal expansion coefficient match with other adjacent cell components such as electrolyte and interconnect materials;
- Good chemical compatibility with the electrolyte and interconnect materials;
- Good stability under an oxidizing atmosphere during fabrication and operation;

- Sufficient porosity for fast oxygen diffusion from cathode to cathode–electrolyte interface;
- Cost effectiveness [2].

In order to optimize the electrochemical properties of the cathode, there is a need to find new compositions, mixed composites, or novel families of materials, and also to modify the morphological aspects of cathode, including active surface area, porosity, and interface area. These two main routes are being intensively studied [186].

Perovskites with a general formula of $ABO_{3-\delta}$ in which A and B are cations with total charge of +6 are the most widely used SOFC cathode materials due to their flexibility in terms of doping. In 1966, $La_xSr_{1-x}CoO_{3-\delta}$ was the first perovskite material used as a cathode for SOFC application [2]. This was followed by the synthesis of several other perovskite materials, which were subsequently tested for the SOFC cathode application. From 1973, strontium-doped lanthanum manganite perovskites ($La_xSr_{1-x}MnO_{3-\delta}$; LSM) were extensively used for the SOFC cathode application. Up until today, it has remained the most investigated and the conventional cathode material for high-temperature SOFCs, however, other perovskite-structured materials have also been widely investigated for lowering the SOFC operating temperature [2]. In the perovskite structure, A is the larger cation with 12-fold oxygen coordination, and B is the smaller cation with 6-fold oxygen coordination, by partial or total substitution of one or both cations, the properties of perovskite such as electronic and ionic conductivity, catalytic activity, and thermal expansion coefficient change to meet the SOFC cathode requirements. The A site is usually occupied by a mixture of rare-earth metals (typically La) and alkaline-earth metals (e.g., Sr, Ca, and Ba). The B site is usually occupied by one or several transition-metal ions (e.g., Mn, Co, Fe, and Ni). The mixed valence can provide a good ORR catalytic activity. The A site and/or B site may also be substituted by aliovalent A′ and/or B′ cations to improve the conductivity [186].

The first materials to be investigated were lanthanum manganite-, cobaltite-, and ferrite-based perovskite materials. In addition, other solid oxides with perovskite structures such as Ruddlesden-popper series (RP) with the general composition of $A_{n+1}B_nO_{3n+1}$ and double perovskites with the general composition of $AA'B_2O_{5+\delta}$ have been investigated as alternative cathode materials [186]. The RP oxides have exhibited a mixed ionic and electronic conductivity, promising electrochemical activity, and appropriate mechanical and chemical compatibilities with other cell components [185,187]. Among this class of materials, the RE_2MO_4 series (RE = rare-earth elements; M = transition metals) have exhibited the highest electrochemical performance [188]. Double perovskites compounds are of alternate layers of single perovskite $ABO_{3-\delta}$. Several manganites, cobaltite or ferrite compounds can adopt this layered structure [185,189]. Although extensive research has been conducted on cathode materials for low- to intermediate-temperature SOFCs, there is still no obvious solution to the quest for optimized cathode [185]. Table 2.7 presents the most relevant properties of the main state-of-the-art SOFC cathode materials.

TABLE 2.7

The Most Relevant Properties of the Main State-of-the-Art SOFC Cathode Materials

Family	Composition	TEC (K^{-1})	σ (S cm^{-1}) 500°C–750°C	D^* (cm^{2}s^{-1}) 600°C	k^* (cm s^{-1}) 600°C
Perovskites	La$_{0.8}$Sr$_{0.2}$MnO$_{3-\delta}$	12.0 × 10^{-6}	120–130	5.2 × 10^{-18}(a)	1.5 × 10^{-11}(a)
	La$_{0.5}$Sr$_{0.5}$CoO$_{3-\delta}$	21.3 × 10^{-6}	1300–1800	2.6 × 10^{-9}	1.3 × 10^{-6}
	La$_{0.6}$Sr$_{0.4}$Co$_{0.2}$Fe$_{0.8}$O$_{3-\delta}$	15.3 × 10^{-6}	300–330	1.7 × 10^{-10}	1.1 × 10^{-7}
	Ba$_{0.5}$Sr$_{0.5}$Co$_{0.8}$Fe$_{0.2}$O$_{3-\delta}$	24.0 × 10^{-6}	30–35	3.3 × 10^{-7}	1.4 × 10^{-5}
Double perovskites	GdBaCo$_2$O$_{5+\delta}$	16.4 × 10^{-6}	550–925	7.0 × 10^{-10}	3.1 × 10^{-7}
	PrBaCo$_2$O$_{5+\delta}$	24.6 × 10^{-6}	400–700	6.4 × 10^{-9}	2.8 × 10^{-7}
Ruddlesden-popper series	La$_2$NiO$_{4+\delta}$	13.0 × 10^{-6}	55–65	8.7 × 10^{-9}	1.7 × 10^{-8}
	Pr$_2$NiO$_{4+\delta}$	13.6 × 10^{-6}	100–120	2.5 × 10^{-8}	5.1 × 10^{-7}
	Nd$_2$NiO$_{4+\delta}$	12.7 × 10^{-6}	35–45	9.7 × 10^{-9}	1.1 × 10^{-7}

Source: Data adapted and modified from Ref. [186].

σ: Overall conductivity; D^*: oxygen self-diffusion coefficient; k^*: oxygen surface exchange coefficient; TEC: thermal expansion coefficient.

(a) For La$_{0.8}$Sr$_{0.2}$MnO$_{3-\delta}$, D^* and k^* were extrapolated from high-temperature values.

2.4.1 MANGANITE-BASED PEROVSKITE CATHODES

Perovskite-structured manganites (Ln, A)MnO$_{3\pm\delta}$ (Ln = La-Yb or Y; A = Ca, Sr, Ba, and Pb) and there derivatives have exhibited high electronic conductivity, moderate thermal expansion coefficient compatibility with conventional solid electrolytes such as YSZ, and sufficient electrocatalytic activity for the ORR at the temperate range of 725°C–825°C. Although the manganites have not necessarily shown the highest total conductivity compared to the other perovskite families such as Co- and Ni- containing perovskites, as shown in Figure 2.53, the latter materials have possessed excessively high thermal expansion coefficients and insufficient thermodynamic stability under oxidizing conditions. So, lanthanum-strontium manganites (La$_{1-x}$Sr$_x$MnO$_{3\pm\delta}$) and their composites are mainly considered as the commonly used cathode materials for high-temperature SOFCs [189].

All perovskite-structured manganites (Ln$_{1-x}$A$_x$MnO$_{3\pm\delta}$) possess a predominant electronic conductivity mixed with relatively low oxygen-ion diffusivity. The electrochemical activity and transport properties of these materials strongly depend on the oxygen non-stoichiometry. Under a high cathodic polarization, the catalytic behavior is usually related to the number of oxygen vacancies generated at the electrode surface. The electronic conductivity of the Ln$_{1-x}$A$_x$MnO$_{3\pm\delta}$ manganites at moderate A^{2+} concentration increases with an increase in x content as Mn^{4+} fraction increases. In general, the maximum conductivity is achieved in the range of x = 0.2–0.5, which shifts toward lower dopant concentrations on heating. The average thermal expansion coefficient of manganite ceramics is typically 6.3–13.0 × 10^{-6} K^{-1}, depending on the material composition. The dopant content for an optimum thermomechanical

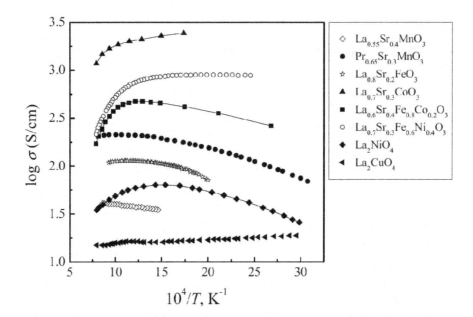

FIGURE 2.53 Comparison of the total conductivities of various perovskite-structured materials at different temperatures in air. (Reprinted with permission from Ref. [189].)

compatibility with zirconia-based solid electrolytes is $x = 0.1$–0.15 for the La-containing manganite cathodes (Ln = La) and is $x = 0.3$–0.5 for Gd-containing manganite cathodes (Ln = Gd). The maximum electrochemical activity at $800°C$–$1000°C$ was obtained in the range of $x = 0.3$–0.7, which shifts toward lower dopant contents with increasing temperature. The manganite electrode performance can increase with introducing Ln-site cation deficiency because of a higher oxygen vacancy content, thus, faster ionic conductivity [189].

Improvement of the manganite electrode performance, optimization of thermal expansion, and an increase in the oxygen permeability can be achieved by partial substitution of manganese with oxidation state stable ions and/or transition-metal cations. A moderate addition of the variable-valence dopants can reduce the polarization resistance of the (Ln, Sr) (Mn, M)$O_{3-\delta}$ cathodes [189].

Lanthanum manganite-based perovskites (La$_{1-x}$A$_x$MnO$_{3\pm\delta}$) are still considered as state-of-the-art cathode materials for SOFC applications. Undoped LaMnO$_3$ has an orthorhombic crystal structure at the room temperature. However, it transforms into an orthorhombic/rhombohedral structure at ~$600°C$. The transformation temperature depends on the Mn^{4+} content and stoichiometry of the material since the transformation occurs as a result of Mn^{3+} to Mn^{4+} ions oxidation. Doping lower-valence cations for the La sites (e.g., Ca^{2+} and Sr^{2+}) can increase the concentration of Mn^{4+} in the LaMnO$_3$ which affects the phase transformation temperature. For example, Sr-doped LaMnO$_3$ perovskites, La$_{1-x}$Sr$_x$MnO$_{3\pm\delta}$, can possess three lattice structures, depending on the doping level: rhombohedral ($0 < x < 5$), tetragonal ($x = 0.5$), and cubic ($x = 0.5$). At room temperature, it also possesses three different phases: orthorhombic $(0 < x < 0.15)$, hexagonal $(0.15 < x < 0.45)$, and cubic $(x > 0.45)$. The volume of the LSM unit cell is reduced with an increase in Sr content [5]. Figure 2.54 represents the general four types of perovskite unit cells.

LaMnO$_3$ is an intrinsic p-type conducting material in which the electronic conductivity increases by substitution of the La^{3+} sites in the perovskite structure with divalent cations such as Ca^{2+} (La$_{1-x}$Ca$_x$MnO$_{3\pm\delta}$) and Sr^{2+} (La$_{1-x}$Sr$_x$MnO$_{3\pm\delta}$). The linear relationship between the electronic conductivity of the LaMnO$_3$ and $1/T$, expressed by the Arrhenius equation (see Equation (2.8)), indicates conduction occurring due to a small polaron hopping mechanism [2]. Conductivity is also a function of non-stoichiometry and oxygen partial pressure. LaMnO$_3$-based oxides can have the oxygen-excess non-stoichiometry (La$_{1-x}$A$_x$MnO$_{3+\delta}$) as well as the oxygen-deficient non-stoichiometry (La$_{1-x}$A$_x$MnO$_{3-\delta}$). As an example, for the La$_{1-x}$Sr$_x$MnO$_{3\pm\delta}$ with $x < 0.4$, it has been shown that the oxide becomes oxygen-excess $(3 + \delta)$ at high oxygen partial pressures (10^{-5} to 10^{-10} Pa), it reveals a stoichiometric composition $(\delta = 0)$ at intermediate oxygen partial pressures, while it becomes oxygen-deficient

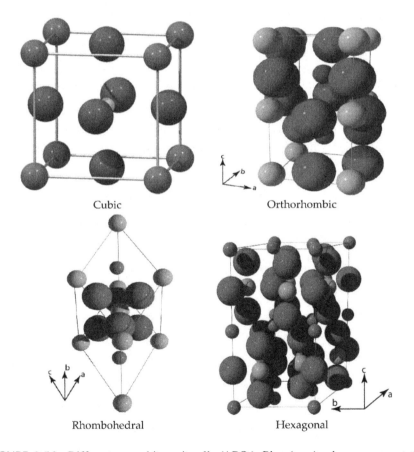

Cubic   Orthorhombic

Rhombohedral   Hexagonal

FIGURE 2.54 Different perovskite unit cells (ABO$_3$). Blue (gray) spheres represent the A cations, yellow (white) spheres represent the B cations and red (black) spheres represent oxygen anions forming an octahedral. (Reprinted with permission from Ref. [190].)

$(3 - \delta)$ at lower oxygen partial pressures ($<10^{-10}$ Pa) where the charge compensation of the positive effective charges of $V_O^{\bullet\bullet}$ is maintained by the oxidation of Mn cations [5]. The oxygen-excess non-stoichiometry of $La_{1-x}Sr_xMnO_{3+\delta}$ is interesting as it is rarely found in other perovskite-structured oxides. However, the oxygen-excess non-stoichiometry of $La_{1-x}Sr_xMnO_{3+\delta}$ disappears with $x > 0.5$.

Among the alkaline-earth dopants, Sr^{2+} dopant is mostly preferred for the $LaMnO_3$-based oxides as the resultant perovskite is more stable and possesses high conductivity under oxidizing conditions. Moreover, there is a small ionic size mismatch between Sr^{2+} and La^{3+} ($r^{2+}_{Sr,XII} = 1.44$ Å and $r^{3+}_{La,XII} = 1.36$ Å). For high-temperature SOFCs (800°C–1000°C), the $La_{1-x}Sr_xMnO_{3-\delta}$ perovskite is the state-of-the-art conventional cathode materials due to its excellent electronic conductivity, high catalytic activity for the ORR, high thermal stability, and high compatibility with GDC, LSGM, and YSZ electrolytes at operating conditions [186]. The $La_{1-x}Sr_xMnO_{3-\delta}$ perovskite with $x = {\sim}0.1$–0.2 exhibits the best performance in terms of both conductivity and chemical and mechanical stabilities at operating temperatures [191,192]. By decreasing the operating temperature, the polarization resistance of the LSM cathode has shown a significant increase from <1 ohm cm^2 at 850°C to $\sim$12 ohm cm^2 at 700°C, as shown in Figure 2.55 [193]. This increase was due to the poor kinetics of the electrode reaction, which makes the LSM unstable for the intermediate-temperature range. Mizusaki et al. [194] have also measured that the electronic conductivity of the $La_{1-x}Sr_xMnO_{3+\delta}$ as a function of temperature (up to 1000°C) and the dopant concentration. As seen in Figure 2.56, a linear increase in electronic conductivity was reported with increasing x and temperature. However, the conductivity reached a maximum value at $x = 0.5$. The same behavior was also reported by Zhong-Tai et al. [195] where a maximum electronic conductivity of 200 S cm^{-1} was achieved for $x = 0.5$ at 1000°C.

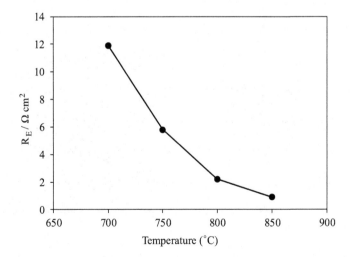

FIGURE 2.55 The polarization resistance of LSM electrodes with respect to the operating temperature. (Data with permission from Ref. [193].)

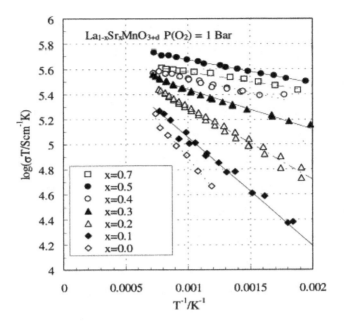

FIGURE 2.56 The electronic conductivity of $La_{1-x}Sr_xMnO_{3+\delta}$ electrodes as a function dopant concentration (x) and temperature (at pure oxygen; $P(O_2) = 1$ bar). (Reprinted with permission from Ref. [194].)

2.4.2 FERRITE-BASED PEROVSKITE CATHODES

The number of ferrite-based oxides with a significant electronic or mixed conductivity and a stable performance under the cathodic conditions of SOFC is even larger than that of the manganite-based systems. These primarily include perovskite-based (Ln, A)FeO$_{3\pm\delta}$ and their derivatives existing in all Ln-A-Fe-O systems, (Ln, A)$_3$FeO$_{12\pm\delta}$ garnets with small Ln^{3+} cations, A$_2$Fe$_2$O$_{5\pm\delta}$ brownmillerites, Ruddlesden-Popper types (Ln, A)$_{n+1}$Fe$_n$O$_z$, and many other intergrowth compounds such as Sr$_4$Fe$_6$O$_{13\pm\delta}$ [189]. However, in most cases, an extensive iron substitution is required to increase the total conductivity to higher values, e.g., 10–30 S cm^{-1} at >450°C due to the structural constraints and defect chemistry features hampering electronic transport of the cathode. The maximum conductivity of the perovskite-based solid solutions under the cathode conditions has been achieved for Ln$_{1-x}$Sr$_x$FeO$_{3-\delta}$ in which Ln = La and $x = 0.5$ [196–199]. The moderate additions of the acceptor-type cations can enhance the concentration of oxygen vacancies, while decreasing the average cation radius in the Ln^{3+} sites and increasing A^{2+} content above 50% promote vacancy-ordering and hole localization processes with a negative impact on the transport properties of the cathode.

In comparison with the manganite-based perovskites, one of the most important problems of the ferrite-based perovskites is related to their high chemical and thermal expansion due to oxygen losses at elevated temperatures, which leads to the thermomechanical incompatibility with conventional solid electrolytes [200–203].

Partial substitution of iron with cations having more stable oxidation such as Ga, Al, Ti, and Cr can partly suppress the expansion due to a decrease in oxygen non-stoichiometry variations [203–206]. However, these dopants may decrease the electronic and ionic transports. With the incorporation of Ni as a dopant, La(Fe, Ni)O_3-based perovskites, the conductivity increases while the thermal expansion remains moderate at temperatures up to 800°C–1000°C. The layered ferrite phases in Ruddlesden-Popper series and brownmillerites exhibit better thermomechanical properties in comparison with perovskites, although the compatibility with electrolyte ceramics is still restricted [189].

Reducing the SOFC operating temperature causes lower oxygen vacancy concentration, thus lowering electrode material ionic conductivity. The content of lower-valence cations strongly affects the oxygen diffusivity; hence, it should be increased up to possible maximum. However, a large amount of acceptor doping usually results in a high thermal and chemical expansion because of the weakening of the metal-oxygen bonds and increasing the atomic vibration inharmonicity [189,203]. For example, in the well-known $La_{1-x}Sr_xCo_{1-y}Fe_yO_{3-\delta}$ (LSCF) system, the moderate additions of dopant significantly increase the total conductivity and electrochemical activity but also increases the thermal expansion coefficients. Figure 2.57 shows the electrochemical performance of $La_{1-x}Sr_xFeO_{3-\delta}$ and $La_{1-x}Sr_xCo_{0.2}Fe_{0.8}O_{3-\delta}$ cathode electrodes with different compositions. The maximum total conductivities reported for the LSCF systems are ~2000 and ~1600 S cm^{-1} at 600°C and 800°C, respectively, in the air for both $La_{0.7}Sr_{0.3}Co_{0.1}Fe_{0.9}O_{3-\delta}$ and $La_{0.6}Sr_{0.4}Co_{0.1}Fe_{0.9}O_{3-\delta}$ ceramics [207].

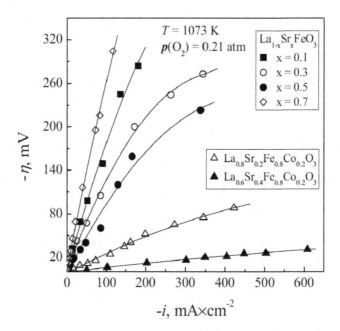

FIGURE 2.57 Current dependence of the cathodic overpotentials of porous $La_{1-x}Sr_xFeO_{3-\delta}$ and $La_{1-x}Sr_xFe_{0.8}Co_{0.2}O_{3-\delta}$ electrodes in contact with YSZ and doped ceria electrolytes, respectively. (Reprinted with permission from Ref. [189].)

Tai et al. [208] showed that the electronic conductivity of the LSCF ceramics with $0.8 \leq y \leq 1$ increases with temperature through a maximum, and then decreases. This may be ascribed to the loss of lattice oxygen at elevated temperatures. The average thermal expansion coefficients measured for the LSCF systems vary from 12.5×10^{-6} to 27.1×10^{-6} K^{-1} in the temperature range of 30°C–1000°C, depending on the ceramic chemical composition [189]. On the creation of La-site cation vacancies in LSCF, the conductivity and thermal expansion both decrease owing to the dominant charge-compensation mechanism with oxygen vacancy formation [189,209]. For $Ln_{0.8}Sr_{0.2}Co_{0.2}Fe_{0.8}O_{3-\delta}$ ceramics (Ln = La-Gd), similar effects on the thermal expansion coefficient and electronic conductivity were observed when Ln^{3+} size becomes smaller. The optimum combination of conductivity, thermomechanical compatibility, and electrochemical behavior in LSCF systems were obtained for the compositions with $x = 0.2$–0.5 and $y = 0.8$, considered as promising cathode materials for the intermediate-temperature SOFC with, particularly ceria-based electrolytes.

The electrochemical performance of the LSCF systems has also extensively been studied [5]. For example, Esquirol et al. [210] studied the electrochemical performance of the $La_{0.6}Sr_{0.4}Co_{0.2}Fe_{0.8}O_3$ cathode. The electrode polarization resistance values of the LSCF sintered at 850°C were 7.5, 0.23, and 0.03 ohm cm^2 at 500°C, 650°C, and 800°C, respectively, which were lower than those of the LSCF cathode sintered at 1000°C. Also, the results showed that the electrochemical activity of the LSCF electrode for the ORR at temperatures below 600°C can be described by a typical TPB model because of the remarkable reduction in the ionic conductivity of the material below 600°C. Tu et al. [211] studied the electrochemical activity of several perovskite oxides $Ln_{0.4}Sr_{0.6}Co_{0.8}Fe_{0.2}O_{3-\delta}$ (Ln = La, Pr, Nd, Sm, Gd) for the ORR. The results exhibited that the $Nd_{0.4}Sr_{0.6}Co_{0.8}Fe_{0.2}O_3$ has the best catalytic activity for the ORR, while $Pd_{0.4}Sr_{0.6}Co_{0.8}Fe_{0.2}O_3$ shows the highest polarization. The catalytic activity for the ORR decreases with the order of dopant at the A-site as:

$$Nd > Sm > La > Gd > Pr$$

However, this is not the same order as the electronic conductivity of the oxides indicating that the catalytic activity of mixed electronic and ionic-conducting oxides depends on other properties of the material such as ionic conductivity and oxygen exchange capability.

Calcium substitution on the La site reduces the sinterability of $LaFeO_3$. A dense $La_{1-x}Ca_xFeO_3$ can be obtained at a sintering temperature of 1320°C. The compositions with $x \leq 0.2$ showed high electronic conductivity of 88 S cm^{-1} at 800°C for $x = 0.15$ and good thermal expansion compatibility with the YSZ electrolyte. Increasing Ca substitution results in material decomposition and the second phase formation [212].

2.4.3 COBALTITE-BASED PEROVSKITE CATHODES

Compared to the ferrite-based oxides, cobaltite-based perovskite materials possess significantly better cathodic and transport properties. However, they have a higher thermal and chemical expansion [189]. The latter feature limits the compatibility of the cobaltite-based perovskite cathodes with solid oxide electrolytes, mainly to derivatives of δ-Bi_2O_3, $Bi_2VO_{5.5-\delta}$, and $La_2Mo_2O_{9-\delta}$. Similar to the manganite- and ferrite-based electrodes, much attention has been paid to perovskite-type (Ln, Sr) $FeO_{3-\delta}$ and their related solid solutions. At the same time, the layered cobaltites with a more stable state of Co cations have significantly been attractive due to relatively high mixed conductivity and fast exchange kinetics. Some important compositional families are: $LnBaCo_2O_{5+\delta}$ (Ln = Pr, Gd-Ho, Y), $LnBaCo_4O_{7+\delta}$ (Ln = Dy-Yb, Y), and also Ruddlesden-Popper type (Ln, $A)_4Co_3O_{10}$ and (Ln, $A)_2CoO_{4\pm\delta}$ (Ln = La-Nd).

Perovskite-type (Ln, $A)CoO_{3-\delta}$ compounds show a greater hole delocalization and mobility compared to Mn- and Fe-based analogues. The maximum total conductivity in $Ln_{1-x}A_xCoO_{3-\delta}$, predominantly p-type electronic has been achieved for Ln = La-Sm, A = Sr, and $x = 0.25$–0.5 at 530°C–1030°C, shifting toward lower x on heating [207,213–216]. Introducing Ln-site vacancies often leads to lower electronic and ionic conductivities. Moreover, the cobaltite-based perovskites have lower thermodynamic stability and are less tolerant to the cation non-stoichiometry compared to manganites and ferrites.

The interaction mechanisms between the cobaltite-based cathodes with the solid electrolytes are similar Mn- and Fe-based cathodes, while the cation inter-diffusion is often enhanced due to lower thermodynamic stability of the cobaltite phase [189]. Especially, the cobaltite cathodes may easily react with zirconia in comparison with the LSM. For the electrochemical cells with CeO_2- or $LaGaO_3$-based solid electrolytes, no blocking layers are formed with the cobaltite-based cathodes, although Co diffusion can have a detrimental effect on cell performance. To avoid this effect, it is necessary to decrease the fabrication temperature of the cell and to utilize sintering additives.

Among the cobaltite-based materials, lanthanum cobaltites have attracted much attention as the manganites and ferrites. However, there are some differences in the characteristics of these materials as the SOFC cathode materials. For instance, the $LaCoO_3$ is much less stable against the reduction compared to the $LaMnO_3$, the thermal expansion coefficient of $LaCoO_3$ is significantly higher than that of the $LaMnO_3$, and the $LaCoO_3$ has a higher electronic conductivity than $LaMnO_3$ under same conditions. Therefore, forming a solid solution of $LaMnO_3$ and $LaCoO_3$ can tailor the cathode thermal expansion and improve the electronic conductivity. Aruna et al. [217] studied the electronic conductivity and thermal expansion coefficient of the solid solution of $LaMn_{1-x}Co_xO_3$. The results showed that substitution of a low concentration of Co into $LaMnO_3$ could not improve the electronic conductivity where the sample with $x = 0.5$ showed the lowest conductivity due to the formation of the cubic phase of the solid solution. Also, the thermal expansion coefficient increases with an increase in Co concentration. The thermal expansion coefficient values of 12.5×10^{-6} and 16.3×10^{-6} K^{-1} were obtained for $LaMn_{0.7}Co_{0.3}O_3$ and $LaMn_{0.4}Co_{0.6}O_3$ solid solutions, respectively. The thermal expansion coefficient for $LaMnO_3$ is 11.33×10^{-6} K^{-1} [218]. Huang et al. [219] studied the electronic conductivity and thermal expansion coefficient of the Sr- and Ni-doped $LaCoO_3$ and $LaFeO_3$ perovskites. The results showed a good electronic conductivity and thermal expansion coefficient values of 15.6×10^{-6} and 12.8×10^{-6} K^{-1} for $La_{0.8}Sr_{0.2}Co_{0.8}Ni_{0.2}O_3$ (LSCN) and $La_{0.8}Sr_{0.2}Fe_{0.8}Ni_{0.2}O_3$

(LSFN), respectively. Hence, LSCN and LSFN are suggested as attractive alternative cathode materials to the conventional cathodes for the intermediate-temperature SOFCs.

Sr-doped $LaCoO_3$ has shown high electronic and ionic conductivities among lanthanum cobaltite materials. Petrov et al. [220] studied the electronic conductivity of $La_{1-x}Sr_xCoO_{3-\delta}$ with different chemical compositions in air, as shown in Figure 2.58. As can be seen, for $x = 0$, 0.1, and 0.2, the conductivity increases with temperature to reach a maximum value, then decreases. For $x \geq 0.3$, the conductivity continuously decreases with increasing temperature. $La_{1-x}Sr_xCoO_{3-\delta}$ exhibits a typical p-type conductivity in which the conductivity decreases with a decrease in oxygen partial pressure at high temperatures. However, the thermal expansion coefficient of Co-rich perovskites is usually very high for both doped CeO_2 and YSZ electrolytes. The average thermal expansion coefficient of $LaCoO_3$ is about 20×10^{-6} K^{-1}, which is significantly higher than those of YSZ and LSM.

Typical examples illustrating the superior performance of the SOFCs using cobaltite cathodes in contact with various solid electrolytes are presented in Table 2.8. Their high electrochemical activity is great interest for SOFC applications, provided that matching of thermal expansion can be achieved by the development of electrolyte-containing composites [221–223]. The maximum power density in LSGM-based SOFCs with $(La, Sr)MO_{3-\delta}$ cathodes at 1000°C increased in the

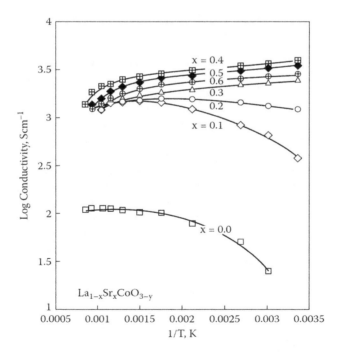

FIGURE 2.58 The electronic conductivity of $La_{1-x}Sr_xCoO_{3-\delta}$ in air as a function of Sr dopant concentration (x content). (Reprinted with permission from Ref. [220].)

TABLE 2.8

Examples of the Maximum Power Density in H_2 Fueled SOFCs with Cobaltite Cathodes

Cathode	Electrolyte	Anode	T (°C)	Maximum Power Density (W cm^{-2})
$La_{0.9}Sr_{0.1}CoO_3$	YSZ	Ni	1000	0.39
			800	0.05
$La_{0.9}Sr_{0.1}CoO_3$	LSGM	Ni	1000	0.93
			800	0.38
$La_{0.6}Sr_{0.4}CoO_3$	LSGM	Ni	1000	0.71
			800	0.21
$La_{0.6}Sr_{0.4}CoO_3$	LSGM	Co	1000	0.53
$La_{0.2}Sr_{0.8}CoO_3$	LSGM	Ni	800	0.44
$La_{0.2}Ca_{0.8}CoO_3$	LSGM	Ni	800	0.46
$La_{0.4}Ba_{0.6}CoO_3$	LSGM	Ni	600	0.12
$La_{0.2}Ba_{0.8}CoO_3$	LSGM	Ni	800	0.52
$Sm_{0.6}Sr_{0.4}CoO_3$	LSGM	Ni	800	0.44
$Sm_{0.5}Sr_{0.5}CoO_3$	LSGM	Ni	800	0.53
			600	0.08
$Sm_{0.5}Sr_{0.5}CoO_3$	CGO	Ni-CGO	600	0.27

Source: Data adapted and modified from Ref. [187].

order of M = Cr < Mn < Fe < Co. Similar trends are also observed for electronic conductivity, oxygen-ion diffusivity, and anion deficiency. In contrast, the lowest polarization resistance in (Ln, Sr)CoO_3 cathodes was repeatedly found for the compositions with Ln^{3+} cation radius smaller than that of the La^{3+}, which includes Sm, Nd, and Pr.

2.4.4 DEGRADATION MECHANISMS OF CONVENTIONAL LANTHANUM-BASED MANGANITE, FERRITE, AND COBALTITE CATHODE MATERIALS

In general, poisoning and corrosion of cathode by the external contaminants are the major issues degrading the SOFC cathodes long-term performance with ambient air supply. For example, CO_2 and humidity in the air during the cell storage and working condition always suffer the cathode, which can result in the cathode delamination from the electrolyte [224–228]. Moreover, Cr released from the commercial Cr-containing interconnect material can easily contaminate the cathode, which leads to the SOFC performance degradation [228]. Silica is also known as a typical contaminant from the sealant materials, which poisons the cathode through reacting with the cathode material, reducing the surface exchange reactions [228–231]. Figure 2.59 shows the possible sources of corrosion and poisoning of the cathode material. Therefore, the investigation of poisoning and corrosion of cathode materials, especially the most

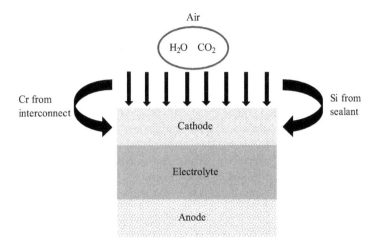

FIGURE 2.59 The possible poisoning and corrosion of cathode materials (drawn not to scale).

common manganite-, ferrite-, and cobaltite-based perovskite materials, is essential for the development of viable commercial SOFCs.

2.4.4.1 Effect of CO_2 on Cathode Performance

The CO_2 adsorption and its effects on oxygen reduction arising from the interaction with the $LaBO_3$ perovskite oxides (B = Cr, Mn, Fe, Co, Ni) have been investigated in several studies [232–234]. The chemical nature of B-site cation in the ABO_3 perovskite structure has been found to affect the adsorption characteristic of CO_2. The CO_2 coverage was shown to follow the order of $LaCrO_3 > LaFeO_3 > LaCoO_3$ [234]. Zhao et al. [235] studied the corrosion of $La_{0.8}Sr_{0.2}MnO_{3+\delta}$ (LSM) and $La_{0.6}Sr_{0.4}CoO_{3-\delta}$ (LSC) cathodes in the presence of CO_2. The results showed cell performance degradation as well as an increase in the polarization resistance with the addition of CO_2 into the O_2 stream. The CO_2 adsorption occurred in the temperature range of 550°C–650°C on the LSC cathode obeyed Temkin model, while it occurred in the temperature range of 650°C–800°C on the LSM cathode obeyed Freundlich model. The difference in CO_2 adsorption mechanism between LSM and LSC cathodes is mainly due to the diversity of the structure and chemical composition. On the LSM cathode, CO_2 can prevent dissociation of the adsorbed oxygen molecule or diffusion of oxygen species, while on the LSC cathode, oxygen dissociative adsorption is impeded because of CO_2 corrosion.

In other studies, the effects of atmospheric CO_2 on surface segregation and phase formation in $La_{0.6}Sr_{0.4}Co_{0.2}Fe_{0.8}O_{3-\delta}$ were investigated [236]. The results revealed that the kinetics of Sr surface segregation increased in the presence of the atmospheric CO_2 due to the increase in thermodynamic driving force for $SrCO_3$ formation. On the other hand, $La_{0.6}Sr_{0.4}FeO_{3-\delta}$ and $La_{0.6}Sr_{0.4}Fe_{0.8}Nb_{0.2}O_{3-\delta}$ cathodes showed a good CO_2 tolerance with a stable crystal structure and performance under CO_2 atmosphere [237,238].

In general, the effect of CO_2 on the cathode material can be expressed as:

- The adsorbed CO_2 on the cathode surface blocks the active sites for the ORR. Moreover, CO_2 tends to react with the alkaline-earth elements and form a metal carbonate blocking layer which has a negative effect on the cathode electrochemical performance;
- CO_2 adsorption on the cathode happens easily at lower temperatures. The stability of formed carbonate is also significantly affected by the acidity of materials;
- The adsorbed CO_2 certainly affect the performance of lanthanum-, strontium- and barium-based cathode materials;
- Introducing Fe or Nb into the B-site of the perovskite oxides can improve the CO_2 resistance;
- The performance degradation of cathode resulted from CO_2 corrosion can be reversible at the certain exposure time and temperature [228].

2.4.4.2 Effect of Humidity on Cathode Performance

Considering humidity effect on the cathode performance, the concentration of water vapor is usually negligible within the cell since water vapor is mainly formed by the electrochemical reactions at the SOFC anode with the oxygen-ion-conducting electrolyte. However, it is still critical to evaluate the humidity impact on cell performance since the SOFC cathode is continuously supplied with the ambient air. Furthermore, since water vapor is generated in the cathode side for the SOFCs with proton-conducting electrolytes, the concentration of moisture level in the cathode may be high, and thus, the performance of cathode may significantly be influenced by humidity.

The effect of humidity on the degradation and performance of LSM and LSM-YSZ materials has been studied [239–243]. The results showed an obvious OCV decrease with 20–40 vol% water steam addition to the cathode stream. The OCV drop was found to correlate with the low-conducting La_2O_3 phase formed on the LSM surface. However, a slight increase in OCV was observed with only 3 vol% water steam in the cathode stream at 800°C. A possible reaction process proposed for the degradation of LSM cathode at high water vapor concentration is described as follows:

$$LaMnO_3 + 1.5H_2O = La(OH)_3 + 0.5Mn_2O_3 \qquad (2.21)$$

$$2La(OH)_3 = La_2O_3 + 3H_2O \qquad (2.22)$$

The degradation of LSM at high water vapor concentrations is also shown in Figure 2.60. LSM cathode surface change includes a decrease in the manganese concentration and an increase in the strontium concentration. A formation of zirconia and zirconate phases was also observed for LSM-YSZ composite cathode materials as a result of humidity and oxygen partial pressure in the cathode [243].

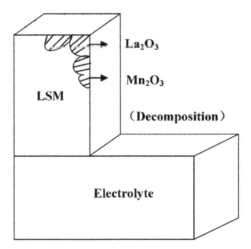

FIGURE 2.60 Degradation mechanism of LSM cathode at very high water vapor concentrations. (Reprinted with permission from Ref. [241].)

The impact of humidity on LSCF materials as one of the promising cathode material has not completely been understood. Liu et al. [241] compared the effect of steam with a different concentration in the air on the LSM and LSCF cathodes. The results showed that both LSM and LSCF suffer degradation under high water vapor concentration due to the formation of Mn_3O_4 phase on the LSM and Sr enrichment as well as SrO formation on the LSCF, respectively. In addition, the LSCF cathode exhibited less tolerance to water vapor compared to the LSM cathode in long-term operation. In the other study, very high tolerance to humidification in the cathode stream was obtained for the LSCF-CGO composite cathode with no detectable degradation [244]. Furthermore, LSCF showed different stability behaviors in low and high temperatures with CO_2/H_2O containing air. At 750°C, the LSCF cathode showed good electrochemical performance and high stability due to the low reactivity of LSCF with CO_2 and H_2O. However, due to $SrCO_3$ formation at the surface, the oxygen reduction activity of LSCF was severely affected below 680°C with O_2-CO_2 (2.83%)-H_2O (2.64%) stream, leading to the degradation of performance. In the CO_2 (5%)-H_2O (2.81%)-He atmosphere, $SrCO_3$ phase is formed at 400°C-680°C as a result of the reaction between LSCF and CO_2 in which the existence of water increases such reactions [225].

In general, the effect of humidity on the cathode material can be expressed as:

- The humidity effect depends on the nature of the cathode material;
- A cell with an improved cathode/electrolyte interface can be more durable in humid air;
- The cathode performance degradation due to humidity can be reversible at a certain exposure time and temperature [228].

2.4.4.3 Effect of Cr on Cathode Performance

The significant tendency to decrease the SOFCs operating temperature greatly increases the feasibility of using low-cost metallic materials for interconnect rather than ceramic materials. Although metallic materials have high thermal and electronic conductivity, low ionic conductivity, good machinability, and low cost, they suffer from oxidation and corrosion, particularly at elevated temperatures, leading to a substantial increase in ohmic resistance of SOFC stack. Among the potential metallic materials, Cr-rich alloys have attracted much attention due to their high resistance to oxidation at high temperatures. However, the volatilization of gaseous Cr^{6+} and formation of gaseous Cr species ($CrO_2(OH)_2$) lead to a drastic degradation of cathode performance, which has become a major problem for SOFC development [228]. The above reactions are presented as follows:

$$1.5O_2(g) + Cr_2O_3(s) = 2CrO_3(g) \tag{2.23}$$

$$O_2(g) + 4H_2O(g) + 2Cr_2O_3(s) = 4CrO(OH)_2(g) \tag{2.24}$$

$$1.5O_2(g) + 2H_2O(g) + Cr_2O_3(s) = 2CrO_2(OH)_2(g) \tag{2.25}$$

The schematic diagram of the effect of gaseous Cr species on the cathode is shown in Figure 2.61.

Jiang and Zhen [246] studied the Cr deposition mechanism in LSM cathode. The results showed that at high temperatures, Mn^{2+} is formed during cathode polarization, which plays as a nucleation agent for Cr_2O_3 formation from Cr-Mn-O nucleus. The formation of stable Cr-Mn-O nuclei accelerates the crystallization and growth of Cr_2O_3 oxide, and (Cr, Mn)$_3$O$_4$ and/or (Cr, Mn)$_2$O$_4$ spinel phases. The Cr deposition process depends on the kinetics of nucleation reaction and the interactions between formed nuclei and the volatile Cr species.

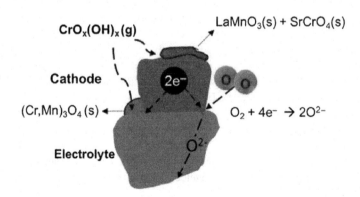

FIGURE 2.61 Cr poisoning mechanism. Volatile Cr species at the TPB are reduced to form a solid compound such as Cr_2O_3, blocking active site for the ORR. (Reprinted with permission from Ref. [245].)

The electrical current is also important in the interaction between the cathode and interconnect, thus, corrosion of Cr from the metallic interconnect (stainless steels) and poisoning the cathode. In addition, it has been shown that Mn is a critical element for Cr deposition under polarization. Mn deficient composition significantly inhibits Cr deposition on the cathode [247].

The LSM/YSZ composite cathodes have shown improved tolerance toward Cr deposition compared to the pure LSM cathodes. In one study, the microstructural analysis showed that there is no preferential deposition of Cr species at the electrode–electrolyte interface for the reaction on the LSM/YSZ composite electrode [248]. Hessler-Wyser et al. [249] also examined the Cr distribution at the LSM/YSZ composite cathode after cell polarization at 800°C for 1900 h. As observed in Figure 2.62, strontium zirconate (SZO) layer was formed at the LSM and YSZ interface. In the region near the LSM/YSZ interface, Cr was not detected (Figure 2.62a);

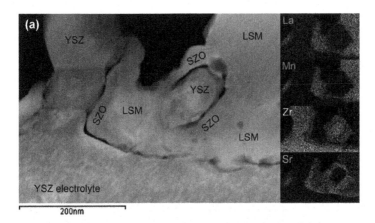

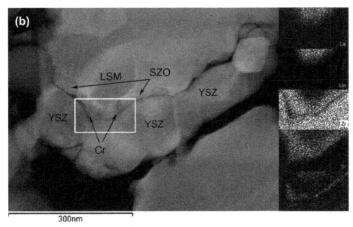

FIGURE 2.62 TEM images and EDS mapping of the LSM/YSZ composite cathode polarized at 800°C for 1900 h. (a) area at the electrode–electrolyte interface; (b) area at the electrode bulk. (Reprinted with permission from Ref. [249].)

however, in the regions in the bulk of the LSM/YSZ composite cathodes, Cr was observed between LSM and YSZ grains and within SZO layer (Figure 2.62b). At the areas other than TPBs such as the interfaces between two LSM grains, Cr_2O_3 was also detected.

The Cr deposition on the mixed ionic and electronic conducting LSCF cathode has also been investigated. The preferential Cr deposition regions at the surface of LSCF electrode were observed rather than LSCF–electrolyte interface under cathodic polarization [250]. The impact of sintering temperature on Cr poisoning of LSCF-GDC electrode was also studied. The results revealed that lowering sintering temperature increases the electrode surface area, which leads to more Sr-Cr-O nucleation sites, and thus, greater Cr degradation [251].

In general, the effect of Cr on the cathode material is complicated and can be expressed as:

- Lowering the operating temperature can effectively mitigate Cr deposition issue;
- Reducing the surface segregation and improving the stability of material structure can significantly increase the resistance of cathode toward Cr poisoning [228].

2.4.4.4   Effect of Si on Cathode Performance

One of the common impurities in the SOFCs systems hampering wide-spread SOFC commercialization to some extent and degrading long-term cell performance is silicon. Si, usually introduced by the glass-based sealing material as well as during cell operation and material fabrication, has attracted extensive attention for a long time due to its detrimental effect on electrolyte materials, thereby improving cell performance. Recently, the effect of Si on cathode materials has also been investigated. The results showed that Si introduced from glass-based sealing materials as well as current collectors, can seriously decrease the oxygen exchange coefficient in the SOFCs. At high temperatures, Si can poison the Sr containing perovskite materials, which deteriorates the cell performance [252–254]. Yokokawa and Bucher conducted the chemical equilibrium calculations for $La_{0.8}Sr_{0.2}MO_3$ (M = Mn, Fe, Co) perovskites and SiO_2. The results demonstrated that stable products such as Sr_2SiO_4 and $LaSiO_7$ are formed at 800°C under oxygen partial pressure of 0.2 atm. For the LSM/YSZ composite cathode, a thin glass film is formed along the pore surfaces of the LSM/YSZ cathode when it is exposed to silicate glass at 750°C for 100h [255,256].

Bucher et al. studied the degradation of oxygen exchange coefficient of LSCF cathode in dry and humidified air at 600°C [257–259]. The results showed Sr enrichment at the LSCF surface, revealing a pronounced Si accumulation in humidified air due to a very thin silicate layer formation at the LSCF surface, which hinders surface oxygen exchange active sites. On the other hand, although Si is mainly considered a detrimental contaminant for SOFC materials, recent studies showed that introducing Si into the cathode perovskite materials such as $La_{0.6}Sr_{0.4}Co_{0.78}Fe_{0.195}Si_{0.025}O_{3-\delta}$ and $Sr_{0.5}Ca_{0.5}Fe_{0.85}Si_{0.15}O_{3-\delta}$ can improve the electronic conductivity and increase the cathode stability against CO_2 poisoning [260].

In general, the effect of Si on the cathode material can be expressed as:

- The effect of Si on the cathode material is influenced by humidity;
- The incorporation of Si into the composition of cathode materials is meaningful [228].

2.4.5 DOUBLE PEROVSKITE MATERIALS ($AA'B_2O_{5+\delta}$)

In the past decades, a new family oxide with the general formula $AA'B_2O_{5+\delta}$ (with A = rare earth, A' = alkaline earth, and B = Co or Mn), namely double perovskites, has been proposed as next-generation cathode materials for intermediate-temperature SOFCs, demonstrating very good electrochemical performance.

Most of the studies carried out for cathode applications have focused on pure or partially substituted $ReBaCo_2O_{5+\delta}$ (RBCO) cobaltites, where rare-earth (Re) ion occupies the A site, barium occupies the A' site, and cobalt occupies the B site. The RBCO cobaltites have shown a high concentration of oxygen vacancies, thus, good oxygen transport properties, high electronic conductivity, and high catalytic activity which make them very attractive mixed ionic and electronic-conducting materials (see Table 2.7). Systematic studies on RBCO oxides have shown a decrease in thermal expansion coefficient as well as electronic and ionic conductivities with the decreasing size of the Re^{3+} ions from Ln = La to Y, as demonstrated for the Re = Nd, Sm, Gd and Y series and for the Re = Pr, Nd, and Sm series, resulting in higher power densities and lower ASR with a decrease in Re^{3+} ionic radii, both for composite and single-phase cathode materials [186]. ASR values reported for $GdBaCo_2O_{5+\delta}$ (GBCO), $PrBaCo_2O_{5+\delta}$ (PBCO), and PBCO/CGO composites were as low as 0.25 ohm cm^2 (at 625°C), 0.213 ohm cm^2 (at 600°C), and 0.15 ohm cm^2 (at 600°C), respectively [261–263]. Moreover, a fuel cell composed of $PBCO/Ce_{0.8}Sm_{0.2}O_{1.9}/Ni-Ce_{0.8}Sm_{0.2}O_{1.9}$ showed a maximum power density of 583 mW cm^{-2} at 600°C [264].

However, for the high-Re ions, the thermal expansion coefficient largely increases with an increase in the size of Re^{3+} (from Y to La), varying from 15.8 to 24.3×10^{-6} K^{-1}. This results in a large mismatch between the cathode and conventional electrolytes [186]. In order to reduce the thermal expansion coefficient while maintaining good electrochemical properties, it has been proposed to lower the Co content (e.g., by partial substitution with Sc or Ni) and use Ba-deficient compounds ($ReBa_{1-x}Co_2O_{5+\delta}$) [265–268]. Regarding the chemical compatibility with common electrolyte materials, although all these double perovskites are unstable in contact with YSZ electrolyte at high temperatures, their stability with CGO and LSGM decreases with a decrease in the Ln^{3+} cation size. In addition, Sr substitution and Co/Sr co-substitution have improved chemical stability and electrochemical performance [186]. In particular, the $PrBa_{0.5}Sr_{0.5}Co_{1.5}Fe_{0.5}O_{5+\delta}$ cathode composition has shown very low ASR and very high peak power density (2160 mW cm^{-2} at 600°C) in the cell using the Ni-CGO anode composite and CGO as the electrolyte [269].

2.4.6 RUDDLESDEN-POPPER SERIES ($A_{n+1}B_nO_{3n+1}$)

Among the alternatives to perovskites, layered A_2BO_4 oxides with a K_2NiF_4-type structure, known as the $n = 1$ Ruddlesden-Popper (RP) phase with a general formula of $A_{n+1}B_nO_{3n+1}$, have been widely investigated. The content of these materials can be either hyper- or hypo-stoichiometric ($A_2BO_{4\pm\delta}$), depending on A-site and B-site doping, temperature, and oxygen partial pressure. Hence, this results in the presence of the oxygen vacancies or interstitials, which are the pathways for oxygen diffusion. The great advantage for this structure is its capability to accommodate the oxygen interstitials in the rock-salt (AO) layers, which provides good ionic conductivity without a need for aliovalent substitution at the A-sites. Sr- and Ba-free (alkaline-earth-free) compositions can be stable for long-term performance since a mixed A-site occupation may decrease cation segregation and may largely reduce Cr and S poisoning [186,270].

So far, the most widely studied RP compounds for the cathode applications are the $Ln_2NiO_{4\pm\delta}$ nickelates, where Ln = La, Nd, or Pr. These compounds not only have high diffusivity of the oxygen interstitial ions and good catalytic properties but also possess a relatively low thermal expansion coefficient which causes a good match with the conventional electrolyte materials (see Table 2.7). However, the moderately low electronic conductivity of these RP compounds limits the electrochemical performance. Previous studies showed low ASR values for $La_2NiO_{4+\delta}$ (~1.2 ohm cm^2) and $Nd_2NiO_{4+\delta}$ (~2 ohm cm^2) cathodes at 600°C in the cells using YSZ together with a yttria-doped ceria backing layer [271,272]. However, the best-performing nickelates are Pr-based nickelates composites, $Pr_2NiO_{4+\delta}$, which have shown the lowest polarization resistance and superior electrochemical performance (see Table 2.7). For this cathode composition in an anode-supported cell using a Co-doped CGO layer between the cathode and a zirconia electrolyte, the peak power density of 400 mW cm^{-2} at 600°C was achieved [186,273].

For these RP phases, a careful selection of electrolyte material, as well as the fabrication and operation temperatures, are important. For instance, the $La_2NiO_{4+\delta}$ cathode is stable with the LSGM electrolyte at temperatures even up to 1000°C, while it reacts with YSZ and CGO electrolytes at temperatures above 900°C. In addition, it has been shown that the $La_2NiO_{4+\delta}$ cathode performs poorly as a single-phase cathode in an anode-supported YSZ cell, while the cell performance is significantly improved with an SDC/$La_2NiO_{4+\delta}$ interlayer, showing the maximum power density of 2200 mW cm^{-2} at 800°C. This indicates a better performance of the $La_2NiO_{4+\delta}$ cathode composite compared to that of the single-phase $La_2NiO_{4+\delta}$ cathode despite its mixed ionic and electronic conductivity. For the $Pr_2NiO_{4+\delta}$ and $Nd_2NiO_{4+\delta}$ cathodes, using a ceria-based interlayer between the YSZ electrolyte and cathode helps to improve the cell performance and to avoid the delamination problems during operation [273–277].

2.4.7 OTHER CATHODE MATERIALS

Phillipps et al. [278] studied the electronic conductivity and thermal expansion coefficient of $Gd_{0.8}Sr_{0.2}Co_{1-y}Mn_yO_3$ (0.1 ≤ y ≤ 1.0) and $Gd_{0.7}Ca_{0.3}Co_{1-y}Mn_yO_3$ (0.3 ≤ y ≤ 1.0) cathode systems. The results showed a higher electronic conductivity for the $Gd_{0.8}Sr_{0.2}Co_{1-y}Mn_yO_3$ system. The highest conductivity was obtained for the composition of $Gd_{0.8}Sr_{0.2}Co_{0.9}Mn_{0.1}O_3$ (271 S cm^{-1} at 800°C). However, its thermal expansion coefficient ($\sim$24 $\times$ 10^{-6} K^{-1}) was much higher than the thermal expansion coefficient of YSZ electrolyte ($\sim$10 $\times$ 10^{-6} K^{-1}). More thermal expansion compatibility was achieved by increasing the Mn content. The polarization performances of both $Gd_{0.8}Sr_{0.2}Co_{1-y}Mn_yO_3$ and $Gd_{0.7}Ca_{0.3}Co_{1-y}Mn_yO_3$ systems were moderate.

Kostogloudis et al. [279–281] investigated electronic conductivity and thermal expansion coefficient of Sr-doped praseodymium manganites and cobaltites systems [i.e. (Pr, Sr)MnO$_3$, (Pr, Sr)CoO$_3$, and (Pr, Sr)(Co, Mn)O$_3$] with the orthorhombic perovskite GdFeO$_3$-type structure. Among the examined oxide cathodes, the composition of $Pr_{0.5}Sr_{0.5}MnO_3$ showed the best properties with the electronic conductivity of 250 S cm^{-2} at 600°C and a thermal expansion coefficient of 12.2 $\times$ 10^{-6} K^{-1}, required for the intermediate-temperature SOFCs.

LaNiO$_3$ compound is known to have very promising electronic conductivity at room temperature, however, it is decomposed to NiO and La$_2$NiO$_4$ phases with poor conductivity at temperatures above 850°C. Previous studies showed that the LaNiO$_3$ material could be stabilized even at high temperatures while possessing both high electronic conductivity and a thermal expansion coefficient close to that of the common zirconia electrolytes if some of the Ni is substituted by Co or Fe ($LaNi_{1-x}Co_xO_3$ and $LaNi_{1-x}Fe_xO_3$ systems) [282–285].

There are also other families of materials that have been proposed as alternative cathode materials for intermediate-temperature SOFCs:

- Layered ferrites, including brownmillerite, RP, and other more complex intergrowth structures such as $Sr_2Fe_2O_5$, Ca_2FeO_4, $Sr_3Fe_2O_{6+\delta}$, (Sr, La)$_3$(Fe, Co)$_2O_{6+\delta}$, $Sr_4Fe_6O_{12+\delta}$, and $Ba_{1.6}Ca_{2.3}Y_{1.1}Fe_5O_{13}$;
- Layered cooperates such as $LaBaCuMO_{5+\delta}$ and $YSr_2Cu_2MO_{7+\delta}$ (M = Fe, Co);
- Pyrochlore-structured ruthenates with the general formula of $A_2B_2O_7$ such as $Bi_2Ru_2O_7$, $Pb_2Ru_2O_7$, and $Gd_{2-x}La_xZr_2O_7$;
- Tetrahedrally coordinated Co compounds such as RBa(Co, M)$_4O_7$ (R = Y, Ca, In and M = Zn, Fe, Al);
- Bismuth oxides based on BIMEVOX family such as $Bi_4V_2O_{11}$;
- Rutile-structured compounds such as $Ir_{0.5}Mn_{0.5}O_2$; and
- Spinel-structured compounds such as $Mn_{1.5}Co_{1.5}O_4$ [186].

2.5 INTERCONNECTS

SOFC single cells basically consist of two porous electrodes separated by a gas-tight oxygen-ion-conducting electrolyte. To build up a useful voltage, cells are connected in series via an interconnect, also known as bipolar plate. As schematically illustrated in Figure 2.63, the multiple cells are stacked in electrical series, requiring an interconnect for connecting the cells. In both tubular and planar stack designs, interconnect physically separates yet electrically connects the anode of one cell to the cathode of the other. The material requirements for the interconnect, either ceramic or metal are the most demanding [286]. Ceramic has traditionally been used as an interconnect material due to its high heat resistance. However, reducing

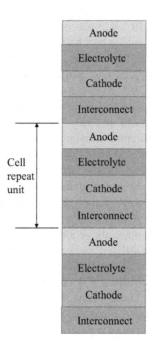

FIGURE 2.63 Schematic of a flat SOFC stack, including three individual fuel cells. (Adapted from Ref. [286].)

the SOFC operating temperature to the range of 600°C–800°C by the development on thinner electrolytes led researchers to pursue opportunities for lowering the interconnect fabrication cost by using metal alloys [287–289]. In comparison with ceramic interconnects, metallic interconnects have shown higher heat and electronic conductivity, higher oxidation resistance, higher mechanical strength, and better manufacturability [290–293].

The criteria for the interconnect materials are the most stringent among all cell components. In general, the interconnect should meet the following requirements:

- Excellent electrical conductivity. The ASR level is accepted to be below 0.1 ohm cm^2;
- Sufficient dimensional, microstructural, and chemical stabilities in both oxidizing and reducing environments at SOFCs operating temperatures for 40,000 h (service lifetime);
- Matched thermal expansion coefficient with those of electrodes and electrolyte to minimize the thermal stresses during start-up and shutdown;
- High oxidation, sulfidation, and carbon cementation resistances;
- No interdiffusion and reaction with the adjoining cell components;
- Sufficient mechanical strength and creep resistance and elevated SOFCs operating temperatures;
- Ease of fabrication and shaping as well as low cost [286,294].

2.5.1 CERAMIC INTERCONNECTS

The traditional ceramic interconnect used in the high-temperature SOFCs (~1000°C) is lanthanum chromite ($LaCrO_3$) with high stability in both cathodic and anodic environments and melting point higher than 2300°C. This material exhibits excellent electronic conductivity under SOFC operating conditions in comparison with typical ceramics [295]. The electronic conductivity of $LaCrO_3$ has shown to be improved by substituting La^{3+} with large cations such as Sr^{2+} and Ca^{2+} and replacing Cr^{3+} with small cations such as Ni^{2+} and Cu^{2+} [286,296,297]. This is associated with lattice distortion arising from a difference in ionic size between the host and the dopant cations, which leads to an increase in charge carrier mobility, and electronic conductivity, subsequently. The thermal expansion coefficient of $LaCrO_3$ is 9.5×10^{-6} K^{-1}, which is compatible with that of the YSZ (10.5×10^{-6} K^{-1}) [298]. Acceptor doping not only alters the electronic conductivity of $LaCrO_3$ but also modifies its thermal expansion coefficient [286]. For instance, Co and Fe dopants both increase the electronic conductivity of $LaCrO_3$ to various levels, but Co doping exhibits a considerable increase in the thermal expansion coefficient of $LaCrO_3$, while Fe lowers the thermal expansion coefficient. Hence, the extent to which acceptor doping improves the properties of $LaCrO_3$ is revealed to differ remarkably.

On the other hand, the $LaCrO_3$ has several weaknesses. The major challenge for the $LaCrO_3$-based materials is their poor sinterability in air due to easy volatilization of Cr(VI) species. It is almost impossible to sinter $LaCrO_3$ to full density in an oxidizing atmosphere. This problem can be addressed by either lowering the vapor pressure of Cr species or sintering under a reducing atmosphere [299–302]. Other certain weaknesses of $LaCrO_3$ are: (1) reduction in conductivity with decreasing oxygen partial pressure as $LaCrO_3$ becomes oxygen-deficient; (2) higher price of lanthanum compared to typical engineering materials; and (3) limitation in the geometry of interconnect made from ceramic $LaCrO_3$ during the fabrication process. In addition, at temperatures below 800°C, the electronic conductivity of doped $LaCrO_3$ has been reported to experience substantial degradation. Thus, this hampers the usage of $LaCrO_3$ interconnects in intermediate-temperature SOFCs operating in the temperature range of 600°C–800°C [286,294]. Due to the abovementioned weaknesses, significant research efforts have been conducted on finding an alternative material to be used for interconnects.

2.5.2 METALLIC INTERCONNECTS

With developing intermediate-temperature SOFCs and reduction of operating temperature to 600°C–800°C, metallic materials have been utilized to replace $LaCrO_3$ interconnects due to their superior advantages in comparison with their high-temperature counterpart [303–308]. Although metallic interconnects were originally developed for the electrolyte-supported planar SOFC, the concept of anode-supported SOFC design progressed substantially over the past years, enabled the replacement of $LaCrO_3$ interconnects with metallic materials at operating temperatures below 800°C. Metallic materials have high thermal and electronic conductivities, eliminating the presence of thermal gradient along the interconnect plane and across the components, and decreasing the cell resistance, thus an increased output. In addition, metallic interconnects have high mechanical strength and have been designed with extra channels to distribute the gas in various flow configurations. This could not be provided by the traditional $LaCrO_3$ interconnects. Last but not least, the fabrication and handling processes of metallic interconnects are easy with relatively low cost [309]. The most popular choices of metallic materials for interconnect applications are Cr-, Fe-, and Ni-based alloys due to their high oxidation and corrosion resistance by forming protective oxide scales of Cr_2O_3 and Al_2O_3. In these alloys, there should be enough Cr content to form a continuous Cr_2O_3 oxide scale to avoid oxidation under SOFC operating conditions. On the other hand, the amount of Al content should be controlled at a minimum to prevent the formation of a continuous Al_2O_3 oxide scale due to the insulating nature of the Al_2O_3 scales [303]. The ideal situation for these interconnects is that they possess sufficient oxidation resistance over the projected service lifetime of the SOFC (40,000 h) with the protective oxide scales exhibiting high electronic conductivity as well. In general, these oxide scales should be not only oxidation and corrosion resistant but also chemically stable and interfacially well-adhered with the substrate as well as exceptionally slow-growing [286]. Table 2.9 compares Cr-, Fe-, and Ni-based interconnects in terms of thermal expansion coefficient match, oxidation resistance, electronic conductivity, mechanical strength, manufacturability, and cost.

2.5.2.1   Cr-Based Alloys

Cr-based alloys are favorable as interconnect materials due to their high oxidation resistance along with good corrosion resistance. Ducrolloy ($Cr-5Fe-1Y_2O_3$), the most representative alloy among Cr-based oxide dispersion strengthened (ODS) alloys used to replace $LaCrO_3$ at elevated SOFC operating temperatures (900°C–1000°C), was first designed by Plansee Company to match the thermal expansion coefficients of other cell components [310]. Ducrolloy also demonstrates excellent oxidation resistance. Other examples for Cr-based ODS alloys are: $Cr-5Fe-0.5CeO_2$, and $Cr-5Fe-1.3La_2O_3$, and $Cr-5Fe0.3Ti-0.5Y_2O_3$.

TABLE 2.9
Comparison Between the Potential Metallic Interconnect Candidates

Interconnect Material	Thermal Expansion Coefficient (K^{-1}; 25°C–800°C)	Oxidation Resistance	Electronic Conductivity	Mechanical Strength	Fabrication	Cost
Cr-based alloy	$11.0–12.5 \times 10^{-6}$	Good	Good	High	Difficult	Very expensive
Fe-based alloy	$11.5–14.0 \times 10^{-6}$	Good	Good	Low	Readily	Inexpensive
Ni-based alloy	$14.0–19.0 \times 10^{-6}$	Good	Good	High	Readily	Expensive

Source: Data adapted and modified from Ref. [309].

The Cr-based alloys are also chosen due to higher electronic conductivity of Cr_2O_3 in comparison with other oxides [286,296]. However, Cr poisoning of the cathode as a result of Cr(VI) species migration/volatilization from the interconnect into the cathode, and excessive Cr_2O_3 growth in these alloys are inevitable due to their high Cr content. The kinetics and mechanism of these phenomena have been well studied elsewhere [311–316]. Excessive grown Cr layer results in spallation after thermal cycling. In addition, ODS alloys are difficult and expensive to be fabricated. In general, the technique used to fabricate near-net-shape ODS interconnects is powder metallurgy due to the sensitivity of oxides dispersion by melting [317].

2.5.2.2 Fe-Cr-Based Alloys

Fe-Cr-based alloys have exhibited better workability along with higher ductility and lower cost in comparison with Cr-based alloys [286]. Fe-Cr-based alloys, specifically ferritic stainless steels with an optimum Cr content of 17%–25%, provide promising oxidation resistance by forming a continuous Cr_2O_3 scale [303,309]. It has been reported that the critical Cr content for the formation of a continuous protective Cr_2O_3 layer on the substrate alloy is approximately 20%–25% [303]. Note that the oxidation resistance is significantly reduced with lowering Cr content to 5%–10% [318,319]. For the very low Cr content (<5%) Fe-Cr-based alloys, the scales formed on the substrate are nearly pure Fe oxide accompanied by $FeCr_2O_4$ spinels and/or internal Cr_2O_3 oxide precipitates where the scales become richer in Cr_2O_3 and spinel with increasing Cr content, accompanied with a decrease in the scale growth rate.

As mentioned, ferritic stainless steels are usually the most promising candidates for SOFC interconnect applications due to their body-centered cubic (BCC) structure, providing quite close thermal expansion coefficient to that of the other SOFC components which is important to avoid spallation or cracks due to mechanical stresses induced by thermal cycling. Furthermore, the processing methods of this type of alloys are quite simple. However, impurities from the steel substrate such as Al and Si can negatively affect the interconnect performance by forming continuous layers of Al_2O_3 and SiO_2, which significantly increases ASR upon oxidation for 4000 h [294,309].

Aside from the advantages mentioned, ferritic stainless steels are exposed to Cr_2O_3 scale growth at elevated temperatures. Also, this scale can grow up to tens of microns after thousands of hours of exposure of interconnect to SOFC environment at intermediate temperature (~800°C) [293,320]. The Cr_2O_3 scale grows continuously with a long exposure of the alloy to a high temperature, which increases the electronic resistance. Further exposure will cause the spallation of the scale and crack formation due to the thermal cycle during SOFC operation. Under cathodic environment, the Cr_2O_3 scale growth also results in thermodynamic instability, leading to generate volatile Cr(VI) species, particularly gaseous chromium oxide (CrO_3) and chromium oxyhydroxide ($CrO_2(OH)_2$), through the following reactions [286,309]:

$$2Cr_2O_3(s) + 3O_2(g) = 4CrO_3(g) \qquad (2.26)$$

$$2Cr_2O_3(s) + 3O_2(g) + 4H_2O(g) = 4CrO_2(OH)_2(g) \qquad (2.27)$$

$$Cr_2O_3(s) + O_2(g) + H_2O(g) = 2CrO_2OH(g) \qquad (2.28)$$

Volatilization of Cr species strongly depends on the water content and the oxygen partial pressure. This process is negligible at the anode side as the highest equilibrium vapor pressure of C–O–H gas species at the anode side (10^{-14} to 10^{-10} Pa) is much lower compared to that of the cathode side (10^{-9} to 10^{-6} Pa) [316]. Hence, Cr gas poisoning is more significant at the cathode side where the Cr(VI) species is formed via a reaction, expressed in Equations (2.26)–(2.28). The mechanism and kinetics of the Cr poisoning at the cathode side is inconsistent and can occur under two theories: (1) volatile Cr(VI) species is either electrochemically reduced or chemically precipitated at the TPB, thus, blocking the active cathode sites; (2) nucleation theory indicating the chemical reduction of high-valence Cr species is facilitated by the nucleation agents on the electrode surface, electrolyte surface, and/or at the interface between these two components [293,317,321,322]. Either way, both theories lead to the prevention of ORR and cell performance degradation. Hence, researchers have focused on developing a suitable coating to overcome this issue and increase the cell lifetime.

2.5.2.3 Ni-Cr-Based Alloys

Compared to the Fe-Cr-based alloys, Ni-Cr-based alloys always exhibit higher oxidation resistance and electronic conductivity. In the Ni-Cr-based alloys, only 15% Cr is needed to obtain a continuous Cr_2O_3 layer and establish a sufficient hot corrosion resistance, which is lower than that of the Fe-Cr-based alloys. In addition, Ni-based alloys have higher mechanical strength [294].

Most Ni-Cr-based alloys demonstrate excellent oxidation resistance in moist hydrogen environment where a thin scale of Cr_2O_3 and $(Mn, Cr, Ni)_3O_4$ spinels are formed on the substrate. Hence, they can be used as clad metal or plated layer at the anode side [323–325]. In air oxidation during high-temperature exposure, high Cr containing alloys such as Haynes 230 and Hastelloy S form a thin layer of Cr_2O_3 and $(Mn, Cr, Ni)_3O_4$ spinel. On the other hand, low Cr containing alloys such as Haynes 242 form a thick double-layer consisting of a NiO layer above a Cr_2O_3 scale, raising concern over its oxidation resistance for the interconnect applications [294,323].

The most important challenge with the Ni-Cr-based alloys is related to their potential thermal expansion coefficient mismatch to cell components. Therefore, a novel design of interconnect is necessary to take full advantages of the Ni-Cr-based alloys. Jablonski and Alman [326] have developed a new series of Ni-Cr-based alloys containing Mo, W, Ti, Al, etc. with thermal expansion coefficient similar to ferritic stainless steels, and oxidation behavior and mechanical properties comparable to that of the commercial alloy Haynes 230 which potentially can be used as SOFC interconnect.

2.5.3 INTERCONNECT PROTECTIVE COATINGS

Recently, in order to mitigate excessive Cr_2O_3 scale growth and Cr poisoning, various coatings have been developed for SOFC metallic interconnects. The coating utilized should have the following characteristics: (1) the diffusion coefficients of

Cr and O in the coating should be minimal so that the transport of Cr and O can effectively be hindered; (2) the coating should be stable and chemically compatible with the substrate and other cell components, particularly electrodes, seal materials and contact pasts; (3) the coating should have low ohmic resistance to maximize the electronic conductivity; (4) it should be thermodynamically stable in both oxidizing and reducing environments at SOFC operating temperature range; and (5) the coating should possess thermal expansion coefficient match with the substrate to be resistant to spallation during thermal cycling [294]. Three types of materials have widely been used for the interconnect coating applications: nitrides, perovskites, and spinels. Also, various deposition techniques such as physical vapor deposition (PVD), chemical vapor deposition (CVD), sol-gel dip coating, magnetron sputtering, plasma spray, electrolytic deposition (ELD), and electrophoretic deposition (EPD) have been used for SOFC interconnect coatings in order to reduce the fabrication cost [293]. Among these techniques, EPD is very well known in academia and industry due to its promising strengths such as various material combinations, low equipment and fabrication costs, and uniform layer deposition on complex shapes. These advantages can lead to developing a cost-effective SOFC interconnect in more complex shapes with better performance.

Nitride coatings have widely been used due to their excellent wear resistance. In addition, they have always been considered as an alternative coating for interconnect applications because of their high-temperature stability and low resistance [327]. PVD has been the method widely used to prepare protective nitride coatings for SOFC interconnects due to its capability to control coating material composition and morphology as well as the versatility of this technique. Previous studies showed that PVD is an effective method to fabricate high-quality Cr-N, Cr-Al-N, and Cr-Al-O-N coatings for the SOFC interconnect applications, reducing Fe and Cr migration from the substrate [328–330]. In addition, TiAlN and SmCoN coatings revealed good stability at 700°C, which is helpful to prevent Fe and Cr migration from the substrate. A quite low ASR has also been obtained for short-term operation [331,332]. The main drawback of the nitride coatings is that they are not quite stable at the temperature above 600°C.

Among perovskite coatings, $LaCrO_3$, traditional interconnect material, as well as Sr-doped lanthanum chromite ($La_{1-x}Sr_xCrO_3$), Sr-doped lanthanum cobaltite ($La_{1-x}Sr_xCoO_3$), Sr-doped lanthanum ferrite ($La_{1-x}Sr_xFeO_3$), and Sr-doped lanthanum manganite ($La_{1-x}Sr_xMnO_3$) on ferritic stainless steels have been extensively studied using radio-frequency magnetron sputtering, sol-gel, and screen printing techniques [309]. The electronic conductivity of these perovskites in oxidizing environments and their stability under low oxygen partial pressures can decrease the electronic resistance and improve the metallic interconnect performance. Despite the improvements achieved, the ionic conductivity of perovskites remains an ineffective barrier for volatile Cr species or oxygen diffusion [333]. Considering inherent porosity of perovskite, it is necessary to deposit a fully dense coating to inhibit Cr migration and increase conductivity.

Compared to perovskite coatings, some spinel coatings have shown better performance in preventing Cr migration and oxygen diffusion. However, Cr volatilization is still a challenge for Cr containing spinel coatings (e.g., $MnCr_2O_4$)

[334]. Among Cr-free spinel interconnect coatings, $(Mn, Co)_3O_4$ spinel is considered the most promising candidate for SOFC interconnect coatings due to its high electronic conductivity, ASR stability, and thermal expansion coefficient match with the substrate as well as its capability in scale growth reduction and Cr diffusion prevention from the metallic substrate during long-term SOFC operation [335–338]. For example, a previous work on Plansee Ducrolloy $(Cr-5\%Fe-1\%Y_2O_3)$ showed that a $(Mn, Co)_3O_4$ spinel coating could reduce Cr migration significantly. The ASR value was also as low as 0.024 mohm cm^2 for 10,000 h when using LSM and LSC contact pastes [339]. Yang et al. [340,341] have evaluated long-term performance of $(Mn, Co)_3O_4$ spinel coating for 6 months and 125 thermal cycling. The results indicated that the coating adhered well to the substrate with no spallation.

The addition of some dopants such as Ni and Cu to the spinel Mn-Co coating composition has been shown to improve the electronic conductivity and adherence of the coating to the interconnect. In addition, dopants, especially Cu, can lower the sintering temperature and further match the thermal expansion coefficient of the coating to the ferritic stainless steel substrate. Cu dopant can also reduce the usage of expensive and toxic Co element in the spinel coating [342–345].

2.6 SEALANTS

Sealants play a critical role in ensuring the proper function of SOFC by preventing gas leakage from the cathode and the anode chambers and separating the air and fuel, as shown in Figure 2.64. Hence, the development of suitable sealant materials has been one of the major challenges for the implementation of SOFCs. In order to achieve mechanical integrity and stability, and the required adherence for sealants, several approaches have been used, including both rigid seals (no applied load during operation) and compressive seals (load applied to seal during operation). A major advantage of compressive seals is that the exact thermal expansion coefficient match is not required with other SOFC components since the sealant is not rigidly fixed to them. On the other hand, rigid seals do not require the applied loading during operation but have more stringent demands for thermal expansion coefficient matching, cracking, and adherence. Using rigid glass and glass-ceramic seals is the most common approach where the properties of sealant can be improved for SOFC applications by variation of the material composition. However, ceramic materials are inherently brittle. In order to address this issue, metallic, metallic-ceramic and ceramic-ceramic composite seals have been developed in both rigid and compressive configurations. Using multiphase seals has shown to improve the wettability, interfaces compliance and stress relief, gas-tightness, and stability of the sealant [346]. The different approaches for developing SOFC sealants are discussed in the following.

2.6.1 GLASS SEALANTS

Historically, glass joining and compressive sealing are the two techniques having been used for sealing a planar SOFC. Glass was originally used due to its simplicity of fabrication and usage. The glass also has to wet and adhere to the components

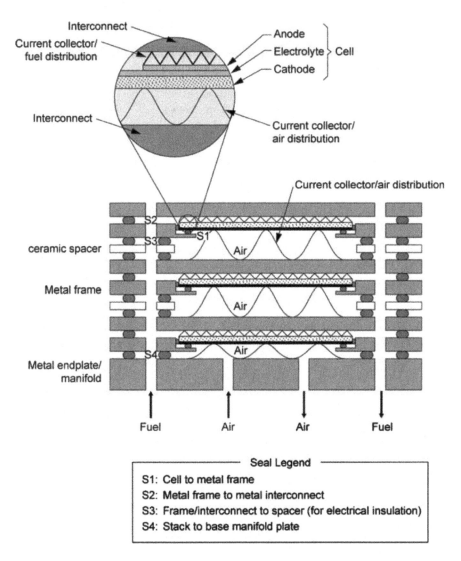

FIGURE 2.64 Schematic of seals typically found in a planar design SOFC stack. (Reprinted with permission from Ref. [347].)

to be sealed. As mentioned, a matched thermal expansion coefficient with other adjoining cell components is the first requirement for a rigid sealant. Thermal expansion mismatch between the glass sealant and other cell materials can result in high stresses and further failure during heating and cooling cycles. The thermal expansion coefficient and softening point of some silica-based glasses are listed in Table 2.10. The softening point and crystallization temperature are important for the selection of a suitable glass sealant since the glass must flow sufficiently to provide an adequate seal while maintaining the sufficient rigidity for mechanical integrity.

TABLE 2.10

Thermal Expansion Coefficient and Softening Point of Selected Glass Compositions

Material	Thermal Expansion Coefficient (K^{-1}; 0°C–300°C)	Softening Point (°C)
Pyrex	3.25×10^{-6}	821
Soda lime	9.35×10^{-6}	696
Potash soda alkali lead	9.70×10^{-6}	658
Potash soda lead	12.4×10^{-6}	500
Lead Zn borate	12.0×10^{-6}	358
Zn boric lead	10.0×10^{-6}	372
Pb Zn borosilicate	10.0×10^{-6}	374
Alkali strontium	9.90×10^{-6}	688

Source: Data adapted and modified from Ref. [347].

Glass seals are very susceptible to cracking caused by tensile stresses after melting and cooling due to their brittle and non-yielding characteristic. The silicate glasses are easily fluxed by the perovskites if they are in physical contact [346,347].

Glass has relatively low strength compared to polycrystalline ceramics. The structure of some amorphous glasses with high thermal expansion coefficients often tends to crystallize at holding for a long time at high temperatures, which can change the thermal expansion coefficient of glass. Therefore, the crystallized glasses, also known as glass-ceramic, with special compositions, have purposefully been used for sealing applications. Glass-ceramic sealants are much stronger than glass sealants with a known thermal expansion coefficient for their crystalline structures [347].

2.6.2 Glass-Ceramic Sealants

Glass-ceramic is formed by the deliberate and controlled partially or fully crystallization of an amorphous glass with special composition when held in high-temperature range followed by a carefully designed cooling profile below its melting/solidification temperature range. The strength and properties of the glass-ceramics are controlled and improved by controlling the amount and nature of the crystalline phase(s). Examples of suitable glass-ceramic materials are boroaluminosilicate, borosilicate, aluminosilicate, silicate, and borate, among which Ba-containing glass-ceramics are the most promising materials because of their relatively high thermal expansion coefficient. For the Ba-containing glass-ceramics, the crystallization increases the thermal expansion coefficient. For instance, it has been shown that the thermal expansion coefficients of $BaO-MgO-SiO_2$ and $BaO-ZnO-SiO_2$ increase with an increase in BaO content for constant SiO_2 content [348]. This is due to the formation of $BaSiO_3$, which has a large thermal expansion coefficient in comparison with, for example, $MgSiO_3$. It has also been shown that the variation of $BaO-Al_2O_3-La_2O_3-B_2O_3-SiO_2$ with suitable ratios of B_2O_3 and SiO_2 results in appropriate binding and wetting properties along with compatible thermal expansion coefficient with YSZ and high chemical stability at 800°C–850°C over 100 h [349]. However, Ba-containing glass-ceramics can react with chromium oxide and form $BaCrO_4$ phase in contact with Cr-containing cell components which leads to an interface with high thermal expansion coefficient and a mechanical weakness with subsequent delamination [350]. Example reactions are shown in Equations (2.29) and (2.30):

$$2Cr_2O_3 + 4BaO + 3O_2 = 4BaCrO_4 \qquad (2.29)$$

$$CrO_2(OH)_2 + BaO = BaCrO_4 + H_2O \qquad (2.30)$$

In addition, compositions containing boron react over time with water (steam) and form $B_2(OH)_2$ or $B(OH)_3$ gas. This can decompose the glass and remarkably limit the sealant lifetime. Hence, many attempts have been made to utilize "no" or "low" boron glass compositions.

2.6.3 Compressive Sealants

Compressive sealants are placed between sealing surfaces, which are then compressed constantly using external forces to the fuel cell stack (e.g., using a load frame and springs or hydraulics) to achieve airtightness. Thus, compressive sealants must be ductile and resistant to oxide formation in the presence of air. The addition of the load frame will cause complexity and higher cost for the SOFC. The best compressive sealant is the one that allows the surfaces to slide past each other while maintaining a gas seal. For compressive sealants, it is not important to match the thermal expansion coefficient with other cell components, as is for glass-based sealants. Ceramic powders have been used to form a compressive sealant in SOFCs due to their high oxidation resistance. However, they typically form a "leaky" seal. Noble metals such as platinum, gold and silver with high oxidation resistance have met the requirements for compressive seals [351,352]. Silver is a more favorable choice because of its lower cost, but it suffers from crack formation along the grain boundaries as well as solubility of both oxygen (from air) and hydrogen (from fuel) where the dissolved oxygen and hydrogen react within the silver to form water, thus, subsequent failure [353,354]. Silver also has a high vapor pressure and a high thermal expansion coefficient (20×10^{-6} K^{-1}). Hence, silver-metal oxides such as Ag-CuO and Ag-V_2O_5 have been used instead to overcome these problems [355]. Another approach to compressive seal is to use mica (although its thermal expansion coefficient is relatively low; 6.9×10^{-6} K^{-1}). The application of pressure to the overlapping plate-like mica crystals/particles can create a seal, as shown in Figure 2.65. However, using the mica sealant has some problems such as leak through the interface between the mica and the metal or ceramic component, interfacial reactions, and crystallization [356]. While mica alone cannot provide proper sealing, using metal/mica composites and or a compliant layer (e.g., glass or metal) at the surface of the mica layer, have been developed as a proposed improvement [357]. An alternative method for combining

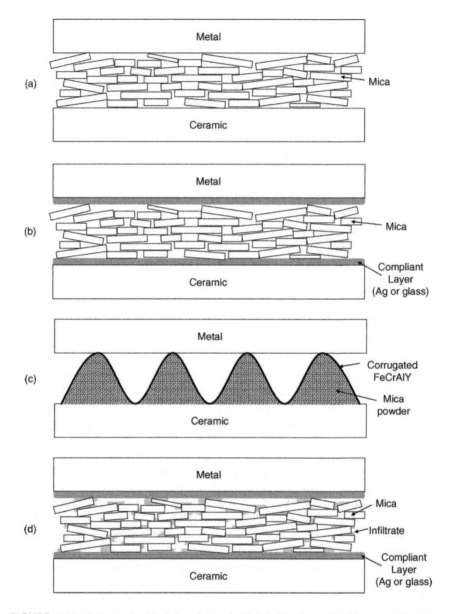

FIGURE 2.65 Mica seals: (a) plain mica seal, (b) hybrid mica seal with a compliant layer (glass or metal), (c) mica powder combined with corrugated alloy, and (d) hybrid mica seal with compliant layer and infiltrated mica. (Reprinted with permission from Ref. [346].)

metal and mica is to place mica powder in the gaps of a corrugated metal seal, as observed in Figure 2.65c (e.g., FeCrAlY alloy is added to increase the thermal expansion coefficient and plasticity of the sealant. This also reduces leak rate after five thermal cycles). The sealing between adjacent mica particles can further be improved by infiltrating mica. The two types of mica-based sealants which have extensively been used are phlogopite ($KMg_3(AlSi_3O_{10})(OH)_2$) and muscovite ($KAl_2(AlSi_3O_{10})(F, OH)_2$) due to their high stability at elevated temperatures up to 800°C. The most significant difference between these two types is that the thermal expansion coefficient of phlogopite is higher than that of the muscovite [356,358,359].

2.6.4 CERAMIC- AND GLASS-COMPOSITE SEALANTS

The ceramic-composite sealant approach is to produce a deformable seal based on using a glass above its transition temperature with control of the viscosity and thermal expansion coefficient modified by using ceramic powder additives [360]. For example, NexTech developed BaO-CaO-Al_2O_3-SiO_2 glass compositions together with mica, talc, alumina powders, or zirconia fibers as the additive to increase the viscosity. This seal was similar to dense and machinable mica-glass-ceramics SOFC seals when sufficient glass content was used in the composition of seal densified upon melting [361]. In another research, Sandia National Laboratory investigated ZnO-CaO-SrO-Al_2O_3-B_2O_3-SiO_2 glasses in conjunction with metallic nickel powder additives (5–30 vol%). The results showed that adding nickel increases the thermal expansion coefficient of the composite [360].

REFERENCES

1. Fabbri, E., Bi, L., Pergolesi, D., and Traversa, E. 2012. Towards the next generation of solid oxide fuel cells operating below 600°C with chemically stable proton-conducting electrolytes. *Advanced Materials* 24:195–208.
2. Mahato, N., Banerjee, A., Gupta, A., Omar, S., and Balani, K. 2015. Progress in material selection for solid oxide fuel cell technology: a review. *Progress in Materials Science* 72:141–337.
3. Wincewicz, K.C., and Cooper, J.S. 2005. Taxonomies of SOFC material and manufacturing alternatives. *Journal of Power Sources* 140:280–296.
4. Tietz, F., Buchkremer, H.-P., and Stovër, D. 2002. Components manufacturing for solid oxide fuel cells. *Solid State Ionics* 152–153:373–381.
5. Fergus, J.W., Hui, R., Li, X., Wilkinson, D.P., and Zhang, J. 2009. *Solid Oxide Fuel Cells: Materials Properties and Performance*. CRC Press, Taylor & Francis Group: Boca Raton, FL.
6. Kharton, V.V., Marques, F.M.B., and Atkinson, A. 2004. Transport properties of solid oxide electrolyte ceramics: a brief review. *Solid State Ionics* 174:135–149.
7. Kaur, G. 2016. *Solid Oxide Fuel Cell: Interfacial Compatibility of SOFC Glass Seals*. Springer: Switzerland.
8. Strickler, D.W., and Carlson, W.G. 1964. Ionic conductivity of cubic solid solutions in the system CaO-Y_2O_3-ZrO_2. *Journal of the American Ceramic Society* 47:122–127.
9. Dixon, J.M., LaGrange, L.D., Merten, U., Miller, C.F., and Porter, J.T. 1963. Electrical resistivity of stabilized zirconia at elevated temperatures. *Journal of the Electrochemical Society* 110:276–280.
10. Guo, X., and Maier, J. 2001. Grain boundary blocking effect in zirconia: a Schottky barrier analysis. *Journal of the Electrochemical Society* 148:E121–E126.
11. Yeh, T.-H., Hsu, W.-C., and Chou, C.-C. 2005. Mechanical and electrical properties of ZrO (3Y) doped with RENbO (RE = Yb, Er, Y, Dy, YNd, Sm, Nd). *Journal de Physique IV EDP Sciences* 128:213–219.

12. Yamamoto, O., Arati, Y., Takeda, Y., Imanishi, N., Mizutani, Y., Kawai, M., and Nakamura, Y. 1995. Electrical conductivity of stabilized zirconia with ytterbia and scandia. *Solid State Ionics* 79:137–142.
13. Johansen, H.A., and Cleary, J.G. 1964. High-temperature electrical conductivity in the systems CaO-ZrO$_2$ and CaO-HfO$_2$. *Journal of the Electrochemical Society* 111:100–103.
14. Hohnke, D.K. 1980. Ionic conductivity of Zr$_{1-x}$In$_{2x}$O$_{2-x}$. *Journal of Physics and Chemistry of Solids* 41:777–784.
15. Hui, S.R., Roller, J., Yick, S., Zhang, X., Decès-Petit, C., Xie, Y., Maric, R., and Ghosh, D. 2007. A brief review of the ionic conductivity enhancement for selected oxide electrolytes. *Journal of Power Sources* 172(2):493–502.
16. Han, Y., and Zhu, J. 2013. Surface science studies on the zirconia-based model catalysts. *Topics in Catalysis* 56:1525–1541.
17. Malavasi, L., Fisher, C.A.J., and Islam, M.S. 2010. Oxide-ion and proton conducting electrolyte materials for clean energy applications: structural and mechanistic features. *Chemical Society Reviews* 39:4370–4387.
18. Biswas, M., and Sadanala, K.C. 2013. Electrolyte materials for solid oxide fuel cell. *Journal of Powder Metallurgy & Mining* 2(117):1–6.
19. Abdalla, A.M., Hossain, S., Azad, A.T., Petra, P.M.I., Begum, F., Eriksson, S.G., and Azad, A.K. 2018. Nanomaterials for solid oxide fuel cells: a review. *Renewable and Sustainable Energy Reviews* 82:353–368.
20. Haering, C., Roosen, A., and Schichl, H. 2005. Degradation of the electrical conductivity in stabilised zirconia systems: Part I: yttria-stabilised zirconia. *Solid State Ionics* 176:253–259.
21. Fergus, J.W. 2006. Electrolytes for solid oxide fuel cells. *Journal of Power Sources* 162:30–40.
22. Vonk, V., Khorshidi, N., Stierle, A., and Dosch, H. 2013. Atomic structure and composition of the yttria-stabilized zirconia (111) surface. *Surface Science* 612(100):69–76.
23. Prakash, B.S., Kumar, S.S., and Aruna, S. 2014. Properties and development of Ni/YSZ as an anode material in solid oxide fuel cell: a review. *Renewable Sustainable Energy Reviews* 36:149–179.
24. Saebea, D., Authayanun, S., Patcharavorachot, Y., Chatrattanawet, N., and Arpornwichanop, A. 2018. Electrochemical performance assessment of low-temperature solid oxide fuel cell with YSZ-based and SDC-based electrolytes. *International Journal of Hydrogen Energy* 43(2):921–931.
25. Zakaria, Z., Hassan, S.H.A., Shaari, N., Yahaya, A.Z., and Kar, Y.B. 2020. A review on recent status and challenges of yttria stabilized zirconia modification to lowering the temperature of solid oxide fuel cells operation. *International Journal of Energy Research* 44:631–650.
26. Koval'chuk, A., Kuz'min, A.V., Osinkin, D.A., Farlenkov, A.S., Solov'ev, A.A., Shipilova, A.V., Ionov, I.V., Bogdanovich, N.M., and Beresnev, S.M. 2018. Single SOFC with supporting Ni-YSZ anode, bilayer YSZ/GDC film electrolyte, and La$_2$NiO$_{4+\delta}$ cathode. *Russian Journal of Electrochemistry* 54(6):541–546.
27. Yu, W., Cho, G.Y., Hong, S., Lee, Y., Kim, Y.B., An, J., and Cha, S.W. 2016. PEALD YSZ-based bilayer electrolyte for thin film-solid oxide fuel cells. *Nanotechnology* 27(41):415402.
28. da Silva, C.A., Ribeiro, N.F., and Souza, M.M. 2009. Effect of the fuel type on the synthesis of yttria stabilized zirconia by combustion method. *Ceramic International* 35(8):3441–3446.
29. Yin, W., Meng, B., Meng, X., and Tan, X. 2009. Highly asymmetric yttria stabilized zirconia hollow fibre membranes. *Journal of Alloys and Compounds* 476(1–2):566–570.

30. Chen, Y., Omar, S., Keshri, A.K., Balani, K., Babu, K., Nino, J.C., Seal, S., and Agarwal, A. 2009. Ionic conductivity of plasma-sprayed nanocrystalline yttria-stabilized zirconia electrolyte for solid oxide fuel cells. *Scripta Materialia* 60(11):1023–1026.

31. Suzuki, T., Zahir, M.H., Yamaguchi, T., Fujishiro, Y., Awano, M., and Sammes, N. 2010. Fabrication of micro-tubular solid oxide fuel cells with a single-grain-thick yttria stabilized zirconia electrolyte. *Journal of Power Sources* 195(23):7825–7828.

32. Smeacetto, F., Salvo, M., Ajitdoss, L.C., Perero, S., Moskalewicz, T., Boldrini, S., Doubova, L., and Ferraris, M. 2010. Yttria-stabilized zirconia thin film electrolyte produced by RF sputtering for solid oxide fuel cell applications. *Materials Letters* 64(22):2450–2453.

33. Jang, D.Y., Kim, H.K., Kim, J.W., Bae, K., Schlupp, M.V.F., Park, S.W., Prestat, M., and Shim, J.H. 2015. Low-temperature performance of yttria-stabilized zirconia prepared by atomic layer deposition. *Journal of Power Sources* 274:611–618.

34. Park, J., Lee, Y., Chang, I., Cho, G.Y., Ji, S., Lee, W., and Cha, S.W. 2016. Atomic layer deposition of yttria-stabilized zirconia thin films for enhanced reactivity and stability of solid oxide fuel cells. *Energy* 116:170–176.

35. Guo, X., and Waser, R. 2006. Electrical properties of the grain boundaries of oxygen ion conductors: acceptor-doped zirconia and ceria. *Progress in Materials Science* 51(2):151–210.

36. Xin, X., Lü, Z., Ding, Z., Huang, X., Liu, Z., Sha, X., Zhang, Y., and Su, W. 2006. Synthesis and characteristics of nanocrystalline YSZ by homogeneous precipitation and its electrical properties. *Journal of Alloys and Compounds* 425(1–2):69–75.

37. Liu, Y., and Lao, L.E. 2006. Structural and electrical properties of ZnO-doped 8 mol% yttria-stabilized zirconia. *Solid State Ionics* 177(1–2):159–163.

38. Badwal, S.P.S., and Rajendran, S. 1994. Effect of micro- and nano-structures on the properties of ionic conductors. *Solid State Ionics* 70–71:83–95.

39. Ji, Y., Liu, J., Lü, Z., Zhao, X., He, T., and Su, W. 1999. Study on the properties of Al_2O_3-doped $(ZrO_2)_{0.92}(Y_2O_3)_{0.08}$ electrolyte. *Solid State Ionics* 126(3–4):277–283.

40. Xiao, J., Han, Q., Yu, F., Zhang, Y., Wu, H., Li, X., Zeng, X., Dong, P., Zhang, Y., and Liu, J. 2018. Co-precipitation synthesis of alumina doped yttria stabilized zirconia. *Journal of Alloys and Compounds* 731:1080–1088.

41. Lee, J.-H., Yoon, S.M., Kim, B.-K., Kim, J., Lee, H.-W., and Song, H.-S. 2001. Electrical conductivity and defect structure of yttria-doped ceria-stabilized zirconia. *Solid State Ionics* 144(1–2):175–184.

42. Schmid, H., Gilardi, E., Gregori, G., Longo, P., Maier, J., and van Aken, P.A. 2018. Structure and chemistry of interfaces between ceria and yttria-stabilized zirconia studied by analytical STEM. *Ultramicroscopy* 188:90–100.

43. Flegler, A.J., Burye, T.E., Yang, Q., and Nicholas, J.D. 2014. Cubic yttria stabilized zirconia sintering additive impacts: a comparative study. *Ceramic International* 40(10):16323–16335.

44. Lee, J.G., Jeon, O.S., Ryu, K.H., Park, M.G., Min, S.H., Hyun, S.H., and Shul, Y.G. 2015. Effects of 8 mol% yttria-stabilized zirconia with copper oxide on solid oxide fuel cell performance. *Ceramic International* 41(6):7982–7988.

45. An, J., Kim, Y.B., Park, J., Gür, T.M., and Prinz, F.B. 2013. Three-dimensional nanostructured bilayer solid oxide fuel cell with 1.3 W/cm^2 at 450°C. *Nano Letters* 13(9):4551–4555.

46. Selvaraj, T., Johar, B., and Khor, S.F. 2019. Iron/zinc doped 8 mol% yttria stabilized zirconia electrolytes for the green fuel cell technology: a comparative study of thermal analysis, crystalline structure, microstructure, mechanical and electrochemical properties. *Materials Chemistry and Physics* 222:309–320.

47. Nomura, K., Mizutani, Y., Kawai, M., Nakamura, Y., and Yamamoto, O. 2000. Aging and Raman scattering study of scandia and yttria doped zirconia. *Solid State Ionics* 132:235–239.

48. Lee, M.-J., Jung, J.-H., Zhao, K., Kim, B.-K., Xu, Q., Ahn, B.-G., Kim, S.S.-H., and Kim, S.-Y. 2014. Fabrication and electrochemical properties of SOFC single cells using porous yttria-stabilized zirconia ceramic support layer coated with Ni. *Journal of the European Ceramic Society* 34(7):1771–1776.

49. Timurkutluk, B., Ciflik, Y., and Korkmaz, H. 2015. Strength evaluation of glass–ceramic composites containing yttria stabilized zirconia after thermal cycling. *Ceramic International* 41(5):6985–6990.

50. Chong, F.D., Tan, C.Y., Singh, R., and Muchtar, A., Somalu, M.R., Ng, C.K., Yap, B.K., Teh, Y.C., and Meng, Y. 2016. Effect of manganese oxide on the sinterability of 8 mol% yttria-stabilized zirconia. *Materials Characterization* 120:331–336.

51. Araki, W., and Arai, Y. 2010. Oxygen diffusion in yttria-stabilized zirconia subjected to uniaxial stress. *Solid State Ionics* 181(8–10):441–446.

52. Badwal, S.P.S., and Ciacchi, F.T. 2000. Oxygen-ion conducting electrolyte materials for solid oxide fuel cells. *Ionics* 6(1–2):1–21.

53. Seabaugh, M.M., Day, M.J., Sabolsky, K., Ibanez, S., and Swartz, S.L. 2005. Materials and components for solid oxide fuel cell systems. *Electrochemical Society Proceedings SOFC IX* 2(7):1037–1044.

54. Yamamoto, O., Arachi, Y., Sakai, H., Takeda, Y., Imanishi, N., Mizutani, Y., Kawai, M., and Nakamura, Y. 1998. Zirconia based oxide ion conductors for solid oxide fuel cells. *Ionics* 4(5–6):403–408.

55. Hirano, M., Oda, T., Ukai, K., and Mizutani, Y. 2003. Effect of Bi_2O_3 additives in Sc stabilized zirconia electrolyte on a stability of crystal phase and electrolyte properties. *Solid State Ionics* 158(3–4):215–223.

56. Sarat, S., Sammes, N., and Smirnova, A. 2006. Bismuth oxide doped scandia-stabilized zirconia electrolyte for the intermediate temperature solid oxide fuel cells. *Journal of Power Sources* 160(2):892–896.

57. Lybye, D., Liu, Y.-L., Mogensen, M., and Linderoth, S. 2005. Effect of impurities on the Conductivity of Sc and Y Co-Doped ZrO_2. *Electrochemical Society Proceedings SOFC IX* 2(7):954–996.

58. Lei, Z., and Zhu, Q. 2007. Phase transformation and low temperature sintering of manganese oxide and scandia co-doped zirconia. *Materials Letters* 61(6):1311–1314.

59. Koteswararao, P., Suresh, B.M., Wani, B.N., and Rao, P.V.B. 2017. Review on different components of solid oxide fuel cells. *Journal of Powder Metallurgy & Mining* 6:181.

60. Oh, S., Park, J., Shin, J.W., Yang, B.C., Zhang, J., Jang, D.J., and An, J. 2018. High performance low-temperature solid oxide fuel cells with atomic layer deposited-yttria stabilized zirconia embedded thin film electrolyte. *Journal of Materials Chemistry A* 6:7401–7408.

61. Hohnke, D.K. 1981. Ionic conduction in doped oxides with the fluorite structure. *Solid State Ionics* 5:531–534.

62. Steele, B.C.H. 2000. Appraisal of $Ce_{1-y}Gd_yO_{2-y/2}$ electrolytes for IT-SOFC operation at 500°C. *Solid State Ionics* 129(1–4):95–110.

63. Zha, S., Xia, C., and Meng, G. 2003. Effect of Gd (Sm) doping on properties of ceria electrolyte for solid oxide fuel cells. *Journal of Power Sources* 115(1):44–48.

64. Zhang, T.S., Ma, J., Cheng, H., and Chan, S.H. 2006. Ionic conductivity of high-purity Gd-doped ceria solid solutions. *Materials Research Bulletin* 41(3):563–568.

65. Babu, A.S., Bauri, R., and Reddy, S. 2016. Processing and conduction behavior of nanocrystalline Gd-doped and rare earth co-doped ceria electrolytes. *Electrochimica Acta* 209:541–550.

66. Zhang, T.S., Ma, J., Chan, S.H., Hing, P., and Kilner, J.A. 2004. Intermediate-temperature ionic conductivity of ceria-based solid solutions as a function of gadolinia and silica contents. *Solid State Sciences* 6(6):565–572,

67. Kharton, V.V., Figueiredo, F.M., Navarro, L., Naumovich, E.N., Kovalevsky, A.V., Yaremchenko, A.A., Viskup, A.P., Carneiro, A., Marques, F.M.B., and Frade, J.R. 2001. Ceria-based materials for solid oxide fuel cells. *Journal of Materials Science* 36(5):1105–1117.

68. Inaba, H., and Tagawa, H. 1996. Review: ceria-based solid electrolytes. *Solid State Sciences* 83:1–16.

69. Jung, G.-B., Huang, T.-J., and Chang, C.-L. 2002. Effect of temperature and dopant concentration on the conductivity of samaria-doped ceria electrolyte. *Journal of Solid State Electrochemistry* 6(4):225–230.

70. Zhan, Z., Wen, T.-L., Tu, H., and Lu, Z.-Y. 2001. AC impedance investigation of samarium-doped ceria. *Journal of the Electrochemical Society* 148(5):A427–A432.

71. Huang, W., Shuk, P., and Greenblatt, M. 1997. Properties of sol-gel prepared $Ce_{1-x}Sm_xO_{2-x/2}$ solid electrolytes. *Solid State Sciences* 100(1–2):23–27.

72. Wang, F.-Y., Chen, S., and Cheng, S. 2004. Gd^{3+} and Sm^{3+} co-doped ceria based electrolytes for intermediate temperature solid oxide fuel cells. *Electrochemistry Communications* 6(8):743–746.

73. Park, J., Yoon, H., and Wachsman, E.D. 2005. Fabrication and characterization of high-conductivity bilayer electrolytes for intermediate-temperature solid oxide fuel cells. *Journal of the American Ceramic Society* 88:2402–2408.

74. Ling, Y., Wang, X., Ma, Z., Wei, K., Wu, Y., Khan, M., Zheng, K., Shen, S., and Wang, S. 2020. Review of experimental and modelling developments for ceria-based solid oxide fuel cells free from internal short circuits. *Journal of Materials Science* 55(1):1–23.

75. Basu, S. 2007. *Recent Trends in Fuel Cell Science and Technology.* Springer and Anamaya Publishers: New Delhi, India.

76. Gao, Z., Mogni, L.V., Miller, E.C., Railsback, J.G., and Barnett, S.A. 2016. A perspective on low-temperature solid oxide fuel cells. *Energy & Environmental Science* 9:1602–1644.

77. Wachsman, E.D., and Lee, K.T. 2011. Lowering the temperature of solid oxide fuel cells. *Science* 33:4935–4939.

78. Skinner, S.J., and Kilner, J.A. 2003. Oxygen ion conductors. *Materials Today* 6(3):30–37.

79. Maric, R., Ohara, S., Fukui, T., Yoshida, H., Nishimura, M., Inagaki, T., and Miura, K. 1999. Solid oxide fuel cells with doped lanthanum gallate electrolyte and $LaSrCoO_3$ cathode, and Ni-samaria-doped ceria cermet anode. *Journal of the Electrochemical Society* 146(6):2006–2010.

80. Zhang, X., Ohara, S., Maric, R., Okawa, H., Fukui, T., Yoshida, H., Inagaki, T., and Miura, K. 2000. Interface reactions in the NiO–SDC–LSGM system. *Solid State Ionics* 133(3–4):153–160.

81. Zhang, X., Ohara, S., Okawa, H., Maric, R., and Fukui, T. 2001. Interactions of a $La_{0.9}Sr_{0.1}Ga_{0.8}Mg_{0.2}O_{3-\delta}$ electrolyte with Fe_2O_3, Co_2O_3 and NiO anode materials. *Solid State Ionics* 139(1–2):145–152.

82. Zhang, X., Ohara, S., Maric, R., Mukai, K., Fukui, T., Yoshida, H., Nishimura, M., Inagaki, T., and Miura, K. 1999. Ni-SDC cermet anode for medium-temperature solid oxide fuel cell with lanthanum gallate electrolyte. *Journal of Power Sources* 83(1–2):170–177.

83. Ohara, S., Maric, R., Zhang, X., Mukai, K., Fukui, T., Yoshida, H., Inagaki, T., and Miura, K. 2000. High performance electrodes for reduced temperature solid oxide fuel cells with doped lanthanum gallate electrolyte: I. Ni–SDC cermet anode. *Journal of Power Sources* 86(1–2):455–458.

84. Inagaki, T., Miura, K., Yoshida, H., Maric, R., Ohara, S., Zhang, X., Mukai, K., and Fukui, T. 2000. High-performance electrodes for reduced temperature solid oxide fuel cells with doped lanthanum gallate electrolyte: II. La(Sr)CoO$_3$ cathode. *Journal of Power Sources* 86(1–2):347–351.

85. Ishihara, T. 2006. Development of new fast oxide ion conductor and application for intermediate temperature solid oxide fuel cells. *Bulletin of the Chemical Society of Japan* 79(8):1155–1166.

86. Petric, A., and Huang, P. 1996. Oxygen conductivity of Nd(SrCa)Ga(Mg)O$_{3-\delta}$ perovskites. *Solid State Sciences* 92(1–2):113–117.

87. Sinha, A., Näfe, H., Sharma, B.P., and Gopalan, P. 2008. Study on ionic and electronic transport properties of calcium-doped GdAlO$_3$. *Journal of the Electrochemical Society* 155(3):B309–B314.

88. Energy Information Administration of the US Department of Energy. 2014. Annual energy outlook 2014 with projections to 2040. https://www.eia.gov/outlooks/aeo/pdf/0383(2014).pdf.

89. Hossain, S., Abdalla, A.M., Jamain, S.N.B., Zaini, J.H., and Azad, A.K. 2017. A review on proton conducting electrolytes for clean energy and intermediate temperature-solid oxide fuel cells. *Renewable and Sustainable Energy Reviews* 79:750–764.

90. Babilo, P., Uda, T., and Haile, S.M. 2007. Processing of yttrium-doped barium zirconate for high proton conductivity. *Journal of Materials Research* 22(5):1322–1330.

91. Akbarzadeh, A.R., Kornev, I., Malibert, C., Bellaiche, L., and Kiat, J.M. 2005. Combined theoretical and experimental study of the low-temperature properties of BaZrO$_3$. *Physical Review B: Condensed Matter and Materials Physics* 72:205104–205111.

92. Yamazaki, Y., Hernandez-Sanchez, R., and Haile, S.M. 2009. High total proton conductivity in large-grained yttrium-doped barium zirconate. *Chemistry of Materials* 21(13):2755–2762.

93. Singhal, S.C., and Kendall, K. 2003. *High-Temperature Solid Oxide Fuel Cells: Fundamentals, Design and Applications.* Elsevier: Oxford, UK.

94. Kakinuma, K., Tomita, A., Yamamura, H., and Atake, T. 2006. Water vapor absorption and proton conductivity of (Ba$_{1-x}$La$_x$)$_2$In$_2$O$_{5+x}$. *Journal of Materials Science* 41(19):6435–6440.

95. Nakayama, S., Aono, H., and Sadaoka, Y. 1995. Ionic conductivity of Ln$_{10}$(SiO$_4$)$_6$O$_3$ (Ln = La, Nd, Sm, Gd and Dy). *Chemistry Letters* 24(6):431–432.

96. Nakayama, S., Kageyama, T., Aono, H., and Sadaoka, Y. 1995. Ionic conductivity of lanthanoid silicates, Ln$_{10}$(SiO$_4$)$_6$O$_3$ (Ln = La, Nd, Sm, Gd, Dy, Y, Ho, Er and Yb). *Journal of Materials Chemistry* 5(11):1801–1805.

97. Kendrick, E., Islam, M.S., and Slater, P.R. 2007. Developing apatites for solid oxide fuel cells: insight into structural, transport and doping properties. *Journal of Materials Chemistry* 17(30):3104–3111.

98. Sansom, J.E.H., Richings, D., and Slater, P.R. 2001. A powder neutron diffraction study of the oxide-ion-conducting apatite-type phases, La$_{9.33}$Si$_6$O$_{26}$ and La$_8$Sr$_2$Si$_6$O$_{26}$. *Solid State Ionics* 139(3–4):205–210.

99. Pramana, S.S., Klooster, W.T., and White, T.J. 2007. Framework 'interstitial' oxygen in La$_{10}$(GeO$_4$)$_5$(GeO$_5$)O$_2$ apatite electrolyte. *Acta Crystallographica Section B: Structural Science* 63(4):597–602.

100. Kendrick, E., and Slater, P.R. 2008. Synthesis of hexagonal lanthanum germanate apatites through site selective isovalent doping with yttrium. *Materials Research Bulletin* 43(8–9):2509–2513.

101. Lacorre, P., Goutenoire, F., Bohnke, O., Retoux, R., and Laligant, Y. 2000. Designing fast oxide-ion conductors based on La$_2$Mo$_2$O$_9$. *Nature* 404:856–858.

102. Tealdi, C., Chiodelli, G., Malavasi, L., and Flor, G. 2004. Effect of alkaline-doping on the properties of $La_2Mo_2O_9$ fast oxygen ion conductor. *Journal of Materials Chemistry* 14(24):3553–3557.

103. Georges, S., Goutenoire, F., Altorfer, F., Sheptyakov, D., Fauth, F., Suard, E., and Lacorre, P. 2003. Thermal, structural and transport properties of the fast oxide-ion conductors $La_{2-x}R_xMo_2O_9$ (R = Nd, Gd, Y). *Solid State Ionics* 161(3–4):231–241.

104. Wang, X.P., Cheng, Z.J., and Fang, Q.F. 2005. Influence of potassium doping on the oxygen-ion diffusion and ionic conduction in the $La_2Mo_2O_9$ oxide-ion conductors. *Solid State Ionics* 176(7–8):761–765.

105. Weber, I.T., Baracho, P.R., Rangel, F., Paris, E.C., and Muccillo, E.N.S. 2009. Pure and Gd doped LAMOX powders and thin films obtained by chemical route. *Materials Science and Technology* 25(11):1346–1350.

106. Kendrick, E., Kendrick, J., Knight, K.S., Islam, M.S., and Slater, P.R. 2007. Cooperative mechanisms of fast-ion conduction in gallium-based oxides with tetrahedral moieties. *Nature Materials* 6(11):871–875.

107. Li, S., Schönberger, F., and Slater, P. 2003. $La_{1-x}Ba_{1+x}GaO_{4-x/2}$: a novel high temperature proton conductor. *Chemical Communications* 21:2694–2695.

108. Thomas, C.I., Kuang, X., Deng, Z., Niu, H., Claridge, J.B., and Rosseinsky, M.J. 2010. Phase stability control of interstitial oxide ion conductivity in the $La_{1+x}Sr_{1-x}Ga_3O_{7+x/2}$ melilite family. *Chemistry of Materials* 22(8):2510–2516.

109. Haugsrud, R., and Norby, T. 2006. Proton conduction in rare-earth ortho-niobates and ortho-tantalates. *Nature Materials* 5(3):193–196.

110. Spacil, H.S. 1970. U.S. Patent 3,503,809.

111. Huang, Y.H., Dass, R.I., Denyszyn, J.C., and Goodenough, J.B. 2006. Synthesis and characterization of $Sr_2MgMoO_{6-\delta}$ an anode material for the solid oxide fuel cell. *Journal of the Electrochemical Society* 153(7):A1266–A1272.

112. Gorte, R.J., and Vohs, J.M. 2009. Nanostructured anodes for solid oxide fuel cells. *Current Opinion in Colloid & Interface Science* 14(4):236–244.

113. Khan, M.S., Lee, S.B., Song, R.H., Lee, J.W., Lim, T.H., and Park, S.J. 2016. Fundamental mechanisms involved in the degradation of nickel–yttria stabilized zirconia (Ni–YSZ) anode during solid oxide fuel cells operation: a review. *Ceramics International* 42(1):35–48.

114. Zhang, X., Robertson, M., Decès-Petit, C., Xie, Y., Hui, R., Yick, S., Staite, M., Styles, E., Roller, J., Marie, R., and Ghosh, D. 2005. Ni-YSZ cermet supported thin ceria-based electrolyte solid oxide fuel cell for reduced temperature (500–600°C) operation. *ECS Proceedings Volumes PV* 2005–2007:1102–1109.

115. Zhang, X., Robertson, M., Deces-Petit, C., Xie, Y., Hui, R., Yick, S., Styles, E., Roller, J., Kesler, O., Maric, R., and Ghosh, D. 2006. NiO–YSZ cermets supported low temperature solid oxide fuel cells. *Journal of Power Sources* 161(1):301–307.

116. Maric, R., Roller, J., and Neagu, R. 2011. Flame-based technologies and reactive spray deposition technology for low-temperature solid oxide fuel cells: technical and economic aspects. *Journal of Thermal Spray Technology* 20(4):696–718.

117. Maric, R., Furusaki, K., Nishijima, D., and Neagu, R. 2011. Thin film low temperature solid oxide fuel cell (LTSOFC) by reactive spray deposition technology (RSDT). *ECS Transactions* 35(1):473–481.

118. Zhu, H., Kee, R.J., Janardhanan, V.M., Deutschmann, O., and Goodwin, D.G. 2005. Modeling elementary heterogeneous chemistry and electrochemistry in solid-oxide fuel cells. *Journal of the Electrochemical Society* 152(12):A2427–A2440.

119. Goodwin, D.G., Zhu, H., Colclasure, A.M., and Kee, R.J. 2009. Modeling electrochemical oxidation of hydrogen on Ni–YSZ pattern anodes. *Journal of the Electrochemical Society* 156(9):B1004–B1021.

120. Bessler, W.G., Vogler, M., Störmer, H., Gerthsen, D., Utz, A., Weber, A., and Ivers-Tiffée, E. 2010. Model anodes and anode models for understanding the mechanism of hydrogen oxidation in solid oxide fuel cells. *Physical Chemistry Chemical Physics* 12(42):13888–13903.

121. Yu, J.H., Park, G.W., Lee, S., and Woo, S.K. 2007. Microstructural effects on the electrical and mechanical properties of Ni–YSZ cermet for SOFC anode. *Journal of Power Sources* 163(2):926–932.

122. Dees, D.W., Claar, T.D., Easler, T.E., Fee, D.C., and Mrazek, F.C. 1987. Conductivity of porous Ni/ZrO_2-Y_2O_3 cermets. *Journal of the Electrochemical Society* 134(9):2141–2146.

123. Grahl-Madsen, L., Larsen, P.H., Bonanos, N., Engell, J., and Linderoth, S. 2006. Mechanical strength and electrical conductivity of Ni-YSZ cermets fabricated by viscous processing. *Journal of Materials Science* 41(4):1097–1107.

124. Wang, Y., Walter, M.E., Sabolsky, K., and Seabaugh, M.M. 2006. Effects of powder sizes and reduction parameters on the strength of Ni–YSZ anodes. *Solid State Ionics* 177(17–18):1517–1527.

125. Kawada, T., Sakai, N., Yokokawa, H., Dokiya, M., Mori, M., and Iwata, T. 1990. Characteristics of slurry-coated nickel zirconia cermet anodes for solid oxide fuel cells. *Journal of the Electrochemical Society* 137(10):3042–3047.

126. Koide, H., Someya, Y., Yoshida, T., and Maruyama, T. 2000. Properties of Ni/YSZ cermet as anode for SOFC. *Solid State Ionics* 132(3–4):253–260.

127. Huebner, W., Reed, D.M., and Anderson, H.U. 1999. Solid oxide fuel cell performance studies: anode development. *Electrochemical Society Proceedings Volumes* 99(19):503–512.

128. Jiang, S.P., and Badwal, S.P.S. 1999. An electrode kinetics study of H_2 oxidation on Ni/Y_2O_3–ZrO_2 cermet electrode of the solid oxide fuel cell. *Solid State Ionics* 123(1–4):209–224.

129. Primdahl, S., Jørgensen, M.J., Bagger, C., and Kindl, B. 1999. Thin anode supported SOFC. *Electrochemical Society Proceedings Volumes* 99(19):773–802.

130. Zhao, F., and Virkar, A.V. 2005. Dependence of polarization in anode-supported solid oxide fuel cells on various cell parameters. *Journal of Power Sources* 141(1):79–95.

131. Iwata, T. 1996. Characterization of Ni-YSZ anode degradation for substrate-type solid oxide fuel cells. *Journal of the Electrochemical Society* 143(5):1521–1525.

132. Khan, M.S., Wahyudi, W., Lee, S.B., Song, R.H., Lee, J.W., Lim, T.H., and Park, S.J. 2015. Effect of various sintering inhibitors on the long term performance of Ni-YSZ anodes used for SOFCs. *International Journal of Hydrogen Energy* 40(35):11968–11975.

133. Gong, M., Liu, X., Trembly, J., and Johnson, C. 2007. Sulfur-tolerant anode materials for solid oxide fuel cell application. *Journal of Power Sources* 168(2):289–298.

134. Sasaki, K., Susuki, K., Iyoshi, A., Uchimura, M., Imamura, N., Kusaba, H., Teraoka, Y., Fuchino, H., Tsujimoto, K., Uchida, Y., and Jingo, N. 2006. H_2S poisoning of solid oxide fuel cells. *Journal of the Electrochemical Society* 153(11):A2023–A2029.

135. Zha, S., Cheng, Z., and Liu, M. 2007. Sulfur poisoning and regeneration of Ni-based anodes in solid oxide fuel cells. *Journal of the Electrochemical Society* 154(2):B201–B206.

136. Li, T.S., Wang, W.G., Chen, T., Miao, H., and Xu, C. 2010. Hydrogen sulfide poisoning in solid oxide fuel cells under accelerated testing conditions. *Journal of Power Sources* 195(20):7025–7032.

137. Sasaki, K., Haga, K., Yoshizumi, T., Minematsu, D., Yuki, E., Liu, R., Uryu, C., Oshima, T., Ogura, T., Shiratori, Y., and Ito, K. 2011. Chemical durability of solid oxide fuel cells: Influence of impurities on long-term performance. *Journal of Power Sources* 196(22):9130–9140.

138. Cheng, Z., and Liu, M. 2007. Characterization of sulfur poisoning of Ni–YSZ anodes for solid oxide fuel cells using in situ Raman microspectroscopy. *Solid State Ionics* 178(13–14):925–935.

139. Brightman, E., Ivey, D.G., Brett, D.J.L., and Brandon, N.P. 2011. The effect of current density on H_2S-poisoning of nickel-based solid oxide fuel cell anodes. *Journal of Power Sources* 196(17):7182–7187.

140. Wang, J.H., and Liu, M. 2008. Surface regeneration of sulfur-poisoned Ni surfaces under SOFC operation conditions predicted by first-principles-based thermodynamic calculations. *Journal of Power Sources* 176(1):23–30.

141. Choi, S., Wang, J., Cheng, Z., and Liu, M. 2008. Surface modification of Ni-YSZ using niobium oxide for sulfur-tolerant anodes in solid oxide fuel cells. *Journal of the Electrochemical Society* 155(5):B449–B454.

142. Kock, A.J.H.M., De Bokx, P.K., Boellaard, E., Klop, W., and Geus, J.W. 1985. The formation of filamentous carbon on iron and nickel catalysts: II. Mechanism. *Journal of Catalysis* 96(2):468–480.

143. Holstein, W.L. 1995. The roles of ordinary and soret diffusion in the metal-catalyzed formation of filamentous carbon. *Journal of Catalysis* 152(1):42–51.

144. He, H., and Hill, J.M. 2007. Carbon deposition on Ni/YSZ composites exposed to humidified methane. *Applied Catalysis A: General* 317(2):284–292.

145. Chen, T., Wang, W.G., Miao, H., Li, T., and Xu, C. 2011. Evaluation of carbon deposition behavior on the nickel/yttrium-stabilized zirconia anode-supported fuel cell fueled with simulated syngas. *Journal of Power Sources* 196(5):2461–2468.

146. Klemensø, T., Chung, C., Larsen, P.H., and Mogensen, M. 2005. The mechanism behind redox instability of anodes in high-temperature SOFCs. *Journal of the Electrochemical Society* 152(11):A2186–A2192.

147. Serra, J.M., Vert, V.B., Buchler, O., Meulenberg, W.A., and Buchkremer, H.P. 2008. IT-SOFC supported on mixed oxygen ionic-electronic conducting composites. *Chemistry of Materials* 20(12):3867–3875.

148. Somalu, M.R., Yufit, V., Cumming, D., Lorente, E., and Brandon, N.P. 2011. Fabrication and characterization of Ni/ScSZ cermet anodes for IT-SOFCs. *International Journal of Hydrogen Energy* 36(9):5557–5566.

149. Zha, S., Rauch, W., and Liu, M. 2004. Ni-$Ce_{0.9}Gd_{0.1}O_{1.95}$ anode for GDC electrolyte-based low-temperature SOFCs. *Solid State Ionics* 166(3–4):241–250.

150. Maric, R., Ohara, S., Fukui, T., Inagaki, T., and Fujita, J.I. 1998. High-performance Ni-SDC cermet anode for solid oxide fuel cells at medium operating temperature. *Electrochemical and Solid-State Letters* 1(5):201–203.

151. Maric, R., Fukui, T., Ohara, S., Yoshida, H., Nishimura, M., Inagaki, T., and Miura, K. 2000. Powder prepared by spray pyrolysis as an electrode material for solid oxide fuel cells. *Journal of Materials Science* 35(6):1397–1404.

152. Hui, S.R., Yang, D., Wang, Z., Yick, S., Deces-Petit, C., Qu, W., Tuck, A., Maric, R., and Ghosh, D. 2007. Metal-supported solid oxide fuel cell operated at 400–600°C. *Journal of Power Sources* 167(2):336–339.

153. Hui, S.R., Yang, D., Wang, Z., Yick, S., Decès-Petit, C., Qu, W., Tuck, A., Maric, R., and Ghosh, D. 2007. Metal-supported solid oxide fuel cell operated at 400–600°C. *ECS Transactions* 7(1):763–769.

154. Wang, Z., Berghaus, J.O., Yick, S., Decès-Petit, C., Qu, W., Hui, R., Maric, R., and Ghosh, D. 2008. Dynamic evaluation of low-temperature metal-supported solid oxide fuel cell oriented to auxiliary power units. *Journal of Power Sources* 176(1):90–95.

155. He, B., Wang, W., Zhao, L., and Xia, C. 2011. Ni-LnO_x (Ln = Dy, Ho, Er, Yb and Tb) cermet anodes for intermediate-temperature solid oxide fuel cells. *Electrochimica Acta* 56(20):7071–7077.

156. Chen, M., Kim, B.H., Xu, Q., Nam, O.J., and Ko, J.H. 2008. Synthesis and performances of Ni–SDC cermets for IT-SOFC anode. *Journal of the European Ceramic Society* 28(15):2947–2953.

157. Zhao, J., Xu, X., Zhou, W., and Zhu, Z. 2017. MnO-Co composite modified Ni-SDC anode for intermediate temperature solid oxide fuel cells. *Fuel Processing Technology* 161:241–247.

158. Da Silva, F.S., and de Souza, T.M. 2017. Novel materials for solid oxide fuel cell technologies: a literature review. *International Journal of Hydrogen Energy* 42(41):26020–26036.

159. Qu, J., Wang, W., Chen, Y., Deng, X., and Shao, Z. 2016. Stable direct-methane solid oxide fuel cells with calcium-oxide-modified nickel-based anodes operating at reduced temperatures. *Applied Energy* 164:563–571.

160. Yamamoto, K., Hashishin, T., Matsuda, M., Qiu, N., Tan, Z., and Ohara, S. 2014. High-performance Ni nanocomposite anode fabricated from Gd-doped ceria nanocubes for low-temperature solid-oxide fuel cells. *Nano Energy* 6:103–108.

161. Liu, Q.L., Khor, K.A., and Chan, S.H. 2006. High-performance low-temperature solid oxide fuel cell with novel BSCF cathode. *Journal of Power Sources* 161(1):123–128.

162. Lomberg, M., Ruiz-Trejo, E., Offer, G., and Brandon, N.P. 2014. Characterization of Ni-infiltrated GDC electrodes for solid oxide cell applications. *Journal of the Electrochemical Society* 161(9):F899–F905.

163. Sin, A., Kopnin, E., Dubitsky, Y., Zaopo, A., Aricò, A.S., La Rosa, D., Gullo, L.R., and Antonucci, V. 2007. Performance and life-time behaviour of NiCu–CGO anodes for the direct electro-oxidation of methane in IT-SOFCs. *Journal of Power Sources* 164(1):300–305.

164. Hussain, A.M., Høgh, J.V., Zhang, W., and Bonanos, N. 2012. Efficient ceramic anodes infiltrated with binary and ternary electrocatalysts for SOFCs operating at low temperatures. *Journal of Power Sources* 216:308–313.

165. Cowin, P.I., Petit, C.T., Lan, R., Irvine, J.T., and Tao, S. 2011. Recent progress in the development of anode materials for solid oxide fuel cells. *Advanced Energy Materials* 1(3):314–332.

166. Li, X., Zhao, H., Zhou, X., Xu, N., Xie, Z., and Chen, N. 2010. Electrical conductivity and structural stability of La-doped $SrTiO_3$ with A-site deficiency as anode materials for solid oxide fuel cells. *International Journal of Hydrogen Energy* 35(15):7913–7918.

167. Savaniu, C.D., and Irvine, J.T. 2009. Reduction studies and evaluation of surface modified A-site deficient La-doped $SrTiO_3$ as anode material for IT-SOFCs. *Journal of Materials Chemistry* 19(43):8119–8128.

168. Yoo, K.B., and Choi, G.M. 2009. Performance of La-doped strontium titanate (LST) anode on $LaGaO_3$-based SOFC. *Solid State Ionics* 11(180):867–871.

169. Ruiz-Morales, J.C., Canales-Vázquez, J., Savaniu, C., Marrero-López, D., Zhou, W., and Irvine, J.T. 2006. Disruption of extended defects in solid oxide fuel cell anodes for methane oxidation. *Nature* 439:568–571.

170. Lu, X.C., Zhu, J.H., Yang, Z., Xia, G., and Stevenson, J.W. 2009. Pd-impregnated SYT/LDC composite as sulfur-tolerant anode for solid oxide fuel cells. *Journal of Power Sources* 192(2):381–384.

171. Kurokawa, H., Yang, L., Jacobson, C.P., De Jonghe, L.C., and Visco, S.J. 2007. Y-doped SrTiO3 based sulfur tolerant anode for solid oxide fuel cells. *Journal of Power Sources* 164(2):510–518.

172. Kharton, V.V., Tsipis, E.V., Marozau, I.P., Viskup, A.P., Frade, J.R., and Irvine, J.T.S. 2007. Mixed conductivity and electrochemical behavior of $(La_{0.75}Sr_{0.25})_{0.95}Cr_{0.5}Mn_{0.5}O_{3-\delta}$. *Solid State Ionics* 178(1–2):101–113.

173. Lay, E., Gauthier, G., Rosini, S., Savaniu, C., and Irvine, J.T. 2008. Ce-substituted LSCM as new anode material for SOFC operating in dry methane. *Solid State Ionics* 179(27–32):1562–1566.

174. Danilovic, N., Vincent, A., Luo, J.L., Chuang, K.T., Hui, R., and Sanger, A.R. 2009. Correlation of fuel cell anode electrocatalytic and ex situ catalytic activity of perovskites $La_{0.75}Sr_{0.25}Cr_{0.5}X_{0.5}O_{3-\delta}$ (X= Ti, Mn, Fe, Co). *Chemistry of Materials* 22(3):957–965.

175. Papazisi, K.M., Balomenou, S., and Tsiplakides, D. 2010. Synthesis and characterization of $La_{0.75}Sr_{0.25}Cr_{0.9}M_{0.1}O_3$ perovskites as anodes for CO-fuelled solid oxide fuel cells. *Journal of Applied Electrochemistry* 40(10):1875–1881.

176. Kobsiriphat, W., Madsen, B.D., Wang, Y., Shah, M., Marks, L.D., and Barnett, S.A. 2010. Nickel-and ruthenium-doped lanthanum chromite anodes: effects of nanoscale metal precipitation on solid oxide fuel cell performance. *Journal of the Electrochemical Society* 157(2):B279–B284.

177. Kobsiriphat, W., Madsen, B.D., Wang, Y., Marks, L.D., and Barnett, S.A. 2009. $La_{0.8}Sr_{0.2}Cr_{1-x}Ru_xO_{3-\delta}$–$Gd_{0.1}Ce_{0.9}O_{1.95}$ solid oxide fuel cell anodes: Ru precipitation and electrochemical performance. *Solid State Ionics* 180(2–3):257–264.

178. Lashtabeg, A., Canales-Vazquez, J., Irvine, J.T., and Bradley, J.L. 2009. Structure, conductivity, and thermal expansion studies of redox stable rutile niobium chromium titanates in oxidizing and reducing conditions. *Chemistry of Materials* 21(15):3549–3561.

179. Flores, J.C.P., and García-Alvarado, F. 2009. Electrical conductivity of the oxygen-deficient rutile $CrNbO_{4-\delta}$. *Solid State Sciences* 11(1):207–213.

180. Danilovic, N., Luo, J.L., Chuang, K.T., and Sanger, A.R. 2009. $Ce_{0.9}Sr_{0.1}VO_x$ (x = 3, 4) as anode materials for H_2S-containing CH_4 fueled solid oxide fuel cells. *Journal of Power Sources* 192(2):247–257.

181. Danilovic, N., Luo, J.L., Chuang, K.T., and Sanger, A.R. 2009. Effect of substitution with Cr^{3+} and addition of Ni on the physical and electrochemical properties of $Ce_{0.9}Sr_{0.1}VO_3$ as a H_2S-active anode for solid oxide fuel cells. *Journal of Power Sources* 194(1):252–262.

182. Cheng, Z., Zha, S., Aguilar, L., Wang, D., Winnick, J., and Liu, M. 2006. A solid oxide fuel cell running on H_2S/CH_4 fuel mixtures. *Electrochemical and Solid-State Letters* 9(1):A31–A33.

183. Petit, C.T., Lan, R., Cowin, P.I., Kraft, A., and Tao, S. 2011. Structure, conductivity and redox stability of solid solution $Ce_{1-x}Ca_xVO_4$ ($0 \leq x \leq 0.4125$). *Journal of Materials Science* 46(2):316–326.

184. Petit, C.T., Lan, R., Cowin, P.I., Irvine, J.T., and Tao, S. 2011. Novel redox reversible oxide, Sr-doped cerium orthovanadate to metavanadate. *Journal of Materials Chemistry* 21(2):525–531.

185. Pelosato, R., Cordaro, G., Stucchi, D., Cristiani, C., and Dotelli, G. 2015. Cobalt based layered perovskites as cathode material for intermediate temperature solid oxide fuel cells: a brief review. *Journal of Power Sources* 298:46–67.

186. Kilner, J.A., and Burriel, M. 2014. Materials for intermediate-temperature solid-oxide fuel cells. *Annual Review of Materials Research* 44:365–393.

187. Singh, D., and Singh, R. 2010. Synthesis and characterization of Ruddlesden-Popper (RP) type phase $LaSr_2MnCrO_7$. *Journal of Chemical Sciences* 122(6):807–811.

188. Hu, Y., Bouffanais, Y., Almar, L., Morata, A., Tarancon, A., and Dezanneau, G. 2013. $La_{2-x}Sr_xCoO_{4-\delta}$ (x = 0.9, 1.0, 1.1) Ruddlesden-Popper-type layered cobaltites as cathode materials for IT-SOFC application. *International Journal of Hydrogen Energy* 38(7):3064–3072.

189. Tsipis, E.V., and Kharton, V.V. 2008. Electrode materials and reaction mechanisms in solid oxide fuel cells: a brief review. *Journal of Solid State Electrochemistry* 12(11):1367–1391.

190. Atta, N.F., Galal, A., and El-Ads, E.H. 2016. Perovskite nanomaterials–synthesis, characterization, and applications. In *Perovskite Materials-Synthesis, Characterization, and Applications*, Edited by Pan, L., and Zhu, G., InTech: Rijeka, Croatia, 107–151.

191. Kuo, J.H., Anderson, H.U., and Sparlin, D.M. 1990. Oxidation-reduction behavior of undoped and Sr-doped $LaMnO_3$: defect structure, electrical conductivity, and thermoelectric power. *Journal of Solid State Chemistry* 87(1):55–63.

192. Yokokawa, H., Sakai, N., Kawada, T., and Dokiya, M. 1990. Thermodynamic analysis on interface between perovskite electrode and YSZ electrolyte. *Solid State Ionics* 40:398–401.

193. Jiang, S.P., and Wang, W. 2005. Fabrication and performance of GDC-impregnated (La, Sr)MnO_3 cathodes for intermediate temperature solid oxide fuel cells. *Journal of the Electrochemical Society* 152(7):A1398–A1408.

194. Mizusaki, J., Yonemura, Y., Kamata, H., Ohyama, K., Mori, N., Takai, H., Tagawa, H., Dokiya, M., Naraya, K., Sasamoto, T., and Inaba, H. 2000. Electronic conductivity, Seebeck coefficient, defect and electronic structure of nonstoichiometric $La_{1-x}Sr_xMnO_3$. *Solid State Ionics* 132(3–4):167–180.

195. Zhong-Tai, Z., Lin, O., and Zi-Long, T. 1995. Synthesis and characteristics of $La_{1-x}Sr_xMnO_3$ ceramics for cathode materials of SOFC. *ECS Proceedings Volumes* 1995–1:502–511.

196. Kovalevsky, A.V., Kharton, V.V., Snijkers, F.M.M., Cooymans, J.F.C., Luyten, J.J., and arques, F.M.B. 2007. Oxygen transport and stability of asymmetric $SrFe(Al)O_{3-\delta}$-$SrAl_2O_4$ composite membranes. *Journal of Membrane Science* 301(1–2): 238–244.

197. Patrakeev, M.V., Bahteeva, J.A., Mitberg, E.B., Leonidov, I.A., Kozhevnikov, V.L., and Poeppelmeier, K.R. 2003. Electron/hole and ion transport in $La_{1-x}Sr_xFeO_{3-\delta}$. *Journal of Solid State Chemistry* 172(1):219–231.

198. Tsipis, E.V., Patrakeev, M.V., Kharton, V.V., Yaremchenko, A.A., Mather, G.C., Shaula, A.L., Leonidov, I.A., Kozhevnikov, V.L., and Frade, J.R. 2005. Transport properties and thermal expansion of Ti-substituted $La_{1-x}Sr_xFeO_{3-\delta}$ ($x = 0.5$–0.7). *Solid State Sciences* 7(4):355–365.

199. Kharton, V.V., Patrakeev, M.V., Waerenborgh, J.C., Kovalevsky, A.V., Pivak, Y.V., Gaczyński, P., Markov, A.A., and Yaremchenko, A.A. 2007. Oxygen nonstoichiometry, Mössbauer spectra and mixed conductivity of $Pr_{0.5}Sr_{0.5}FeO_{3-\delta}$. *Journal of Physics and Chemistry of Solids* 68(3):355–366.

200. Lein, H.L., Wiik, K., and Grande, T. 2006. Thermal and chemical expansion of mixed-conducting $La_{0.5}Sr_{0.5}Fe_{1-x}Co_xO_{3-\delta}$ materials. *Solid State Ionics* 177(19–25):1795–1798.

201. Søgaard, M., Hendriksen, P.V., and Mogensen, M. 2007. Oxygen nonstoichiometry and transport properties of strontium substituted lanthanum ferrite. *Journal of Solid State Chemistry* 180(4):1489–1503.

202. Zając, W., Świerczek, K., and Molenda, J. 2007. Thermochemical compatibility between selected (La, Sr)(Co, Fe, Ni)O_3 cathodes and rare earth doped ceria electrolytes. *Journal of Power Sources* 173(2):675–680.

203. Kharton, V.V., Yaremchenko, A.A., Patrakeev, M.V., Naumovich, E.N., and Marques, F.M.B. 2003. Thermal and chemical induced expansion of $La_{0.3}Sr_{0.7}$(Fe, Ga)$O_{3-\delta}$ ceramics. *Journal of the European Ceramic Society* 23(9):1417–1426.

204. Kharton, V.V., Shaulo, A.L., Viskup, A.P., Avdeev, M., Yaremchenko, A.A., Patrakeev, M.V., Kurbakov, A.I., Naumovich, E.N., and Marques, F.M.B. 2002. Perovskite-like system (Sr, La)(Fe, Ga)$O_{3-\delta}$: structure and ionic transport under oxidizing conditions. *Solid State Ionics* 150(3–4):229–243.

205. Park, C.Y., and Jacobson, A.J. 2005. Thermal and chemical expansion properties of $La_{0.2}Sr_{0.8}Fe_{0.55}Ti_{0.45}O_{3-x}$. *Solid State Ionics* 176(35–36):2671–2676.

206. Tsipis, E.V., Kharton, V.V., Vyshatko, N.V., Frade, J.R., and Marques, F.M.B. 2005. Ion transport properties and Seebeck coefficient of Fe-doped La(Sr)Al(Mg)O$_{3-\delta}$. *Solid State Sciences* 7(3):257–267.

207. Petric, A., Huang, P., and Tietz, F. 2000. Evaluation of La–Sr–Co–Fe–O perovskites for solid oxide fuel cells and gas separation membranes. *Solid State Ionics* 135(1–4):719–725.

208. Tai, L.W., Nasrallah, M.M., Anderson, H.U., Sparlin, D.M., and Sehlin, S.R. 1995. Structure and electrical properties of La$_{1-x}$Sr$_x$Co$_{1-y}$Fe$_y$O$_3$. Part 1. The system La$_{0.8}$Sr$_{0.2}$Co$_{1-y}$Fe$_y$O$_3$. *Solid State Ionics* 76(3–4):259–271.

209. Kostogloudis, G.C., and Ftikos, C. 1999. Properties of A-site-deficient La$_{0.6}$Sr$_{0.4}$Co$_{0.2}$Fe$_{0.8}$O$_{3-\delta}$-based perovskite oxides. *Solid State Ionics* 126(1–2):143–151.

210. Esquirol, A., Brandon, N.P., Kilner, J.A., and Mogensen, M. 2004. Electrochemical characterization of La$_{0.6}$Sr$_{0.4}$Co$_{0.2}$Fe$_{0.8}$O$_3$ cathodes for intermediate-temperature SOFCs. *Journal of The Electrochemical Society* 151(11):A1847–A1855.

211. Tu, H.Y., Takeda, Y., Imanishi, N., and Yamamoto, O. 1999. Ln$_{0.4}$Sr$_{0.6}$Co$_{0.8}$Fe$_{0.2}$O$_{3-\delta}$ (Ln = La, Pr, Nd, Sm, Gd) for the electrode in solid oxide fuel cells. *Solid State Ionics* 117(3–4):277–281.

212. Hung, M.H., Rao, M.M., and Tsai, D.S. 2007. Microstructures and electrical properties of calcium substituted LaFeO$_3$ as SOFC cathode. *Materials Chemistry and Physics* 101(2–3):297–302.

213. Kharton, V.V., Naumovich, E.N., Vecher, A.A., and Nikolaev, A.V. 1995. Oxide ion conduction in solid solutions Ln$_{1-x}$Sr$_x$CoO$_{3-\delta}$ (Ln = La, Pr, Nd). *Journal of Solid State Chemistry* 120(1):128–136.

214. Ishihara, T., Fukui, S., Nishiguchi, H., and Takita, Y. 2002. Mixed electronic-oxide ionic conductor of BaCoO$_3$ doped with La for cathode of intermediate-temperature-operating solid oxide fuel cell. *Solid State Ionics* 152:609–613.

215. Ohbayashi, H., Kudo, T., and Gejo, T. 1974. Crystallographic, electric and thermochemical properties of the perovskite-type Ln$_{1-x}$Sr$_x$CoO$_3$ (Ln: Lanthanoid element). *Japanese Journal of Applied Physics* 13(1):1.

216. Kharton, V.V., Naumovich, E.N., Zhuk, P.P., Demin, A.K., and Nikolaev, A.V. 1992. Physicochemical and electrochemical properties of Ln(Sr)CoO$_3$ electrode materials. *Soviet Electrochemistry* 28(11):1376–1384.

217. Aruna, S.T., Muthuraman, M., and Patil, K.C. 2000. Studies on combustion synthesized LaMnO$_3$–LaCoO$_3$ solid solutions. *Materials Research Bulletin* 35(2):289–296.

218. Aruna, S., and Patil, K. 1997. Combustion synthesis and properties of strontium substituted lanthanum manganites La$_{1-x}$Sr$_x$MnO$_3$ ($0 \le x \le 0.3$). *Journal of Materials Chemistry* 7(12):2499–2503.

219. Huang, K., Lee, H.Y., and Goodenough, J.B. 1998. Sr-and Ni-doped LaCoO$_3$ and LaFeO$_3$ perovskites new cathode materials for solid-oxide fuel cells. *Journal of the Electrochemical Society* 145(9):3220–3227.

220. Petrov, A.N., Kononchuk, O.F., Andreev, A.V., Cherepanov, V.A., and Kofstad, P. 1995. Crystal structure, electrical and magnetic properties of La$_{1-x}$Sr$_x$CoO$_{3-y}$. *Solid State Ionics* 80(3–4):189–199.

221. Zhu, X.D., Sun, K.N., Zhang, N.Q., Chen, X.B., Wu, L.J., and Jia, D.C. 2007. Improved electrochemical performance of SrCo$_{0.8}$Fe$_{0.2}$O$_{3-\delta}$–La$_{0.45}$Ce$_{0.55}$O$_{2-\delta}$ composite cathodes for IT-SOFC. *Electrochemistry Communications* 9(3):431–435.

222. Zhou, W., Shao, Z., Ran, R., Chen, Z., Zeng, P., Gu, H., Jin, W., and Xu, N. 2007. High performance electrode for electrochemical oxygen generator cell based on solid electrolyte ion transport membrane. *Electrochimica Acta* 52(22):6297–6303.

223. Zhang, X., Robertson, M., Yick, S., Deĉes-Petit, C., Styles, E., Qu, W., Xie, Y., Hui, R., Roller, J., Kesler, O., and Maric, R. 2006. Sm$_{0.5}$Sr$_{0.5}$CoO$_3$ + Sm$_{0.2}$Ce$_{0.8}$O$_{1.9}$ composite cathode for cermet supported thin Sm$_{0.2}$Ce$_{0.8}$O$_{1.9}$ electrolyte SOFC operating below 600°C. *Journal of Power Sources* 160(2):1211–1216.

224. Lee, K.C., Choi, M.B., Lim, D.K., Singh, B., and Song, S.J. 2013. Effect of humidification on the performance of intermediate-temperature proton conducting ceramic fuel cells with ceramic composite cathodes. *Journal of Power Sources* 232:224–233.

225. Zhao, Z., Liu, L., Zhang, X., Wu, W., Tu, B., Cui, D., Ou, D., and Cheng, M. 2013. High-and low-temperature behaviors of $La_{0.6}Sr_{0.4}Co_{0.2}Fe_{0.8}O_{3-\delta}$ cathode operating under CO_2/H_2O-containing atmosphere. *International Journal of Hydrogen Energy* 38(35):15361–15370.

226. Hu, B., Mahapatra, M.K., Keane, M., Zhang, H., and Singh, P. 2014. Effect of CO_2 on the stability of strontium doped lanthanum manganite cathode. *Journal of Power Sources* 268:404–413.

227. Yang, Q., and Lin, Y.S. 2006. Kinetics of carbon dioxide sorption on perovskite-type metal oxides. *Industrial & Engineering Chemistry Research* 45(18):6302–6310.

228. Yang, Z., Guo, M., Wang, N., Ma, C., Wang, J., and Han, M. 2017. A short review of cathode poisoning and corrosion in solid oxide fuel cell. *International Journal of Hydrogen Energy* 42(39):24948–24959.

229. Porras-Vazquez, J.M., Smith, R.I., and Slater, P.R. 2014. Investigation into the effect of Si doping on the cell symmetry and performance of $Sr_{1-y}Ca_yFeO_{3-\delta}$ SOFC cathode materials. *Journal of Solid State Chemistry* 213:132–137.

230. Schuler, J.A., Gehrig, C., Wuillemin, Z., Schuler, A.J., Wochele, J., Ludwig, C., and Hessler-Wyser, A. 2011. Air side contamination in solid oxide fuel cell stack testing. *Journal of Power Sources* 196(17):7225–7231.

231. Perz, M., Bucher, E., Gspan, C., Waldhäusl, J., Hofer, F., and Sitte, W. 2016. Long-term degradation of $La_{0.6}Sr_{0.4}Co_{0.2}Fe_{0.8}O_{3-\delta}$ IT-SOFC cathodes due to silicon poisoning. *Solid State Ionics* 288:22–27.

232. Tejuca, L.G., Rochester, C.H., Fierro, J.L.G., and Tascón, J.M. 1984. Infrared spectroscopic study of the adsorption of pyridine, carbon monoxide and carbon dioxide on the perovskite-type oxides $LaMO_3$. *Journal of the Chemical Society, Faraday Transactions 1: Physical Chemistry in Condensed Phases* 80(5):1089–1099.

233. Hammami, R., Batis, H., and Minot, C. 2009. Combined experimental and theoretical investigation of the CO_2 adsorption on $LaMnO_{3+y}$ perovskite oxide. *Surface Science* 603(20):3057–3067.

234. Pena, M.A., and Fierro, J.L.G. 2001. Chemical structures and performance of perovskite oxides. *Chemical Reviews* 101(7):1981–2018.

235. Zhao, Z., Liu, L., Zhang, X., Wu, W., Tu, B., Ou, D., and Cheng, M. 2013. A comparison on effects of CO_2 on $La_{0.8}Sr_{0.2}MnO_{3+\delta}$ and $La_{0.6}Sr_{0.4}CoO_{3-\delta}$ cathodes. *Journal of Power Sources* 222:542–553.

236. Yu, Y., Luo, H., Cetin, D., Lin, X., Ludwig, K., Pal, U., Gopalan, S., and Basu, S. 2014. Effect of atmospheric CO_2 on surface segregation and phase formation in $La_{0.6}Sr_{0.4}Co_{0.2}Fe_{0.8}O_{3-\delta}$ thin films. *Applied Surface Science* 323:71–77.

237. Zhang, K., Meng, B., Tan, X., Liu, L., Wang, S., and Liu, S. 2014. CO_2-Tolerant ceramic membrane driven by electrical current for oxygen production at intermediate temperatures. *Journal of the American Ceramic Society* 97(1):120–126.

238. Gui, L., Wan, Y., Wang, R., Wang, Z., He, B., and Zhao, L. 2015. A comparison of oxygen permeation and CO_2 tolerance of $La_{0.6}Sr_{0.4}Co_{0.2}Fe_{0.6}Nb_{0.2}O_{3-\delta}$ and $La_{0.6}Sr_{0.4}Fe_{0.8}Nb_{0.2}O_{3-\delta}$ ceramic membranes. *Journal of Alloys and Compounds* 644:788–792.

239. Kim, S.H., Ohshima, T., Shiratori, Y., Itoh, K., and Sasaki, K. 2007. Effect of water vapor and SO_x in air on the cathodes of solid oxide fuel cells. *MRS Online Proceedings Library Archive* 1041-R03-10. Symposium R – Life-Cycle Analysis for New Energy Conversion and Storage Systems. doi:10.1557/PROC-1041-R03-10.

240. Hagen, A., Neufeld, K., and Liu, Y.L. 2010. Effect of humidity in air on performance and long-term durability of SOFCs. *Journal of the Electrochemical Society* 157(10):B1343–B1348.

241. Liu, R.R., Kim, S.H., Taniguchi, S., Oshima, T., Shiratori, Y., Ito, K., and Sasaki, K. 2011. Influence of water vapor on long-term performance and accelerated degradation of solid oxide fuel cell cathodes. *Journal of Power Sources* 196(17):7090–7096.

242. Chen, X., Zhen, Y., Li, J., and Jiang, S.P. 2010. Chromium deposition and poisoning in dry and humidified air at $(La_{0.8}Sr_{0.2})_{0.9}MnO_{3+\delta}$ cathodes of solid oxide fuel cells. *International Journal of Hydrogen Energy* 35(6):2477–2485.

243. Knöfel, C., Chen, M., and Mogensen, M. 2011. The effect of humidity and oxygen partial pressure on LSM–YSZ cathode. *Fuel Cells* 11(5):669–677.

244. Nielsen, J., Hagen, A., and Liu, Y.L. 2010. Effect of cathode gas humidification on performance and durability of solid oxide fuel cells. *Solid State Ionics* 181(11–12): 517–524.

245. Aphale, A., Liang, C., Hu, B., and Singh, P. 2017. Cathode degradation from airborne contaminants in solid oxide fuel cells: a review. In *Solid Oxide Fuel Cell Lifetime and Reliability*, Edited by Brandon, N.P., Ruiz-Trejo, E., and Boldrin, P., Academic Press, Elsevier: London, UK, 101–119.

246. Jiang, S.P., and Zhen, Y. 2008. Mechanism of Cr deposition and its application in the development of Cr-tolerant cathodes of solid oxide fuel cells. *Solid State Ionics* 179(27–32):1459–1464.

247. Jin, T., and Lu, K. 2012. Chromium deposition and interfacial interactions of an electrolyte–air electrode-interconnect tri-layer for solid oxide fuel cells. *Journal of Power Sources* 202:143–148.

248. Jiang, S.P., Zhen, Y.D., and Zhang, S. 2006. Interaction Between Fe–Cr metallic interconnect and (La, Sr)MnO_3/YSZ composite cathode of solid oxide fuel cells. *Journal of the Electrochemical Society* 153(8):A1511–A1517.

249. Hessler-Wyser, A., Wuillemin, Z., Schuler, J.A., and Faes, A. 2011. TEM investigation on zirconate formation and chromium poisoning in LSM/YSZ cathode. *Journal of Materials Science* 46(13):4532–4539.

250. Matsuzaki, Y., and Yasuda, I. 2001. Dependence of SOFC cathode degradation by chromium-containing alloy on compositions of electrodes and electrolytes. *Journal of the Electrochemical Society* 148(2):A126–A131.

251. Xiong, C., Taillon, J.A., Pellegrinelli, C., Huang, Y.L., Salamanca-Riba, L.G., Chi, B., Jian, L., Pu, J., and Wachsman, E.D. 2016. Long-term Cr poisoning effect on LSCF-GDC composite cathodes sintered at different temperatures. *Journal of the Electrochemical Society* 163(9):F1091–F1099.

252. Schrödl, N., Bucher, E., Egger, A., Kreiml, P., Teichert, C., Höschen, T., and Sitte, W. 2015. Long-term stability of the IT-SOFC cathode materials $La_{0.6}Sr_{0.4}CoO_{3-\delta}$ and $La_2NiO_{4+\delta}$ against combined chromium and silicon poisoning. *Solid State Ionics* 276:62–71.

253. Viitanen, M.M., Welzenis, R.V., Brongersma, H.H., and Van Berkel, F.P.F. 2002. Silica poisoning of oxygen membranes. *Solid State Ionics* 150(3–4):223–228.

254. Zhao, L., Perry, N.H., Daio, T., Sasaki, K., and Bishop, S.R. 2015. Improving the Si impurity tolerance of $Pr_{0.1}Ce_{0.9}O_{2-\delta}$ SOFC electrodes with reactive surface additives. *Chemistry of Materials* 27(8):3065–3070.

255. Yokokawa, H., Horita, T., Yamaji, K., Kishimoto, H., and Brito, M.E. 2012. Degradation of SOFC cell/stack performance in relation to materials deterioration. *Journal of the Korean Ceramic Society* 49(1):11–18.

256. Bucher, E., Sitte, W., Klauser, F., and Bertel, E. 2011. Oxygen exchange kinetics of $La_{0.58}Sr_{0.4}Co_{0.2}Fe_{0.8}O_3$ at 600°C in dry and humid atmospheres. *Solid State Ionics* 191(1):61–67.

257. Bucher, E., Sitte, W., Klauser, F., and Bertel, E. 2012. Impact of humid atmospheres on oxygen exchange properties, surface-near elemental composition, and surface morphology of $La_{0.6}Sr_{0.4}CoO_{3-\delta}$. *Solid State Ionics* 208:43–51.

258. Bucher, E., Sitte, W., Caraman, G.B., Cherepanov, V.A., Aksenova, T.V., and Ananyev, M.V. 2006. Defect equilibria and partial molar properties of (La, Sr)(Co, Fe)$O_{3-\delta}$. *Solid State Ionics* 177(35–36):3109–3115.

259. Bucher, E., and Sitte, W. 2011. Long-term stability of the oxygen exchange properties of (La, Sr)$_{1-z}$(Co, Fe)$O_{3-\delta}$ in dry and wet atmospheres. *Solid State Ionics* 192(1):480–482.

260. Porras-Vazquez, J.M., and Slater, P.R. 2012. Synthesis and characterization of oxyanion-doped cobalt containing perovskites. *Fuel Cells* 12(6):1056–1063.

261. Tarancón, A., Skinner, S.J., Chater, R.J., Hernandez-Ramirez, F., and Kilner, J.A. 2007. Layered perovskites as promising cathodes for intermediate temperature solid oxide fuel cells. *Journal of Materials Chemistry* 17(30):3175–3181.

262. Zhang, K., Ge, L., Ran, R., Shao, Z., and Liu, S. 2008. Synthesis, characterization and evaluation of cation-ordered $LnBaCo_2O_{5+\delta}$ as materials of oxygen permeation membranes and cathodes of SOFCs. *Acta Materialia* 56(17):4876–4889.

263. Kim, G., Wang, S., Jacobson, A.J., Reimus, L., Brodersen, P., and Mims, C.A. 2007. Rapid oxygen ion diffusion and surface exchange kinetics in $PrBaCo_2O_{5+x}$ with a perovskite related structure and ordered A cations. *Journal of Materials Chemistry* 17(24):2500–2505.

264. Zhu, C., Liu, X., Yi, C., Yan, D., and Su, W. 2008. Electrochemical performance of $PrBaCo_2O_{5+\delta}$ layered perovskite as an intermediate-temperature solid oxide fuel cell cathode. *Journal of Power Sources* 185(1):193–196.

265. Che, X., Shen, Y., Li, H., and He, T. 2013. Assessment of $LnBaCo_{1.6}Ni_{0.4}O_{5+\delta}$ (Ln = Pr, Nd, and Sm) double-perovskites as cathodes for intermediate-temperature solid-oxide fuel cells. *Journal of Power Sources* 222:288–293.

266. Li, X., Jiang, X., Xu, H., Xu, Q., Jiang, L., Shi, Y., and Zhang, Q. 2013. Scandium-doped $PrBaCo_{2-x}Sc_xO_{6-\delta}$ oxides as cathode material for intermediate-temperature solid oxide fuel cells. *International Journal of Hydrogen Energy* 38(27):12035–12042.

267. Pang, S., Jiang, X., Li, X., Wang, Q., and Su, Z. 2012. Characterization of Ba-deficient $PrBa_{1-x}Co_2O_{5+\delta}$ as cathode material for intermediate temperature solid oxide fuel cells. *Journal of Power Sources* 204:53–59.

268. Pang, S.L., Jiang, X.N., Li, X.N., Xu, H.X., Jiang, L., Xu, Q.L., Shi, Y.C., and Zhang, Q.Y. 2013. Structure and properties of layered-perovskite $LaBa_{1-x}Co_2O_{5+\delta}$ ($x = 0$–0.15) as intermediate-temperature cathode material. *Journal of Power Sources* 240:54–59.

269. Choi, S., Yoo, S., Kim, J., Park, S., Jun, A., Sengodan, S., Kim, J., Shin, J., Jeong, H.Y., Choi, Y., and Kim, G. 2013. Highly efficient and robust cathode materials for low-temperature solid oxide fuel cells: $PrBa_{0.5}Sr_{0.5}Co_{2-x}Fe_xO_{5+\delta}$. *Scientific Reports* 3:2426.

270. Skinner, S.J. 2001. Recent advances in perovskite-type materials for solid oxide fuel cell cathodes. *International Journal of Inorganic Materials* 3(2):113–121.

271. Hildenbrand, N., Nammensma, P., Blank, D.H., Bouwmeester, H.J., and Boukamp, B.A. 2013. Influence of configuration and microstructure on performance of $La_2NiO_{4+\delta}$ intermediate-temperature solid oxide fuel cells cathodes. *Journal of Power Sources* 238:442–453.

272. Mesguich, D., Bassat, J.M., Aymonier, C., Brüll, A., Dessemond, L., and Djurado, E. 2013. Influence of crystallinity and particle size on the electrochemical properties of spray pyrolyzed $Nd_2NiO_{4+\delta}$ powders. *Electrochimica Acta* 87:330–335.

273. Ferchaud, C., Grenier, J.C., Zhang-Steenwinkel, Y., Van Tuel, M.M., Van Berkel, F.P., and Bassat, J.M. 2011. High performance praseodymium nickelate oxide cathode for low temperature solid oxide fuel cell. *Journal of Power Sources* 196(4):1872–1879.

274. Hernández, A.M., Mogni, L., and Caneiro, A. 2010. $La_2NiO_{4+\delta}$ as cathode for SOFC: reactivity study with YSZ and CGO electrolytes. *International Journal of Hydrogen Energy* 35(11):6031–6036.

275. Sayers, R., Liu, J., Rustumji, B., and Skinner, S.J. 2008. Novel K_2NiF_4-type materials for solid oxide fuel cells: compatibility with electrolytes in the intermediate temperature range. *Fuel Cells* 8(5):338–343.

276. Lalanne, C., Prosperi, G., Bassat, J.M., Mauvy, F., Fourcade, S., Stevens, P., Zahid, M., Diethelm, S., and Grenier, J.C. 2008. Neodymium-deficient nickelate oxide $Nd_{1.95}NiO_{4+\delta}$ as cathode material for anode-supported intermediate temperature solid oxide fuel cells. *Journal of Power Sources* 185(2):1218–1224.

277. Laberty, C., Zhao, F., Swider-Lyons, K.E., and Virkar, A.V. 2007. High-performance solid oxide fuel cell cathodes with lanthanum-nickelate-based composites. *Electrochemical and Solid-State Letters* 10(10):B170–B174.

278. Phillipps, M.B., Sammes, N.M., and Yamamoto, O. 1999. $Gd_{1-x}A_xCo_{1-y}Mn_yO_3$ (A = Sr, Ca) as a cathode for the SOFC. *Solid State Ionics* 123(1–4):131–138.

279. Kostogloudis, G.C., Vasilakos, N., and Ftikos, C. 1997. Preparation and characterization of $Pr_{1-x}Sr_xMnO_{3\pm\delta}$ ($x = 0$, 0.15, 0.3, 0.4, 0.5) as a potential SOFC cathode material operating at intermediate temperatures (500–700°C). *Journal of the European Ceramic Society* 17(12):1513–1521.

280. Kostogloudis, G.C., Vasilakos, N., and Ftikos, C. 1998. Crystal structure, thermal and electrical properties of $Pr_{1-x}Sr_xCoO_{3-\delta}$ ($x = 0$, 0.15, 0.3, 0.4, 0.5) perovskite oxides. *Solid State Ionics* 106(3–4):207–218.

281. Kostogloudis, G.C., Fertis, P., and Ftikos, C. 1998. The perovskite oxide system $Pr_{1-x}Sr_xCo_{1-y}Mn_yO_{3-\delta}$: crystal structure and thermal expansion. *Journal of the European Ceramic Society* 18(14):2209–2215.

282. Rajeev, K.P., and Raychaudhuri, A.K. 1992. Quantum corrections to the conductivity in a perovskite oxide: a low-temperature study of $LaNi_{1-x}Co_xO_3$ ($0 \leq x \leq 0.75$). *Physical Review B* 46(3):1309–1320.

283. Drennan, J., Tavares, C.P., and Steele, B.C.H. 1982. An electron microscope investigation of phases in the system La-Ni-O. *Materials Research Bulletin* 17(5):621–626.

284. Hrovat, M., Katsarakis, N., Reichmann, K., Bernik, S., Kuščer, D., and Holc, J. 1996. Characterisation of $LaNi_{1-x}Co_xO_3$ as a possible SOFC cathode material. *Solid State Ionics* 83(1–2):99–105.

285. Chiba, R., Yoshimura, F., and Sakurai, Y. 1999. An investigation of $LaNi_{1-x}Fe_xO_3$ as a cathode material for solid oxide fuel cells. *Solid State Ionics* 124(3–4):281–288.

286. Zhu, W.Z., and Deevi, S.C. 2003. Development of interconnect materials for solid oxide fuel cells. *Materials Science and Engineering: A* 348(1–2):227–243.

287. Anwar, M., Muchtar, A., and Somalu, M.R. 2016. Effects of various co-dopants and carbonates on the properties of doped ceria-based electrolytes: a brief review. *International Journal of Applied Engineering Research* 11(19):9921–9928.

288. Evans, A., Bieberle-Hütter, A., Galinski, H., Rupp, J.L., Ryll, T., Scherrer, B., Tölke, R., and Gauckler, L.J. 2009. Micro-solid oxide fuel cells: status, challenges, and chances. *Monatshefte für Chemie-Chemical Monthly* 140(9):975–983.

289. Kalinina, E.G., Pikalova, E.Y., Menshikova, A.V., and Nikolaenko, I.V. 2016. Electrophoretic deposition of a self-stabilizing suspension based on a nanosized multi-component electrolyte powder prepared by the laser evaporation method. *Solid State Ionics* 288:110–114.

290. Brylewski, T., Kruk, A., Adamczyk, A., Kucza, W., Stygar, M., and Przybylski, K. 2012. Synthesis and characterization of the manganese cobaltite spinel prepared using two "soft chemical" methods. *Materials Chemistry and Physics* 137(1):310–316.

291. Park, B.K., Kim, D.W., Song, R.H., Lee, S.B., Lim, T.H., Park, S.J., Park, C.O., and Lee, J.W. 2015. Design of a dual-layer ceramic interconnect based on perovskite oxides for segmented-in-series solid oxide fuel cells. *Journal of Power Sources* 300:318–324.

292. Antepara, I., Villarreal, I., Rodríguez-Martínez, L.M., Lecanda, N., Castro, U., and Laresgoiti, A. 2005. Evaluation of ferritic steels for use as interconnects and porous metal supports in IT-SOFCs. *Journal of Power Sources* 151:103–107.
293. Aznam, I., Muchtar, A., Somalu, M.R., Ghazali, M.J., Mah, J.C.W., and Baharuddin, N.A. 2018. Interconnect development for solid oxide fuel cell application. *Journal of Advanced Research in Fluid Mechanics and Thermal Sciences* 51(2):227–233.
294. Wu, J., and Liu, X. 2010. Recent development of SOFC metallic interconnect. *Journal of Materials Science & Technology* 26(4):293–305.
295. Minh, N.Q. 1993. Ceramic fuel cells. *Journal of the American Ceramic Society* 76(3):563–588.
296. Fergus, J.W. 2004. Lanthanum chromite-based materials for solid oxide fuel cell interconnects. *Solid State Ionics* 171(1–2):1–15.
297. Sharifzadeh, M. 2019. *Design and Operation of Solid Oxide Fuel Cells: The Systems Engineering Vision for Industrial Application.* Academic Press, Elsevier: London, UK.
298. Tietz, F. 1999. Thermal expansion of SOFC materials. *Ionics* 5(1–2):129–139.
299. Anderson, H.U. 1978. Fabrication and property control of $LaCrO_3$ based oxides. In *Processing of Crystalline Ceramics*, Edited by Palmour III, H., Davis, R.F., Hare, T.M., Springer: Boston, MA, 469–477.
300. Groupp, L., and Anderson, H.U. 1976. Densification of $La_{1-x}Si_xCrO_3$. *Journal of the American Ceramic Society* 59(9–10):449–450.
301. Sakai, N., Kawada, T., Yokokawa, H., Dokiya, M., and Iwata, T. 1990. Sinterability and electrical conductivity of calcium-doped lanthanum chromites. *Journal of Materials Science* 25(10):4531–4534.
302. Tai, L.W., and Lessing, P.A. 1991. Tape casting and sintering of strontium-doped lanthanum chromite for a planar solid oxide fuel cell bipolar plate. *Journal of the American Ceramic Society* 74(1):155–160.
303. Yang, Z., Weil, K.S., Paxton, D.M., and Stevenson, J.W. 2003. Selection and evaluation of heat-resistant alloys for SOFC interconnect applications. *Journal of the Electrochemical Society* 150(9):A1188–A1201.
304. Yang, Z., Walker, M.S., Singh, P., and Stevenson, J.W. 2003. Anomalous corrosion behavior of stainless steels under SOFC interconnect exposure conditions. *Electrochemical and Solid-State Letters* 6(10):B35–B37.
305. Yang, Z., Xia, G., Singh, P., and Stevenson, J.W. 2006. Electrical contacts between cathodes and metallic interconnects in solid oxide fuel cells. *Journal of Power Sources* 155(2):246–252.
306. Yang, Z., Xia, G.G., Wang, C.M., Nie, Z., Templeton, J., Stevenson, J.W., and Singh, P. 2008. Investigation of iron–chromium–niobium–titanium ferritic stainless steel for solid oxide fuel cell interconnect applications. *Journal of Power Sources* 183(2):660–667.
307. Yang, Z., Xia, G., Singh, P., and Stevenson, J.W. 2005. Effects of water vapor on oxidation behavior of ferritic stainless steels under solid oxide fuel cell interconnect exposure conditions. *Solid State Ionics* 176(17–18):1495–1503.
308. Yang, Z., Singh, P., Stevenson, J.W., and Xia, G.G. 2006. Investigation of modified Ni–Cr–Mn base alloys for SOFC interconnect applications. *Journal of the Electrochemical Society* 153(10):A1873–A1879.
309. Mah, J.C., Muchtar, A., Somalu, M.R., and Ghazali, M.J. 2017. Metallic interconnects for solid oxide fuel cell: a review on protective coating and deposition techniques. *International Journal of Hydrogen Energy* 42(14):9219–9229.
310. Zhu, W.Z., and Deevi, S.C. 2003. Opportunity of metallic interconnects for solid oxide fuel cells: a status on contact resistance. *Materials Research Bulletin* 38(6):957–972.
311. Sachitanand, R., Sattari, M., Svensson, J.E., and Froitzheim, J. 2013. Evaluation of the oxidation and Cr evaporation properties of selected FeCr alloys used as SOFC interconnects. *International Journal of Hydrogen Energy* 38(35):15328–15334.

312. Seo, H.S., Yun, D.W., and Kim, K.Y. 2013. Oxidation behavior of ferritic stainless steel containing Nb, Nb–Si and Nb–Ti for SOFC interconnect. *International Journal of Hydrogen Energy* 38(5):2432–2442.

313. Seo, H.S., Yun, D.W., and Kim, K.Y. 2012. Effect of Ti addition on the electric and ionic property of the oxide scale formed on the ferritic stainless steel for SOFC interconnect. *International Journal of Hydrogen Energy* 37(21):16151–16160.

314. Chandra-Ambhorn, S., Wouters, Y., Antoni, L., Toscan, F., and Galerie, A. 2007. Adhesion of oxide scales grown on ferritic stainless steels in solid oxide fuel cells temperature and atmosphere conditions. *Journal of Power Sources* 171(2):688–695.

315. Wu, J., Yan, D., Pu, J., Chi, B., and Jian, L. 2012. The investigation of interaction between $La_{0.9}Sr_{0.1}MnO_3$ cathode and metallic interconnect for solid oxide fuel cell at reduced temperature. *Journal of Power Sources* 202:166–174.

316. Ebbinghaus, B.B. 1993. Thermodynamics of gas phase chromium species: the chromium oxides, the chromium oxyhydroxides, and volatility calculations in waste incineration processes. *Combustion and Flame* 93(1–2):119–137.

317. Fergus, J.W. 2005. Metallic interconnects for solid oxide fuel cells. *Materials Science and Engineering: A* 397(1–2):271–283.

318. Geng, S., Zhu, J., Brady, M.P., Anderson, H.U., Zhou, X.D., and Yang, Z. 2007. A low-Cr metallic interconnect for intermediate-temperature solid oxide fuel cells. *Journal of Power Sources* 172(2):775–781.

319. Zhu, J.H., Geng, S.J., and Ballard, D.A. 2007. Evaluation of several low thermal expansion Fe–Co–Ni alloys as interconnect for reduced-temperature solid oxide fuel cell. *International Journal of Hydrogen Energy* 32(16):3682–3688.

320. Wei, P., Deng, X., Bateni, M.R., and Petric, A. 2007. Oxidation and electrical conductivity behavior of spinel coatings for metallic interconnects of solid oxide fuel cells. *Corrosion* 63(6):529–536.

321. Hua, B., Pu, J., Lu, F., Zhang, J., Chi, B., and Jian, L. 2010. Development of a Fe–Cr alloy for interconnect application in intermediate temperature solid oxide fuel cells. *Journal of Power Sources* 195(9):2782–2788.

322. Jiang, S.P., and Chen, X. 2014. Chromium deposition and poisoning of cathodes of solid oxide fuel cells–a review. *International Journal of Hydrogen Energy* 39(1):505–531.

323. Yang, Z., Xia, G.G., and Stevenson, J.W. 2006. Evaluation of Ni–Cr-base alloys for SOFC interconnect applications. *Journal of Power Sources* 160(2):1104–1110.

324. Chen, L., Yang, Z., Jha, B., Xia, G., and Stevenson, J.W. 2005. Clad metals, roll bonding and their applications for SOFC interconnects. *Journal of Power Sources* 152:40–45.

325. Nielsen, K.A., Dinesen, A.R., Korcakova, L., Mikkelsen, L., Hendriksen, P.V., and Poulsen, F.W. 2006. Testing of Ni-plated ferritic steel interconnect in SOFC stacks. *Fuel Cells* 6(2):100–106.

326. Jablonski, P.D., and Alman, D.E. 2007. Oxidation resistance and mechanical properties of experimental low coefficient of thermal expansion (CTE) Ni-base alloys. *International Journal of Hydrogen Energy* 32(16):3705–3712.

327. PalDey, S.C.D.S., and Deevi, S.C. 2003. Single layer and multilayer wear resistant coatings of (Ti, Al)N: a review. *Materials Science and Engineering: A* 342(1–2):58–79.

328. Pederson, L.R., Singh, P., and Zhou, X.D. 2006. Application of vacuum deposition methods to solid oxide fuel cells. *Vacuum* 80(10):1066–1083.

329. Gannon, P.E., Tripp, C.T., Knospe, A.K., Ramana, C.V., Deibert, M., Smith, R.J., Gorokhovsky, V.I., Shutthanandan, V., and Gelles, D. 2004. High-temperature oxidation resistance and surface electrical conductivity of stainless steels with filtered arc Cr–Al–N multilayer and/or superlattice coatings. *Surface and Coatings Technology* 188:55–61.

330. Kayani, A., Smith, R.J., Teintze, S., Kopczyk, M., Gannon, P.E., Deibert, M.C., Gorokhovsky, V.I., and Shutthanandan, V. 2006. Oxidation studies of CrAlON nanolayered coatings on steel plates. *Surface and Coatings Technology* 201(3–4): 1685–1694.
331. Liu, X., Johnson, C., Li, C., Xu, J., and Cross, C. 2008. Developing TiAlN coatings for intermediate temperature solid oxide fuel cell interconnect applications. *International Journal of Hydrogen Energy* 33(1):189–196.
332. Wu, J., Li, C., Johnson, C., and Liu, X. 2008. Evaluation of SmCo and SmCoN magnetron sputtering coatings for SOFC interconnect applications. *Journal of Power Sources* 175(2):833–840.
333. Shaigan, N., Qu, W., Ivey, D.G., and Chen, W. 2010. A review of recent progress in coatings, surface modifications and alloy developments for solid oxide fuel cell ferritic stainless steel interconnects. *Journal of Power Sources* 195(6):1529–1542.
334. Johnson, C., Wu, J., Liu, X., and Gemmen, R.S. 2007. Solid oxide fuel cell performance using metallic interconnects coated by electroplating methods (No. DOE/NETL-IR-2007–171). National Energy Technology Laboratory (NETL), Pittsburgh, PA, Morgantown, WV, and Albany, OR.
335. Piccardo, P., Gannon, P., Chevalier, S., Viviani, M., Barbucci, A., Caboche, G., Amendola, R., and Fontana, S. 2007. ASR evaluation of different kinds of coatings on a ferritic stainless steel as SOFC interconnects. *Surface and Coatings Technology* 202(4–7):1221–1225.
336. Stevenson, J.W., Yang, Z.G., Xia, G.G., Nie, Z., and Templeton, J.D. 2013. Long-term oxidation behavior of spinel-coated ferritic stainless steel for solid oxide fuel cell interconnect applications. *Journal of Power Sources* 231:256–263.
337. Collins, C., Lucas, J., Buchanan, T.L., Kopczyk, M., Kayani, A., Gannon, P.E., Deibert, M.C., Smith, R.J., Choi, D.S., and Gorokhovsky, V.I. 2006. Chromium volatility of coated and uncoated steel interconnects for SOFCs. *Surface and Coatings Technology* 201(7):4467–4470.
338. Bateni, M.R., Wei, P., Deng, X., and Petric, A. 2007. Spinel coatings for UNS 430 stainless steel interconnects. *Surface and Coatings Technology* 201(8):4677–4684.
339. Larring, Y., and Norby, T. 2000. Spinel and perovskite functional layers between plansee metallic interconnect (Cr-5wt%Fe-1wt%Y_2O_3) and ceramic $(La_{0.85}Sr_{0.15})_{0.91}MnO_3$ cathode materials for solid oxide fuel cells. *Journal of the Electrochemical Society* 147(9):3251–3256.
340. Yang, Z., Xia, G., Simner, S.P., and Stevenson, J.W. 2005. Thermal growth and performance of manganese cobaltite spinel protection layers on ferritic stainless steel SOFC interconnects. *Journal of the Electrochemical Society* 152(9):A1896–A1901.
341. Yang, Z., Xia, G.G., Li, X.H., and Stevenson, J.W. 2007. $(Mn,Co)_3O_4$ spinel coatings on ferritic stainless steels for SOFC interconnect applications. *International Journal of Hydrogen Energy* 32(16):3648–3654.
342. Xiao, J., Zhang, W., Xiong, C., Chi, B., Pu, J., and Jian, L. 2016. Oxidation behavior of Cu-doped $MnCo_2O_4$ spinel coating on ferritic stainless steels for solid oxide fuel cell interconnects. *International Journal of Hydrogen Energy* 41(22):9611–9618.
343. Kumar, C.D., Dekich, A., Wang, H., Liu, Y., Tilson, W., Ganley, J., and Fergus, J.W. 2013. Transition metal doping of manganese cobalt spinel oxides for coating SOFC interconnects. *Journal of the Electrochemical Society* 161(1):F47.
344. Xu, Y., Wen, Z., Wang, S., and Wen, T. 2011. Cu doped Mn–Co spinel protective coating on ferritic stainless steels for SOFC interconnect applications. *Solid State Ionics* 192(1):561–564.

345. Masi, A., Bellusci, M., McPhail, S.J., Padella, F., Reale, P., Hong, J.E., Steinberger-Wilckens, R., and Carlini, M. 2017. Cu-Mn-Co oxides as protective materials in SOFC technology: the effect of chemical composition on mechanochemical synthesis, sintering behaviour, thermal expansion and electrical conductivity. *Journal of the European Ceramic Society* 37(2):661–669.

346. Fergus, J.W. 2005. Sealants for solid oxide fuel cells. *Journal of Power Sources* 147(1–2):46–57.

347. Lessing, P.A. 2007. A review of sealing technologies applicable to solid oxide electrolysis cells. *Journal of Materials Science* 42(10):3465–3476.

348. Lara, C., Pascual, M.J., and Duran, A. 2004. Glass-forming ability, sinterability and thermal properties in the systems RO–BaO–SiO$_2$ (R = Mg, Zn). *Journal of Non-Crystalline Solids* 348:149–155.

349. Sohn, S.B., Choi, S.Y., Kim, G.H., Song, H.S., and Kim, G.D. 2002. Stable sealing glass for planar solid oxide fuel cell. *Journal of Non-Crystalline Solids* 297(2–3):103–112.

350. Batfalsky, P., Haanappel, V.A.C., Malzbender, J., Menzler, N.H., Shemet, V., Vinke, I.C., and Steinbrech, R.W. 2006. Chemical interaction between glass–ceramic sealants and interconnect steels in SOFC stacks. *Journal of Power Sources* 155(2):128–137.

351. Weber, A., Müller, A., Herbstritt, D., and Ivers-Tiffée, E. 2001. Characterization of SOFC single cells. *ECS Proceedings Volumes* 16:952–962.

352. Duquette, J., and Petric, A. 2004. Silver wire seal design for planar solid oxide fuel cell stack. *Journal of Power Sources* 137(1):71–75.

353. Chou, Y.S., and Stevenson, J.W. 2003. Novel silver/mica multilayer compressive seals for solid-oxide fuel cells: the effect of thermal cycling and material degradation on leak behavior. *Journal of Materials Research* 18(9):2243–2250.

354. Singh, P., Yang, Z., Viswanathan, V., and Stevenson, J.W. 2004. Observations on the structural degradation of silver during simultaneous exposure to oxidizing and reducing environments. *Journal of Materials Engineering and Performance* 13(3):287–294.

355. Weil, K.S., Coyle, C.A., Hardy, J.S., Kim, J.Y., and Xia, G.G. 2004. Alternative planar SOFC sealing concepts. *Fuel Cells Bulletin* 2004(5):11–16.

356. Chou, Y.S., and Stevenson, J.W. 2003. Phlogopite mica-based compressive seals for solid oxide fuel cells: effect of mica thickness. *Journal of Power Sources* 124(2):473–478.

357. Bram, M., Reckers, S., Drinovac, P., Mönch, J., Steinbrech, R.W., Buchkremer, H.P., and Stöver, D. 2004. Deformation behavior and leakage tests of alternate sealing materials for SOFC stacks. *Journal of Power Sources* 138(1–2):111–119.

358. Chou, Y.S., and Stevenson, J.W. 2002. Thermal cycling and degradation mechanisms of compressive mica-based seals for solid oxide fuel cells. *Journal of Power Sources* 112(2):376–383.

359. Chou, Y.S., and Stevenson, J.W. 2003. Mid-term stability of novel mica-based compressive seals for solid oxide fuel cells. *Journal of Power Sources* 115(2):274–278.

360. Loehman, R., Gauntt, B., and Shah, R. 2005. *Development of Reliable Methods for Sealing Solid Oxide Fuel Cell Stacks*. Sandia National Laboratories: Albuquerque, NM.

361. Yamamoto, T., Hibiki, I.T.O.H., Masashi, M.O.R.I., Noriyuki, M.O.R.I., and Toshio, A.B.E. 1995. Application of mica glass-ceramics as gas-sealing materials for SOFC. *ECS Proceedings Volumes* 1995(1):245–254.

3 Processing

3.1 DIFFERENT CELL CONCEPTS

The two main concepts are based on the tubular and the planar design, and the main commercial manufacturer of the planar configuration is Bloom Energy; and tubular, the Kyocera Corporation [1]. In 2017, Kyocera Corporation launched the industry's first 3 kW solid oxide fuel cell (SOFC) co-generation system that utilizes Kyocera's proprietary ceramic technologies to deliver 52% generation efficiency. The performance of the Kyocera unit is the highest of any comparable SOFC system currently on the market in the 1–3 kW range, with an overall efficiency of 90% with exhaust heat recovery [2]. The benefit of this system design is the use of exhaust heat from the power generation process to heat water. These characteristics make the system well-suited for larger houses, smaller hotels, hospitals, police stations, fire stations, retail establishments, and other commercial enterprises, including small restaurants. The Bloom Energy system is based on the electrolyte-based planar design, and a 250 kW system is about the size of the shipping container. Bloom Energy commercializes SOFC systems with over 50% net electrical efficiencies. Bloom shipped its first energy server in 2006 and in a report published in 2019 stated that over 350 megawatts (MW) of capacity has been installed at more than 600 sites worldwide [3]. Bloom's customers are renowned, such as Google, Coca-Cola, Ebay, Walmart, and Bank of America, just to name a few. Each energy server can be connected, remotely managed, and monitored by Bloom Energy. The system can be grid-connected or stand-alone configuration with cost predictability and overall power quality. In the last 20 years, the different types of the tubular designs have been developed: (1) flatten tube-type cells made by Kyocera, (2) segment-in-series cells on flatten tubes made by Tokyo gas, (3) sealless tubular cells made by TOTO, (4) disk-type planer cells made by Mitsubishi Materials Corporation and Kansai Electric Power Corporation, (5) segment-in-series cells on tubes made by Mitsubishi Heavy Industry, and (6) TOTO small tubular type [4].

The first tubular design was developed by Siemens Westinghouse, where the cathode acts as the carrier. This tubular SOFC design is schematically presented in Figure 3.1. The tubes are closed at one end and consist mainly of a cathode of perovskite-type material. Then, the electrolyte is directly deposited on the cathode, followed by the anode deposition. In this design, the anode side is in the fuel surrounding, whereas the air is fed to the tube inside. The main advantage of such cells is that the tubular design does not demand a complex sealing, and the separation of air and fuel volumes is easy to manage, but it is more expensive to manufacture than the planar type [5,6].

Mechanically, tubular designs can be cathode-supported, anode-supported, or electrolyte-supported. Tubular configurations allow for fabricating thin-layer structures that can significantly improve power density. Many techniques for tubular

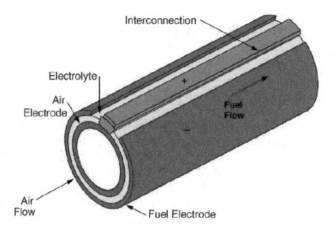

FIGURE 3.1   Cathode-supported tubular SOFC (Simens Westinghouse concept). (Reprinted with permission from Ref. [7].)

cell fabrication have been reported, such as commonly adopted extrusion [8–11], iso-pressing [12], slip casting [13], and dip-coating [14]. The extrusion process is used for a large scale commercial development and started by the production of the cathode-supported tubular SOFC [8]. During the extrusion process, the parameters such as the viscosity of the ceramic slurry, the extrusion speed, and the die temperature need to be optimized for a high-performance cell. The most important process parameter is a die temperature, and the suitable die temperature is ranging from 40°C to 70°C depending on the slurry content [15]. After the extrusion, the next step is the drying process, and the optimization of drying parameters is crucial to avoid cracks [16]. The drying and de-binding of water and the binder have to be done under the slow rate in the specific temperature range that depends on the evaporation rate of the organic components. Many times the pre-sintering process is applied before applying electrolyte to increase the strength of the tube and to address the thermal expansion mismatch [17]. The next generation of the tubular cells are flat tubes. The flat-tube SOFC design is developed by Siemens Westinghouse based on their formerly tubular SOFC, and Kyocera and Korean KIER developed the flat-tubular architecture as well. It was demonstrated that the flat tubular SOFCs improves the power density by reducing the internal ohmic loss [18]. The flat tubular design is shown in Figure 3.2. In this type of architecture, the cells are generally constructed on an extruded anode support with multiple gas channels. Depending on the application, the tubular SOFCs have a dimension from microtubular SOFC to lengths of about 1.5–2 m for rapid start-up times and large gross power, respectively. The advantages of the extrusion process for manufacturing tubular cells are low-cost productions, the ability to produce long tubes, and a high fabrication capacity [19].

Many developers aiming at the kW range prefer planar anode-supported SOFCs due to their potential of lowering the operating temperature below 800°C. Many companies and research organizations have focused on the porous anode substrate and a thin electrolyte membrane. Besides the frequently used ceramic processing

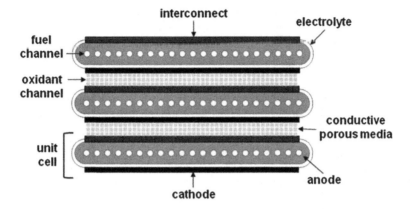

FIGURE 3.2 Schematic diagram of anode-supported flat-tubular SOFCs connected in series. (Reprinted with permission from Ref. [20].)

techniques, i.e., tape casting and screen printing, also alternative methods like warm pressing, tape calendaring, powder spraying, plasma spray process, spray pyrolysis (SP), and flame processes had been explored. The main selection criteria for the ability to make a thin film of electrolytes and electrodes, adoption of the fabrication route are the cost aspects, the potential for automation, reproducibility, and precision of the different techniques – the widely adopted concept in the support where the anode must, therefore, display sufficient mechanical stability. In mass production, the anode support is produced by tape casting.

Planar SOFCs have mainly two design variants, metals- and ceramic-supported SOFC. Ceramic-supported SOFCs are electrolytes based on anode-based SOFCs. To lower the operating temperature, SOFCs with thin YSZ electrolytes on Ni-YSZ anode supports received considerable attention because they can produce power densities of $1 \ W \ cm^{-2}$ or higher at operating temperatures of below 700°C. Both the high power densities and the reduced operating temperatures are advantageous for making practical SOFC power plants. NGK Spark Plug is a leader in manufacturing SOFC around 1 kW with SOFC that demonstrated operation over 10,000 h [21]. The system capacity in NGK Spark Plug unit is 0.5–1.5 kW per unit, with a total efficiency of 80% or larger for co-generation unit, and aiming for a power generation efficiency of 60% or more when only supplying electricity [22]. The NGK planar stack is presented in Figure 3.3a. NGK Spark Plug has a state of art manufacturing facility for the development of SOFC co-power generation (CHP) by conventional ceramic processing such as screen printing and tape casting. The NGK Spark Plug design is made of the Ni-YSZ anode/YSZ electrolyte and $La_{0.6}Sr_{0.4}Co_{0.2}Fe_{0.8}O_3$ (LSCF) cathode. The anode-supported electrolyte consisting of NiO-8YSZ was made by tape casting, and the resulting green tape was cut into squares. The cell consists of the Ni-8YSZ cermet anode/8YSZ electrolyte/gadolinia-doped ceria barrier layer and LSCF cathode layer, as schematically shown in Figure 3.3b.

However, the performance of these types of cells at lower temperatures strongly depends on their cathode–electrolyte interface since interfacial polarization

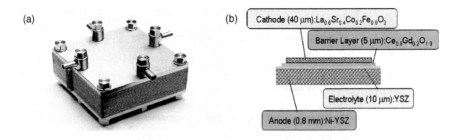

FIGURE 3.3 (a) NGK Spark Plug SOFC [23], and (b) schematic illustration of NGK cell. (Reprinted with permission from Ref. [24].)

resistance increases rapidly as temperature is decreased. Extensive interfacial reactions are known to occur between the Sr-containing perovskite LSCF cathode and the standard YSZ electrolyte even at low temperature and in order to prevent these unwanted reactions between YSZ electrolytes and cathode materials, it is necessary to introduce a thin diffusion barrier layer such as GDC. According to the New Energy and Industrial Technology Development Organization, the Japanese domestic residential micro-CHP has become the first mass production application for household applications [25] and in novel micro grid deployments [26].

Metal-supported SOFCs are considered to have a high potential for use in mobile applications and will lead to a significant cost reduction. Alloys such as ferritic steels are the excellent metal-support, since they have a very low cost and are very similar in their coefficient of thermal expansion with ceramic components, nickel oxide doped with YSZ or GDC. Metal-supported cells could provide excellent thermal cyclability while maintaining a strong interlayer bond. In addition, metallic substrates enable well-established joining techniques such as brazing and welding to be used for the integration of cells into a stack design. The coating of the functional SOFC layers can either be done by thermal spraying processes, SP, flame pyrolysis [27–29], or conventional sintering techniques [30–32]. The metal-supported cell exhibits good mechanical and thermal properties and helps the ceramic functional layers to overcome their weakness to mechanical and thermal shocks [33]. UK-based Ceres Power Holdings developed metal-supported Steel Cell technology for a range of potential power equipment applications [34].

Ceres Power became the first company to effectively make a metal-supported SOFC stack product [35–37]. Ceres Power cell technology is based on a patented steel cell using cerium gadolinium oxide (CGO) as the electrolyte, thus permitting operating temperatures of 500°C–620°C. The ability of the ferritic steel to bond with the ceramic presents an issue. The complexity in the fabrication of metal-supported cells arises from the fact that multiple layers must be sintered at much lower temperatures, as per the sintering requirements of the metal support. In the case of ferritic steels, temperatures higher than 1200°C enhance the possibility of Ni (from the anode) and Fe, Cr (from the support) counter-diffusion, and the approach toward the melting point of the alloy (1400°C–1500°C range is the melting region for SS430 and SS446). As mentioned earlier by Tucker, over-densification in order to sinter the electrolyte

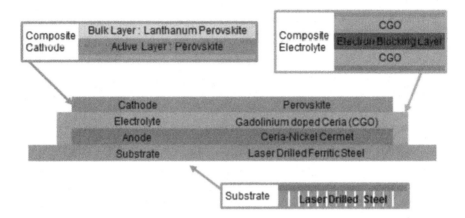

FIGURE 3.4 Schematic representation of a Ceres Power Steel Cell. (Reprinted with permission from Ref. [37].)

layer will result in the collapse of the pore structure and a drastic reduction in porosity, thereby causing major gas diffusional resistances in the anode [33]. Therefore, the different processing techniques have been used to control the size of the pores in the metal support, and some powder metallurgy techniques use metal meshes and laser-drilled halls in supports.

The design of Ceres Power cell is shown in Figure 3.4. Ceres Power fabricates the individual Steel Cells by screen-printing functional layers of ceramic ink onto a drilled sheet of steel. For the commercial-scale market, Ceres has developed a modular stack concept, which in initial testing has achieved 55% net electrical efficiency. For the residential market, Ceres has developed a 1 kW class power system which achieves 50% net electrical efficiency in one of the most compact fuel cell system designs available. These fuel cell power systems are running on natural gas and achieve power only efficiencies in the home or business, which are higher than state-of-the-art centralized generating plants, and an overall efficiency up to 90% when also capturing the heat [37–39].

3.2 METHODS OF PROCESSING CELL COMPONENTS

Methods for the manufacturing of the substrates and those of functional cell components are distinguished. Figure 3.5 describes the current conventional processing routes for the mass production for the different cells, i.e., the tubular, the planar electrolyte-supported, and the planar electrode-supported concepts.

3.2.1 SUBSTRATES

3.2.1.1 Extrusion

Extrusion is a traditional plastic-forming method for the manufacturing of ceramic parts over the large length and is perfectly suited for the tubular-type SOFC. The tubular type of SOFC requires one of the cell components to act as a mechanical support,

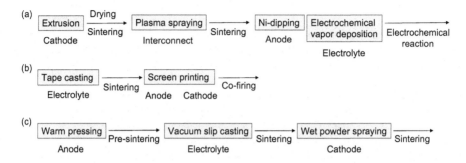

FIGURE 3.5 Ceramic processing routes to (a) the tubular design, (b) the electrolyte-supported design, and (c) the anode-supported design. (Adapted and modified from Ref. [40].)

and only a small amount of liquid is added to the solid, providing flexibility in shaping very hard powders and producing thin sections with support for deposition of the remaining cell layers. The supporting structure can be anode, cathode, or electrolyte depending on the system design, available technologies, and operating conditions that manufactures are envisioning. Extrusion offers several advantages such as uniform density distribution and cross-section with complex structures, greater length of the tube, and economic benefits in terms of capital and operating costs and ability to scale up the process [11,41]. In the field of SOFC, Siemens Westinghouse used extrusion for production of the air-electrode-supported cathode tubes and demonstrated the world's first highly efficient, longest-running 100 kW class SOFC/combined heat and power (SOFC/CHP) system and the first highest efficiency, 220 kW class pressurized SOFC/gas turbine (PSOFC/GT) hybrid system based on this technology [42]. Regardless of the supporting structure, the extrusion principles are that the ceramic slurry with the plasticizer incorporated is pushed or drawn through a die with the desired annular cross-section at the temperature when the plasticizer induces the easy slipping of the suspension. The suspension is then extruded into an external coagulation bath. The most critical component is the suspension stability with a solid charge as high as possible to induce large density values of the green tube. Usually, water-based plastic raw material compounds with water values between 15% and 20% are used [18,43]. Critical extrusion parameters are the screw speed, the temperature of the die, and the temperature profile to obtain a self-sustaining plastic rod. Table 3.1 summarizes the development of single SOFC fabrication using conventional extrusion [44]. Important information, such as the support configuration, materials selection, and cell performance, are presented. Regardless of the raw material, the important parameter is the rheological behavior of the compound, which needs to demonstrate nearly Newtonian flow behavior. To avoid the crack formation and deformation of the tubes, the heating rates during the de-binding must be selected in accordance with water movement from inside of the tube to the surface and the transportation rate of the gaseous species. The crack formations are usually formed during the middle-to-late stages of drying and propagated throughout the process. Nowadays, the research trend is to reduce the SOFC operating temperature down to 400°C–700°C (low- to intermediate-temperature SOFCs) to avoid many of

TABLE 3.1

The Development of Single SOFC Fabrication Using Conventional Extrusion

Configuration	Anode	Electrolyte	Cathode	Performance
Anode-supported	Ni-YSZ Extrusion, 700 μm	YSZ Dip-coating, 20 μm	LSM-YSZ/LSM Dip-coating 30–50 μm	600 mW cm^{-2} at 0.7 V, 850°C
Anode-supported	Ni-ScSZ Extrusion, 400 μm	ScSZ Dip-coating, 3 μm	GDC/LSCF-GDC Dip-coating 6 μm	800 mW cm^{-2} at 0.7 V, 600°C
Anode-supported	Ni-YSZ Extrusion	YSZ Dip-coating, <1 μm	GDC/LSCF-GDC Dip-coating 6 μm	300 mW cm^{-2} at 0.7 V, 600°C
Anode-supported	Ni-YSZ Extrusion	Ni-ScSZ/ScSZ Dip-coating, 10 μm	GDC/LSCF-GDC Dip-coating 6 μm	600 mW cm^{-2} at 0.7 V, 780°C
Anode-supported	Ni-GDC Extrusion	GDC Dip-coating, 10 μm	LSCF-GDC Dip-coating	350 mW cm^{-2} at 0.7 V, 550°C
Electrolyte-supported	Ni-YSZ	YSZ Extrusion, 100–200 μm	LSM	-
Anode-supported	Ni-YSZ Extrusion, 300 μm	YSZ, 15 μm	LSM	350 mW cm^{-2} at 0.5 V, 800°C, H$_2$ and CH$_4$
Anode-supported	Ni-YSZ Extrusion, 300 μm	YSZ, 10–50 μm	LSM	700 mW cm^{-2} at 0.7 V, 800°C, H$_2$ and CH$_4$
Cathode-supported	Ni-YSZ Coating, 50 μm	ScSZ, 20 μm	LSM/LSM-GDC Extrusion, 300 μm	75.6 mW cm^{-2} at 0.7 V, 650°C
Anode-supported	Ni-YSZ Extrusion, 1.5 mm	Ni/GDC/LSGM, 30 μm	LSCF, 20 μm	~160 mW cm^{-2} at 0.8 V, 690°C, town gas

Source: Data adapted and modified from Ref. [44].

the inconveniences related to a high operating temperature and reduce the cost and degradation rate of the material. Therefore, more effort has been invested in the study of anode-supported tube. Thin-film SOFCs with a traditional 100 nm YSZ electrolyte have been reported to be operative at temperatures as low as 350°C with an output power density of 400 mW cm^{-2} at 400°C [45].

3.2.1.2 Tape Casting

Tape casting is the most efficient way to manufacture large areas, thin, and flat parts that are impossible to press or to extrude. The technique can be used to produce flat ceramic or powder. Most manufacturers like Bloom Energy, Fuel Cell Energy, NGK Spark Plugs, and Fuel Cell Limited, just to name a few, are using the so-called doctor blade techniques to produce electrolyte-supported cells (100–200 μm) or anode/cathode-supported cells [46–51]. In the last 20 years, the focus on tape casting has been enhanced by colloidal shaping large-area, thin, and flat ceramic anode-supported tapes [52]. Current SOFCs using tape casting fabrication technology are mostly solvent-based, though the process originated as an aqueous processing method [53,54].

The doctor blade technique is schematically shown in Figure 3.6. The slurry contains different ingredients, and each of them has a different impact on the rheological behavior of the slurry. Usually, the tape is produced by mixing the starting powder with the dispersant in a suitable solvent [55]. The binder, surfactant, and plasticizers are added to provide slurry with adequate properties of the support. For example, plasticizers or sintering agents are usually added to improve either the mechanical properties of the green tape (flexibility, plasticity, etc.) or the physical properties of the sintered tape (density, microstructure). Surfactant is used to actively modify the particle surface to obtain a desired characteristic, such as lower surface charge, higher surface charge, high/low surface energy, or specific surface chemistry. The presence of different ingredients influences the rheology of the ceramic slurry, and this is important when it comes to conducting a fluid flow analysis to the flow of ceramic slurries during tape casting. The well-mixed slurry is then casted and dried to make green tapes. The green tapes are laser cut or punched to a reunited shape and de-bonded and sintered to produce ceramic support. The main property of the support tested and measured after sintering is the bending strength that needs to be up to 80 MPa. For the electrolyte-supported, the density has to be close to 100%, and for the anode- and cathode-supported, the porosity has to be between 26 and 48 vol%. The gas permeability and electrical conductivity of the tapes are measured before functional layers are printed or deposited on the tapes.

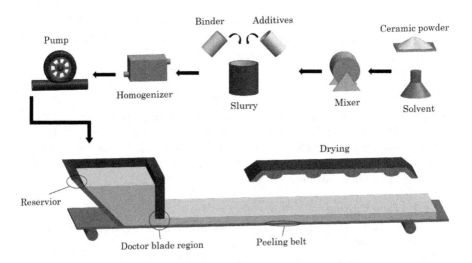

FIGURE 3.6 Schematic illustration of the tape casting process. (Reprinted with permission from Ref. [55].)

3.2.2 COATING METHODS

Manufactures of SOFCs are using different processing techniques to apply the functional layers on the support. The most commonly used process in manufacturing is screen printing. This is still considered a thick-film technology. For printing a film with a thickness less than 1 µm, other thin-film technologies should be used, such as chemical and physical processing. Each of these processes will be briefly described in the following sections.

3.2.2.1 Screen Printing

Screen printing is widely used in the electronic industry and is one of the key processes for multilayer components printing. In addition, this method is widely used to fabricate SOFC components having a thickness in the range of 10–100 µm. In the screen-printing process, a highly viscous paste consisting of a mixture of ceramic powder, an organic binder, and a plasticizer is printed on the substrate forced through the open meshes of a screen. The effect of powder particle size, solid loading, binder, solvent, and dispersant on the ink rheological properties and performance of resultant films must be deeply understood for the fabrication of optimized screen-printing inks. All these components will have a significant influence on the morphology as well as the conducting properties of the film. The structure can be optimized by measuring the rheological properties of inks such as viscosity, yield stress, thixotropy, and viscoelasticity for the application at a specific printer setting. Understanding the relationship between the composition, rheology of the ink, and sintering temperature after screen printing may enhance the properties and performance of the resultant screen-printed films. In the fabrication of SOFCs, thick films of anode, electrolyte, or cathode can be fabricated by screen printing on the support structure [56–60]. Thick films of electrolytes such as YSZ [61,62], SDC [63–65], ScSZ [66], and CGO or GDC [67] have been successfully fabricated by screen printing. There are many parameters that will influence the film density; however, in the ink preparation, the combination of a high solid content (>30 vol%) and a low binder content (<1 wt% of powder) is generally used to produce a dense film with minimum film defects. For the thick films of anode preparation (e.g., Ni/ScSZ, Ni/YSZ, Ni/CGO, and Ni/SDC) [68,69] and cathode preparation (e.g., $La(Sr)CoO_3$, $Sm_{0.5}Sr_{0.5}CoO_3 + Sm_{0.2}Ce_{0.8}O_{1.9}$, and LSCF) [70,71], a higher binder content (>1 wt% of powder) is preferred in order to improve the porosity of the resultant films after sintering. In the case of a porous substrate-supported cell, either metal or ceramic support, all the principal components of the cells can be fabricated by the cost-effective screen-printing technique. The quality and performance of cells are strongly dependent on the rheology of the ink, the screen-printer parameters, and sintering conditions.

3.2.2.2 Thin-Film Technologies

3.2.2.2.1 Chemical Vapor Deposition

There are two main chemical deposition techniques that are suitable for making gas-tight electrolyte films such as chemical vapor deposition (CVD) and electrochemical vapor deposition (EVD). They are also known to be suitable for mass production. The choice of an appropriate thin-film deposition technique is strongly influenced by available infrastructure, cost, the selection of material to be deposited, and the desired film quality. The Siemens Westinghouse had used a CVD technique to deposit YSZ electrolyte layers with a thickness of ~40 μm at a temperature of 950°C, in which zirconium and yttrium chlorides were the starting compounds [72]. A schematic diagram of a typical CVD apparatus is shown in Figure 3.7.

The main disadvantages of the CVD process are the high reaction temperature, the presence of corrosive gases, and the relatively low deposition rates. The process is of interest for producing uniform, pure, reproducible, and adherent films at low or high rates. It is particularly useful in the deposition of coating in the tubes that are difficult to reach by other deposition techniques. In order to lower the cost of deposition as well as deposition temperature, other variables of the CVD process had been developed, such as CVD methods in which precursors are volatile complexes of metals with organic ligands. The choice of mass transfer conditions (coordination of temperatures in desiccator chambers, carrier gas flow rate, total reactor pressure, the surface area of starting compounds, their dispersiveness, etc.) and the temperature of the substrate is instrumental to control in order to obtain a dense film. Few main parameters influence the density of the film: the flow rate, evaporation rate, and substrate temperature. The growth rates of the film thickness are in the range of 1–10 μm h^{-1}. Another process developed by the Siemens Westinghouse is EVD [74]. EVD is a two-step process where the pores closing in the substrate is done by the CVD reaction between the reactant metal chloride vapors and water vapor (or oxygen). EVD is modified from the CVD process, and in the EVD process, film growth occurs due to the presence of an electrochemical potential gradient across the deposited film. The primary application of EVD at Westinghouse was to fabricate a thin layer of the solid electrolyte and interconnection material used in SOFC. In this step, oxygen ions formed on the water vapor side of the substrate diffuse through the thin metal oxide layer to the metal chloride side. The solid film is spreading over the internal pore surface in the desired region across the substrate [75].

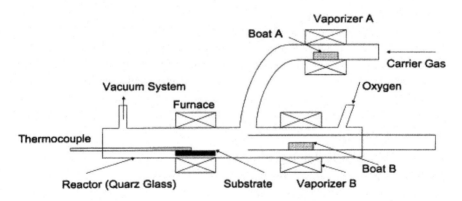

FIGURE 3.7 Schematic diagram of a CVD apparatus for preparation of YSZ films. (Reprinted with permission from Ref. [73].)

3.2.2.2.2 Wet Powder and Suspension Spraying

Spraying of suspension of fine dispersed ceramic or metallic powders enables the production of the uniform coating with thin-film electrolytes and a thick layer onto porous or dense metal or ceramic substrates [76,77]. Both planar [78,79] and tubular [80] SOFCs had been coated by wet powder spraying. The Korean Institute of Energy Research designed and manufactured 5k SOFC gas turbine hybrid system incorporating Julich's 5 kW stack that was manufactured by wet powder spraying [81]. The Julich design is based on the anode-supported SOFC, and the output of the SOFC stack was 8.1 kW for the H_2 and 4.7 kW for the pre-reformed gas under atmospheric pressure. In the hybrid configuration, when the stack was fueled by liquefied natural gas, the pre-reformer was operated in combination with a micro-gas turbine with an output of 5.1 kW at about 3.5 bar. The nominal power of the micro-gas turbine was 25 kW. The data presented lead to the conclusion that Julich stack operated successfully in a pressurized hybrid system with a micro-gas turbine. The processing steps of wet powder spraying are presented in Figure 3.8.

For the anode production, the two powders of NiO and YSZ are ball milled to ensure a good homogeneity of the powder. In the suspension, the powder is mixed with ethanol or isopropanol and a binder. The most important property of the suspension is viscosity and drying rate, and it is important to control that the ethanol does not completely evaporate during the spraying process. In some suspension, the water is added to assist with the issue of evaporation. The suspension is ball milled to create a homogeneous mixture. Before spraying, one more step of steering to homogenize suspension is applied. After that, the spraying gun is used to spray the solution. The spray gun is fixed on a computer-controlled motion system. The most important parameters to control for high-quality and homogenous film deposition are spray velocity, spray distance, and the angle between the nozzle and the substrate. Julich team used the wet powder spraying for the cathode layer deposition with

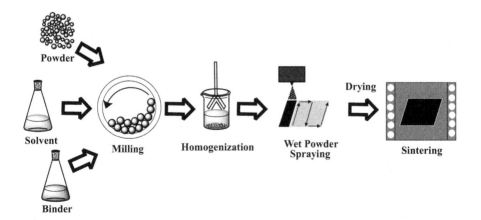

FIGURE 3.8 Processing steps of the wet powder spraying technique. (Adapted and modified from Ref. [40].)

layer thicknesses of more than 50 μm for the anode-supported fuel cell concept [82]. The flow rate and spraying patterns are the two main parameters that influence film thickness. The suspension spraying technique is a cost-effective technology and has a high potential for up-scaling from laboratory to industrial size.

3.2.2.2.3 SP and Plasma Spraying

The driving force in SOFC material development is lowering the temperature of SOFC operation, a high performance of the cell at low temperature below 600°C, low-cost, a rapid processing method that can be done in one continuous process without the need for long sintering times at elevated temperatures. The choice of an appropriate thin-film deposition technique is strongly influenced by the material to be deposited and the desired film quality and capabilities and availabilities of the techniques at the manufacture's side. There is a broad spectrum of spray deposition techniques, and authors use different terminologies, but the commonality in all methods involves the generation of fine aerosol droplets of a liquid precursor solution, which is then deposited onto a heated substrate surface with or without the aid of an external electric field [83–86]. The main advantages of SP deposition are the simplicity of the apparatus, the low cost of the process, and the good reproducibility of the high-quality, dense electrolyte. Regarding the anode, the benefit of the SP is that for cermet anodes, which are composed of Ni-YSZ or Ni-SDC and prepared by SP, the size and distribution spherical particles can be precisely controlled [68,69,87,88]. The anode prepared with the composite particles showed higher SOFC cell performance than that with the mixed ones. In the SP process, the precursor materials are dissolved in a solvent (usually aqueous or alcoholic) to form a precursor solution. The precursor solution is then atomized into discrete droplets, and there are many tools to atomize droplets such as ultrasonication, nebulizer, just to mention two. Then, the solution is sprayed onto a heated substrate, and a solid oxide film is formed on the substrate surface. Unlike CVD, the SP deposition does not require high vapor pressure precursors or vacuum at any stage, which is a great advantage if the technique is to be scaled up for industrial applications. The deposition rate is still low comparing to plasma spray and flame-based technologies; however, the thickness of the films can be easily controlled over a wide range by changing the spray parameters, thus eliminating the major drawbacks of other chemical methods such as spin coating, which produces films with limited thickness [85]. Either dense or porous film can be deposited when using different spray parameters and different heating temperatures of the substrates. By changing the composition of the spray solution during the spray process, the process can be optimized to make gradient films with a gradient porosity as well as composition [89]. Lately, reactive spray deposition technology (RSDT) has been introduced, which is a high deposition rate, high-volume, and low-cost flame SP process [90–93].

In the RSDT process developed by Maric et al. [91,92], the layers are deposited directly onto an anode or metal support. The deposition temperature is in the range of 500°C–1200°C, which addresses sintering issues by removing the necessity of co-firing the cathode and electrolyte. The substrate is heated from the back to prevent thermal shock in the ceramic layer and to assist sintering. The RSDT has the potential to penetrate the SOFC market because of the enormous cost savings advantage, coupled with the ability to coat large surface areas, both metal and ceramic, with multi-element compounds. The cost saving compared to plasma spray and CVD processes are the result of two factors: the RSDT process is an atmospheric deposition process that uses air as oxidant and uses low vapor pressure precursors that are generally 100 times less expensive than those used by currently available technologies like CVD. The capital equipment costs for a functional RSDT system is also ten times less expensive.

The results confirm that the RSDT technique is able to produce different microstructures in one processing step just by adequately tuning the instrument. The controlling parameters that will influence the structure density and homogeneity are the solution concentration, deposition rate, deposition temperature, spraying pattern, and the substrate porosity and temperature. For the high-density film, the lower concentration, lower deposition rate, and higher substrate temperature are required. The technique has been used for deposition of both ion-conducting and proton-conducting electrolyte Yttrium-doped Barium Zirconate (BZY20) with additional dopants on the metal-supported substrates [94]. As mentioned earlier for the large scale production, the thermal and plasma spraying are common techniques used to produce films of SOFC electrode/electrolyte materials. In the plasma spraying, the solid powder is sprayed from the plasma torches into current or alternating current plasma, followed by melting of the solid particles, and formation of splats of material on the substrate. The length of time in the plasma depends on the type of torch, gas flows, and plasma shaping devices (i.e., cooling shrouds). The plasma spray process, especially low-pressure plasma spraying, is a promising process for cell manufacturing because of its fast deposition rate. Atmospheric plasma spraying (APS) as a more cost-effective process has been investigated for the production of dense electrolyte layers [95–97]. There is still difficulty in making a full density of electrolytes, and additional means have to be developed to make the electrolyte layer as dense as possible either by process improvement or by post-processing. For the metal-supported SOFC, another variation of the plasma spray process has been used, a high-velocity oxy-fuel spraying of liquid suspension feedstock [98]. Metal-supported SOFCs composed of a $Ce_{0.8}Sm_{0.2}O_{2-\delta}$ (SDC) electrolyte layer and $Ni-Ce_{0.8}Sm_{0.2}O_{2-\delta}$ (Ni-SDC) cermet anode and $Sm_{0.5}Sr_{0.5}CoO_3$ (SSCo)-SDC cathode were fabricated by suspension thermal spraying on a metal substrate. The gun operating conditions were optimized, and in-flight particle temperature and velocity were measured for a number of different gun operating conditions. All these parameters, including a standoff distance, have an influence on microstructures.

3.2.2.2.4 Other Thin-Film Techniques

Different spattering techniques can be used to deposit thin films with different structures and porosity using different sputtering conditions, such as the target to substrate distance, substrate temperature, deposition pressure, and the deposition power [99].

Laser deposition and variations of laser deposition, such as pulsed laser deposition (PLD), have become an important technique for depositing thin-film and pinhole-free electrolytes [100,101]. The main advantage of the PLD process is tied control over the film properties, such as microstructures, chemical composition, density, and interfacial properties. Grain size from a few nanometers to a few hundred nanometers can be easily prepared by selecting appropriate processing parameters, and it can operate at low processing temperatures, which can suppress grain growth to achieve nanocrystalline materials. The PLD process belongs to the category of physical vapor deposition (PVD). The PVD process is a generic term used for the number of sputtering techniques. Radio frequency sputtering has most widely been used for YSZ low-defect-density films that can be grown on the cold substrate as the process is kinetically driven [102]. The ability to significantly decrease the thickness of the YSZ thickness from 10 µm to 400 nm, which can reduce the operating temperature from 800°C to 450°C, has significant importance in reducing the cost of operation and durability of the cell [103].

3.3 SINTERING/CO-FIRING

Most of the processing techniques, in particular tape casting, extrusion, and screen printing, produce films and substrates that require additional heat treatment (sintering) to obtain the mechanical strength as well as desired electronic conductivity, electrochemical activity, and gas tied properties for the electrolyte. Sintering is widely used as a cost-effective method to tailor the properties for both ceramic- and metal-supported SOFCs. In the manufacturing of the electrolyte-supported SOFC like a Bloom Energy design, the typical manufacturing is a tape casting of the green tape, followed by sintering at a temperature of 1300°C–1400°C, and then, screen printing of the anode and cathode layer with co-firing at temperatures about 1200°C. The process has to be optimized to fabricate the electrolyte with about 100% density, while electrodes porosity should be kept higher than 30%. For the anode-supported cells, a co-firing process and sintering of a multilayer structure is more difficult to control and required that we are able to minimize the distortion of the structure due to mismatch in thermo-mechanical and physical properties of the individual layers [104,105]. Figure 3.9 shows the morphology of the anode-supported electrolyte, as a function of the sintering temperature. The sintering temperature of 1350°C–1400°C yielded a fully dense electrolyte with a porous anode structure [106]. In order to bring SOFCs to the lower temperature (400°C–500°C), the proton-conducting oxide ceramics on the metal support are widely explored as alternatives to conventional oxide conductors, primarily because the proton conductors display higher conductivity at intermediate temperatures. Lower operating temperatures lead to reducing thermal stress and allow the use of less expensive stack materials and balance-of-plant components. The metal support will dictate the sintering conditions and determine the cost of the process. Ferritic stainless steel is a typical choice for the metal support as it displays good oxidation resistance below about 800°C and has a thermal expansion coefficient that is similar to common SOFC ceramic materials. The limited studies on co-sintering BZCY with the stainless steel support have been reported, however, that significant challenges exist for this approach. The sintering temperature needs to occur in the controlled environments to avoid oxidation of the metal support and has to be below 800°C. It was reported in the literature that the inter-diffusion between the anode and steel layers was a significant issue and resulted in contamination of the Ni catalyst and melting of the stainless steel [33]. Hence, the RSDT was used to

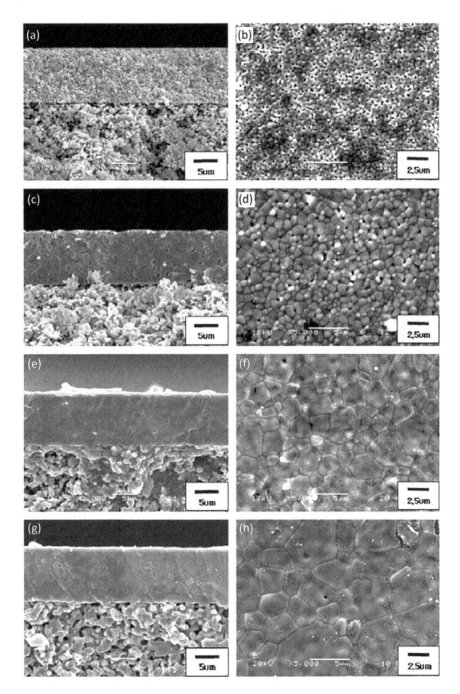

FIGURE 3.9 Scanning electron microscopy of anode-supported microstructure electrolytes depending on the sintering temperatures: fracture surfaces of (a) 1200°C, (c) 1300°C, (e) 1350°C, and (g) 1400°C and electrolyte surfaces of (b) 1200°C, (d) 1300°C, (f) 1350°C, and (h) 1400°C. (Reprinted with permission from Ref. [106].)

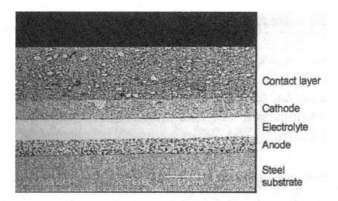

FIGURE 3.10 Metal-supported SOFC, Lit.: Ceres Power. (Reprinted with permission from Ref. [32].)

successfully apply multiple ceramic cell layers, including dense BYZ electrolytes on the pre-sintered ferritic stainless steel support, but cell performance was very low due to the high porosity of the metal-supported mash (larger than 500 µm) and difficulty to deposit the crack-free electrolyte [107].

The most progress in the manufacturing and sintering of SOFCs has been made by Ceres Power. They used the wet processing for CGO electrolyte deposition on the stainless steel support. The dense CGO was obtained after sintering at 1000°C in controlled environments [35,37]. The Ceres stack operates at 500°C–600°C. The Ceres powder cell is shown in Figure 3.10. Ceres Power is developing a platform technology for applications in the 1–25 kW range. The metal-supported SOFC is a promising technology for the fast start-up and has been explored for remote power, auxiliary power units, uninterruptible power supplies, and heat and/or electrical load led micro-CHP applications. The interest in metal-supported SOFCs remains strong. In 2018, Ceres Power entered into a new partnership with Nissan to further explore the concept of metal-supported SOFCs with different fuels. Doosan will develop a 5–30 kW power system with Ceres stack technology. Co-firing of the cell components is a promising path to reduce the number of the processing steps, thereby saving cost for cell processing.

REFERENCES

1. First commercially available SOFC co-generation system in the 3–5 kW class for institutional applications, based on research by Kyocera (as of June 1, 2017). https://global.kyocera.com/news-archive/2017/0702_bnfo.html.
2. Highest efficiency of any 3–5 kW class SOFC system currently on the market. Based on research by Kyocera (as of June 1, 2017), allowing approx. −2.0% of margins for the absolute value due to the progress of use. https://global.kyocera.com/news-archive/2017/0702_bnfo.html.
3. Synapse Energy Economics, Inc. 2019. Bloom energy fuel cells: a cost-effectiveness brief. https://www.synapse-energy.com/sites/default/files/Bloom-Energy-Fuel-Cell-Brief-18-105.pdf.

4. Yokokawa, H. 2013. Report of five-year NEDO project on durability/reliability of SOFC stacks. *ECS Transactions* 57(1):299–308.
5. Stolten, D. 2012. *Fuel Cell Science and Engineering: Materials, Processes, Systems and Technology*. John Wiley & Sons: Hoboken, NJ.
6. Singhal, S.C., and Kendall, K. 2003. *High-Temperature Solid Oxide Fuel Cells: Fundamentals, Design and Applications*, Elsevier: London, UK.
7. Williams, M.C., Strakey, J.P., and Singhal, S.C. 2004. US distributed generation fuel cell program. *Journal of Power Sources* 131(1–2):79–85.
8. Dias, F.J., Kampel, M., Koch, F.J., and Nickel, H. 1994. Ceramic in energy application. In *Proceedings of the 2nd International Conference*, London, The Institute of Energy, Pergamon.
9. Sammes, N.M., Du, Y., and Bove, R. 2005. Design and fabrication of a 100 W anode supported micro-tubular SOFC stack. *Journal of Power Sources* 145(2):428–434.
10. Lee, S.B., Lim, T.H., Song, R.H., Shin, D.R., and Dong, S.K. 2008. Development of a 700 W anode-supported micro-tubular SOFC stack for APU applications. *International Journal of Hydrogen Energy* 33(9):2330–2336.
11. Sammes, N.M., and Du, Y. 2007. Fabrication and characterization of tubular solid oxide fuel cells. *International Journal of Applied Ceramic Technology* 4(2):89–102.
12. Mahata, T., Nair, S.R., Lenka, R.K., and Sinha, P.K. 2012. Fabrication of Ni-YSZ anode supported tubular SOFC through iso-pressing and co-firing route. *International Journal of Hydrogen Energy* 37(4):3874–3882.
13. Zhang, L., He, H.Q., Kwek, W.R., Ma, J., Tang, E.H., and Jiang, S.P. 2009. Fabrication and characterization of anode-supported tubular solid-oxide fuel cells by slip casting and dip coating techniques. *Journal of the American Ceramic Society* 92(2):302–310.
14. Liu, R.Z., Wang, S.R., Huang, B., Zhao, C.H., Li, J.L., Wang, Z.R., Wen, Z.Y., and Wen, T.L. 2009. Dip-coating and co-sintering technologies for fabricating tubular solid oxide fuel cells. *Journal of Solid State Electrochemistry* 13(12):1905–1911.
15. Du, Y., Sammes, N.M., and Tompsett, G.A. 2000. Optimisation parameters for the extrusion of thin YSZ tubes for SOFC electrolytes. *Journal of the European Ceramic Society* 20(7):959–965.
16. Du, Y., and Sammes, N.M. 2001. Fabrication of tubular electrolytes for solid oxide fuel cells using strontium-and magnesium-doped LaGaO3 materials. *Journal of the European Ceramic Society* 21(6):727–735.
17. Son, H.J., Song, R.H., Lim, T.H., Lee, S.B., Kim, S.H., and Shin, D.R. 2010. Effect of fabrication parameters on coating properties of tubular solid oxide fuel cell electrolyte prepared by vacuum slurry coating. *Journal of Power Sources* 195(7):1779–1785.
18. Hassmann, K. 2001. SOFC power plants, the Siemens-Westinghouse approach. *Fuel Cells* 1(1):78–84.
19. Timurkutluk, B., Timurkutluk, C., Mat, M.D., and Kaplan, Y. 2016. A review on cell/stack designs for high performance solid oxide fuel cells. *Renewable and Sustainable Energy Reviews* 56:1101–1121.
20. Park, B.K., Lee, J.W., Lee, S.B., Lim, T.H., Park, S.J., Song, R.H., and Shin, D.R. 2012. A flat-tubular solid oxide fuel cell with a dense interconnect film coated on the porous anode support. *Journal of Power Sources* 213:218–222.
21. NEDO. 2019. Practical application of solid oxide fuel cells. NEDO report 2019, in Japanese, https://www.nedo.go.jp/content/100899256.pdf, https://www.nedo.go.jp/events/report/ZZHY_00005.html.
22. NEDO's presentation at 4th IPHE Workshop on March 1, 2011. https://www.iphe.net/iphe-workshop-on-stationary-applica
23. NGK Spark Plug Co., Ltd. Our R&D activities. https://www.ngkntk.co.jp/english/rd/activities.html.

24. Maric, R., Furusaki, K., Nishijima, D., and Neagu, R. 2011. Thin film low temperature solid oxide fuel cell (LTSOFC) by reactive spray deposition technology (RSDT). *ECS Transactions* 35(1):473–481.

25. Barbieri, E.S., Spina, P.R., and Venturini, M. 2012. Analysis of innovative micro-CHP systems to meet household energy demands. *Applied Energy* 97:723–733.

26. Kopanos, G.M., Georgiadis, M.C., and Pistikopoulos, E.N. 2013. Energy production planning of a network of micro combined heat and power generators. *Applied Energy* 102:1522–1534.

27. Maric, R., Roller, J.M., Neagu, R., Fatih, K., and Tuck, A. 2008. Low Pt loading thin cathode catalyst layer by reactive spray deposition technology. *ECS Transactions* 12(1):59–63.

28. Berghaus, J.O., Legoux, J.G., Moreau, C., Hui, R., Decès-Petit, C., Qu, W., Yick, S., Wang, Z., Maric, R., and Ghosh, D. 2008. Suspension HVOF spraying of reduced temperature solid oxide fuel cell electrolytes. *Journal of Thermal Spray Technology* 17(5–6):700–707.

29. Wang, Z., Berghaus, J.O., Yick, S., Decès-Petit, C., Qu, W., Hui, R., Maric, R., and Ghosh, D. 2008. Dynamic evaluation of low-temperature metal-supported solid oxide fuel cell oriented to auxiliary power units. *Journal of Power Sources* 176(1):90–95.

30. Singhal, S.C., and Dokiya, M. 2003. Solid oxide fuel cells VIII:(SOFC VIII). In *Proceedings of the International Symposium*, The Electrochemical Society, Pennington, NJ, USA.

31. Brandon, N.P., Corcoran, D., Cumming, D., Duckett, A., El-Koury, K., Haigh, D., Leah, R., Lewis, G., Maynard, N., McColm, T., Trezona, R., Selcuk, A., and Schmidt, M. 2004. Development of metal supported solid oxide fuel cells for operation at 500–600°C. *Journal of Materials Engineering and Performance* 13:253–256.

32. Attryde, P., Baker, A., Baron, S., Blake, A., Brandon, N.P., Corcoran, D., Cumming, D., Duckett, A., El-Koury, K., Haigh, D., and Harrington, M. 2005. Stacks and systems based around metal supported SOFCs operating at 500–600°C. *ECS Proceedings Volumes* 2005:113–122.

33. Tucker, M.C. 2010. Progress in metal-supported solid oxide fuel cells: a review. *Journal of Power Sources* 195(15):4570–4582.

34. Green Car Congress. 2016. Ceres power scales up "steel cell" SOFC fuel cell production capability with innovate UK funding. https://www.greencarcongress.com/2016/03/20160306-cerespower.html.

35. Bance, P., Brandon, N.P., Girvan, B., Holbeche, P., O'dea, S., and Steele, B.C.H. 2004. Spinning-out a fuel cell company from a UK University—2 years of progress at Ceres Power. *Journal of Power Sources* 131(1–2):86–90.

36. Leah, R.T., Bone, A., Selcuk, A., Corcoran, D., Lankin, M., Dehaney-Steven, Z., Selby, M., and Whalen, P. 2011. Development of highly robust, volume-manufacturable metal-supported SOFCs for operation below 600°. *ECS Transactions* 35(1):351–367.

37. Leah, R.T., Bone, A., Lankin, M., Selcuk, A., Rahman, M., Clare, A., Rees, L., Phillip, S., Mukerjee, S., and Selby, M. 2015. Ceres power steel cell technology: rapid progress towards a truly commercially viable SOFC. *ECS Transactions* 68(1):95–107.

38. Mukerjee, S., Leah, R., Selby, M., Stevenson, G., and Brandon, N.P. 2017. Life and reliability of solid oxide fuel cell-based products: a review. In *Solid Oxide Fuel Cell Lifetime and Reliability*, Edited by Brandon, N.P., Ruiz-Trejo, E., and Boldrin, P. Academic Press, Elsevier: London, UK, 173–191.

39. Leah, R., Selby, M., Bone, A., McNicol, A., Rees, L., and Mukerjee, S. 2014. In *Proceedings of the 11th European SOFC and SOEC forum A0911*, Chapter 5:62., International Solid Oxide Fuel Cell and Electrolyser Conference with Exhibition and Tutorial, Lucern, Switzerland.

40. Vielstich, W., Lamm, A., and Gasteiger, H.A. 2003. *Handbook of Fuel Cells: Fundamentals Technology and Applications*, Volume 4: Fuel Cell Technology and Applications. John Wiley & Sons: Hoboken, NJ.

41. Mogensen, M., Sammes, N.M., and Tompsett, G.A. 2000. Physical, chemical and electrochemical properties of pure and doped ceria. *Solid State Ionics* 129(1–4):63–94.

42. Huang, K., and Singhal, S.C. 2013. Cathode-supported tubular solid oxide fuel cell technology: a critical review. *Journal of Power Sources* 237:84–97.

43. George, R.A. 2000. Status of tubular SOFC field unit demonstrations. *Journal of Power Sources* 86(1–2):134–139.

44. Orera, V.M., Laguna-Bercero, M.A., and Larrea, A. 2014. Fabrication methods and performance in fuel cell and steam electrolysis operation modes of small tubular solid oxide fuel cells: a review. *Frontiers in Energy Research* 2:22.

45. Huang, H., Nakamura, M., Su, P., Fasching, R., Saito, Y., and Prinz, F.B. 2007. High-performance ultrathin solid oxide fuel cells for low-temperature operation. *Journal of the Electrochemical Society* 154(1):B20–B24.

46. Love, J., Amarasinghe, S., Selvey, D., Zheng, X., and Christiansen, L. 2009. Development of SOFC stacks at ceramic fuel cells limited. *ECS Transactions* 25(2):115–124.

47. Wang, W.G., Guan, W., Li, H., Wang, Z., Wang, J.X., Wu, Y., Zhou, S., and Zuo, G. 2009. Solid oxide fuel cell development at NIMTE. *ECS Transactions* 25(2):85–90.

48. Christiansen, N., Hansen, J.B., Larsen, H.H., Linderoth, S., Larsen, P.H., Hendriksen, P.V., and Hagen, A. 2007. Solid oxide fuel Cell development at topsoe fuel cell A/S and Risø National Laboratory. *ECS Transactions* 7(1):31–38.

49. Borglum, B.P., Tang, E., and Pastula, M. 2009. The status of SOFC development at versa power systems. *ECS Transactions* 25(2):65–70.

50. Mistler, R.E., and Twiname, E.R. 2000. *Tape Casting: Theory and Practice*. American Ceramic Society: Westerville, OH.

51. Mertens, J., Haanappel, V.A., Wedershoven, C., and Buchkremer, H.P. 2006. Sintering behavior of (La, Sr) MnO_3 type cathodes for planar anode-supported SOFCs. *Journal of Fuel Cell Science and Technology* 3(4):415–421.

52. Schafbauer, W., Menzler, N.H., and Buchkremer, H.P. 2014. Tape casting of anode supports for solid oxide fuel cells at Forschungszentrum Jülich. *International Journal of Applied Ceramic Technology* 11(1):125–135.

53. Cable, T.L., and Sofie, S.W. 2007. A symmetrical, planar SOFC design for NASA's high specific power density requirements. *Journal of Power Sources* 174(1):221–227.

54. Linderoth, S. 2009. Solid oxide cell R&D at Risø National Laboratory—and its transfer to technology. *Journal of Electroceramics* 22(1–3):61–66.

55. Jabbari, M., Bulatova, R., Tok, A.I.Y., Bahl, C.R.H., Mitsoulis, E., and Hattel, J.H. 2016. Ceramic tape casting: a review of current methods and trends with emphasis on rheological behaviour and flow analysis. *Materials Science and Engineering: B* 212:39–61.

56. Wang, J., Yan, D., Pu, J., Chi, B., and Jian, L. 2011. Fabrication and performance evaluation of planar solid oxide fuel cell with large active reaction area. *International Journal of Hydrogen Energy* 36(12):7234–7239.

57. Chang, Y.C., Lee, M.C., Kao, W.X., Wang, C.H., Lin, T.N., Chang, J.C., and Yang, R.J. 2011. Characterization of anode-supported solid oxide fuel cells with composite LSM-YSZ and LSM-GDC cathodes. *Journal of The Electrochemical Society* 158(3):B259–B265.

58. Jiang, S.P., Callus, P.J., and Badwal, S.P.S. 2000. Fabrication and performance of Ni/3 mol% Y_2O_3–ZrO_2 cermet anodes for solid oxide fuel cells. *Solid State Ionics* 132(1–2):1–14.

59. Guillodo, M., Vernoux, P., and Fouletier, J. 2000. Electrochemical properties of Ni–YSZ cermet in solid oxide fuel cells: effect of current collecting. *Solid State Ionics* 127(1–2):99–107.

60. Rotureau, D., Viricelle, J.P., Pijolat, C., Caillol, N., and Pijolat, M. 2005. Development of a planar SOFC device using screen-printing technology. *Journal of the European Ceramic Society* 25(12):2633–2636.

61. Ried, P., Lorenz, C., Brönstrup, A., Graule, T., Menzler, N.H., Sitte, W., and Holtappels, P. 2008. Processing of YSZ screen printing pastes and the characterization of the electrolyte layers for anode supported SOFC. *Journal of the European Ceramic Society* 28(9):1801–1808.

62. Phair, J.W., Lundberg, M., and Kaiser, A. 2009. Leveling and thixotropic characteristics of concentrated zirconia inks for screen-printing. *Rheologica Acta* 48(2):121–133.

63. Zhang, X., Robertson, M., Decès-Petit, C., Xie, Y., Hui, R., Yick, S., Staite, M., Styles, E., Roller, J., Marie, R., and Ghosh, D. 2005. Ni-YSZ cermet supported thin ceria-based electrolyte solid oxide fuel cell for reduced temperature (500–600°C) operation. *ECS Proceedings Volumes* 2005:1102–1109.

64. Zhao, L., Huang, X., Zhu, R., Lu, Z., Sun, W., Zhang, Y., Ge, X., Liu, Z., and Su, W. 2008. Optimization on technical parameters for fabrication of SDC film by screen-printing used as electrolyte in IT-SOFC. *Journal of Physics and Chemistry of Solids* 69(8):2019–2024.

65. Zhang, X., Deces-Petit, C., Yick, S., Robertson, M., Kesler, O., Maric, R., and Ghosh, D. 2006. A study on sintering aids for $Sm_{0.2}Ce_{0.8}O_{1.9}$ electrolyte. *Journal of Power Sources* 162(1):480–485.

66. Zhang, X., Robertson, M., Deces-Petit, C., Xie, Y., Hui, R., Qu, W., Kesler, O., Maric, R., and Ghosh, D. 2008. Solid oxide fuel cells with bi-layered electrolyte structure. *Journal of Power Sources* 175(2):800–805.

67. Sanson, A., Roncari, E., Boldrini, S., Mangifesta, P., and Doubova, L. 2010. Eco-friendly screen-printing inks of gadolinia doped ceria. *Journal of Fuel Cell Science and Technology* 7(5):051013.

68. Zhang, X., Ohara, S., Maric, R., Mukai, K., Fukui, T., Yoshida, H., Nishimura, M., Inagaki, T., and Miura, K. 1999. Ni-SDC cermet anode for medium-temperature solid oxide fuel cell with lanthanum gallate electrolyte. *Journal of Power Sources* 83(1–2):170–177.

69. Maric, R., Ohara, S., Fukui, T., Yoshida, H., Nishimura, M., Inagaki, T., and Miura, K. 1999. Solid oxide fuel cells with doped lanthanum gallate electrolyte and $LaSrCoO_3$ cathode, and Ni-samaria-doped ceria cermet anode. *Journal of the Electrochemical Society* 146(6):2006–2010.

70. Inagaki, T., Miura, K., Yoshida, H., Maric, R., Ohara, S., Zhang, X., Mukai, K., and Fukui, T. 2000. High-performance electrodes for reduced temperature solid oxide fuel cells with doped lanthanum gallate electrolyte: II. $La(Sr)CoO_3$ cathode. *Journal of Power Sources* 86(1–2):347–351.

71. Zhang, X., Robertson, M., Yick, S., Deĉes-Petit, C., Styles, E., Qu, W., Xie, Y., Hui, R., Roller, J., Kesler, O., and Maric, R. 2006. $Sm_{0.5}Sr_{0.5}CoO_3 + Sm_{0.2}Ce_{0.8}O_{1.9}$ composite cathode for cermet supported thin $Sm_{0.2}Ce_{0.8}O_{1.9}$ electrolyte SOFC operating below 600°C. *Journal of Power Sources* 160(2):1211–1216.

72. Lin, J., Meijerink, J., Brinkman, H.W., de Vries, K.J., and Burggraa, A.J. 1991. Microporous and dense ceramic membranes prepared by CVD and EVD. In *Proceedings of the 2nd International Conference on Inorganic Membranes-ICIM2–91*, Trans Tech Publications, 465–469, Montpellier, France.

73. Aizawa, M., Kobayashi, C., Yamane, H., and Hirai, T. 1993. Preparation of ZrO_2-$Y2O_3$ films by CVD using β-diketone metal chelates. *Journal of the Ceramic Society of Japan* 101(1171):291–294.

74. Pal, U.B., and Singhal, S.C. 1990. Electrochemical vapor deposition of yttria-stabilized zirconia films. *Journal of the Electrochemical Society* 137(9):2937.

75. Lin, Y.S. 1992. Chemical and electrochemical vapor deposition of zirconia–yttria solid solutions in porous ceramic media, Ph.D. Thesis, University of Twente, Netherlands.
76. Schüller, E., Vaßen, R., and Stöver, D. 2002. Thin electrolyte layers for SOFC via wet powder spraying (WPS). *Advanced Engineering Materials* 4(9):659–662.
77. Liu, Q.L., Chan, S.H., Fu, C.J., and Pasciak, G. 2009. Fabrication and characterization of large-size electrolyte/anode bilayer structures for low-temperature solid oxide fuel cell stack based on gadolinia-doped ceria electrolyte. *Electrochemistry Communications* 11(4):871–874.
78. de Haart, L.G.J., Hauber, T., Mayer, K., and Stimming, U. 1996. Electrochemical performance of an anode supported planar SOFC system. In *Proceedings of the 2nd European Solid Oxide Fuel Cells Forum*, European Fuel Cell Forum, 1:229–235, Oslo, Norway.
79. Wain-Martin, A., Morán-Ruiz, A., Laguna-Bercero, M.A., Campana, R., Larrañaga, A., Slater, P.R., and Arriortua, M.I. 2019. SOFC cathodic layers using wet powder spraying technique with self synthesized nanopowders. *International Journal of Hydrogen Energy* 44(14):7555–7563.
80. Bai, Y., Wang, C., Jin, C., and Liu, J. 2011. Anode current collecting efficiency of tubular anode-supported solid oxide fuel cells. *Fuel Cells* 11(3):465–468.
81. Lim, T.H., Song, R.H., Shin, D.R., Yang, J.I., Jung, H., Vinke, I.N.C., and Yang, S.S. 2008. Operating characteristics of a 5 kW class anode-supported planar SOFC stack for a fuel cell/gas turbine hybrid system. *International Journal of Hydrogen Energy* 33(3):1076–1083.
82. Ruder, A., Buchkremer, H.P., Jansen, H., Mallener, W., and Stöver, D. 1992. Wet powder spraying—a process for the production of coatings. *Surface and Coatings Technology* 53(1):71–74.
83. Stoermer, A.O., Rupp, J.L., and Gauckler, L.J. 2006. Spray pyrolysis of electrolyte interlayers for vacuum plasma-sprayed SOFC. *Solid State Ionics* 177(19–25):2075–2079.
84. Rupp, J.L., Infortuna, A., and Gauckler, L.J. 2006. Microstrain and self-limited grain growth in nanocrystalline ceria ceramics. *Acta Materialia* 54(7):1721–1730.
85. Perednis, D., and Gauckler, L.J. 2004. Solid oxide fuel cells with electrolytes prepared via spray pyrolysis. *Solid State Ionics* 166(3–4):229–239.
86. Perednis, D., Xie, Y., Zhang, X., and Ghosh, D. 2005. Deposition of samaria-doped ceria electrolyte using spray pyrolysis. *ECS Proceedings Volumes* 2005:1087–1092.
87. Maric, R., Ohara, S., Mukai, K., Fukui, T., Yoshida, H., Inagaki, T., and Miura, K. 1999. Performance of a Ni-SDC/La(Sr)Ga(Mg)O$_3$/La(Sr)CoO$_3$ single fuel cell. *ECS Proceedings Volumes* 1999:938–944.
88. Ohara, S., Maric, R., Zhang, X., Mukai, K., Fukui, T., Yoshida, H., Inagaki, T., and Miura, K. 2000. High performance electrodes for reduced temperature solid oxide fuel cells with doped lanthanum gallate electrolyte: I. Ni–SDC cermet anode. *Journal of Power Sources* 86(1–2):455–458.
89. Xie, Y., Neagu, R., Hsu, C.S., Zhang, X., Decès-Petit, C., Qu, W., Hui, R., Yick, S., Robertson, M., Maric, R., and Ghosh, D. 2010. Thin film solid oxide fuel cells deposited by spray pyrolysis. *Journal of Fuel Cell Science and Technology* 7(2):021007.
90. Wegner, K., and Pratsinis, S.E. 2005. Gas-phase synthesis of nanoparticles: scale-up and design of flame reactors. *Powder Technology* 150(2):117–122.
91. Maric, R., Roller, J., and Neagu, R. 2011. Flame-based technologies and reactive spray deposition technology for low-temperature solid oxide fuel cells: technical and economic aspects. *Journal of Thermal Spray Technology* 20(4):696–718.
92. Maric, R., Neagu, R., Zhang-Steenwinkel, Y., Van Berkel, F.P., and Rietveld, B. 2010. Reactive spray deposition technology–an one-step deposition technique for solid oxide fuel cell barrier layers. *Journal of Power Sources* 195(24):8198–8201.

93. Nédélec, R., Neagu, R., Uhlenbruck, S., Maric, R., Sebold, D., Buchkremer, H.P., and Stöver, D. 2011. Gas phase deposition of diffusion barriers for metal substrates in solid oxide fuel cells. *Surface and Coatings Technology* 205(16):3999–4004.
94. Ouimet, R., and Maric, R. 2018. Proton conducting SOFC. In *15th International Symposium on Solid Oxide Fuel Cells: Materials, Science and Technology*, Daytona Beach, FL.
95. Hui, R., Wang, Z., Kesler, O., Rose, L., Jankovic, J., Yick, S., Maric, R., and Ghosh, D. 2007. Thermal plasma spraying for SOFCs: applications, potential advantages, and challenges. *Journal of Power Sources* 170(2):308–323.
96. Mauer, G., Vaßen, R., and Stöver, D. 2009. Atmospheric plasma spraying of yttria-stabilized zirconia coatings with specific porosity. *Surface and Coatings Technology* 204(1–2):172–179.
97. Hui, R., Berghaus, J.O., Decès-Petit, C., Qu, W., Yick, S., Legoux, J.G., and Moreau, C. 2009. High performance metal-supported solid oxide fuel cells fabricated by thermal spray. *Journal of Power Sources* 191(2):371–376.
98. Berghaus, J.O., Legoux, J.G., Moreau, C., Hui, R., Decès-Petit, C., Qu, W., Yick, S., Wang, Z., Maric, R., and Ghosh, D. 2008. Suspension HVOF spraying of reduced temperature solid oxide fuel cell electrolytes. *Journal of Thermal Spray Technology* 17(5–6):700–707.
99. Mattox, D.M. 2010. *Handbook of Physical Vapor Deposition (PVD) Processing*. William Andrew: Norwich, NY.
100. Gruner, H.R., and Tannenberger, H. 1994. SOFC elements by vacuum-plasma-spraying (VPS). In *Proceedings of 1st European Solid Oxide Fuel Cell Conference*, 2:611–616, Lucerne, Switzerland.
101. Yang, D., Zhang, X., Nikumb, S., Decès-Petit, C., Hui, R., Maric, R., and Ghosh, D. 2007. Low temperature solid oxide fuel cells with pulsed laser deposited bi-layer electrolyte. *Journal of Power Sources* 164(1):182–188.
102. La O, G.J., Hertz, J., Tuller, H., and Shao-Horn, Y. 2004. Microstructural features of RF-sputtered SOFC anode and electrolyte materials. *Journal of Electroceramics* 13(1–3):691–695.
103. Hill, T., and Huang, H. 2011. Fabricating pinhole-free YSZ sub-microthin films by magnetron sputtering for micro-SOFCs. *International Journal of Electrochemistry* 2011:1–8.
104. Savo, G., Rainer, A., D'Epifanio, A., Licoccia, S., and Traversa, E. 2005. Co-Sintering of dense YSZ electrolyte films on porous NiO-YSZ supporting anodes for IT-SOFCs. *ECS Proceedings Volumes* 2005:1031–1036.
105. Ye, G., Ju, F., Lin, C., Gopalan, S., Pal, U., and Seccombe, D.A. 2005. Low-cost single-step co-firing technique for SOFC manufacturing. *ECS Proceedings Volumes* 2005:451–459.
106. Moon, H., Kim, S.D., Hyun, S.H., and Kim, H.S. 2008. Development of IT-SOFC unit cells with anode-supported thin electrolytes via tape casting and co-firing. *International Journal of Hydrogen Energy* 33(6):1758–1768.
107. Myles, T.D., Ouimet, R., Kwak, D., and Maric, R. 2016. Characterization and performance of proton conducting solid oxide fuel cells manufactured using reactive spray deposition technology. *ECS Transactions* 72(25):17–23.

4 Cell and Stack Configuration

4.1 GENERAL REQUIREMENTS FOR SOLID OXIDE FUEL CELL (SOFC) DESIGNS

The SOFC, either in a single-cell configuration or in a stack configuration, as shown in Figure 4.1, must deliver the desired electrochemical performance as well as sufficient mechanical stability to achieve long-term performance goals. Therefore, the following key requirements must be fulfilled for choosing a specific SOFC design:

- **Minimum polarization losses**: The cell/stack should be designed in a way that minimizes the polarization losses, gas leakage, short-circuiting, and cross-leakage gases. The polarization losses usually cause a major voltage drop in SOFC.
- **Minimum ohmic losses**: The components of SOFC should be kept as short as possible to minimize the ohmic losses in the stack.
- **Uniform temperature and gas distribution**: During the SOFC operation, the SOFC design should provide a uniform temperature distribution along with proper cooling as well as uniform fuel gas and oxidant distributions over the entire cell/stack. The uniform gas distribution efficiently reduces the mass transport limitation.

(a)   (b)

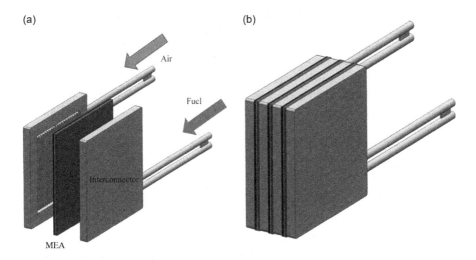

FIGURE 4.1 (a) SOFC single cell and (b) SOFC stack with three cells. (Reprinted with permission from Ref. [2].)

161

- **Good electrochemical contacts**: The contact surface area of the SOFC components should be high. Also, the desired current flow must be regulated by the current collectors.
- **Good mechanical/structural stability**: The mechanical strength of the SOFC cell/stack should be sufficient to sustain the thermal stresses or gas pressure during the operation. Basically, the SOFC stack should possess high resistance to thermal shock conditions resulted from cold start-up, sudden power change, installation and vibration loading conditions, and off-design temperature gradients [1].

4.2   SINGLE-CELL CONFIGURATION

The cell designs can be flexible owing to the solid nature of the SOFC components. The individual cells can be stacked together to obtain a particular stack configuration. SOFC cells can be categorized according to not only the operating temperature (i.e., high, intermediate, and low temperatures) but also the type of the cell support (i.e., self-supported and external supported: anode-, cathode-, electrolyte-, interconnect- or porous substrate-supported). For the self-supported configuration, the structural support of the cell is provided by the thickest layer of the cell components. The self-support systems can be anode-supported, cathode-supported, and electrolyte-supported cell configurations. For the external supported configuration, a porous substrate or interconnect supports the thin layers of the cell [1,2]. Figure 4.2 shows the different single-cell configurations based on the type of cell support along with their advantages and disadvantages.

For the electrode-supported cells, they can provide relatively high performance at lower operating temperatures due to reduced electrolyte thickness, although they exhibit mass transport limitations arising because of thick electrode layers. The most common and preferred design in this respect is the anode-supported cell design as the anodic polarization is quite less than the cathodic polarization. For the electrolyte-supported cells, they have relatively strong structures and are less susceptible to mechanical failures, although the higher operating temperature is required (900°C–1000°C) to reduce the ohmic losses. YSZ is the most commonly used electrolyte for this configuration. The thickness of the electrolyte for the electrolyte-supported configuration is typically >100 μm. However, the cell operating temperature can be reduced to <800°C when the thickness of the electrolyte is 5–20 μm, providing more choice of materials. The use of new material to support the anode, cathode, and electrolyte increases the complexity in the design of the porous supported cell. In addition, interconnect/metallic supports can be used. This type has shown the structure as strong as the electrolyte-supported configuration, even though the design of flow fields is problematic and restricted by the cell support requirements [1,2].

4.3   DESIGN OF SOFCS AND STACKS

So far, there have been many designs proposed for SOFCs and stacks, as shown in Figure 4.3. However, there are two major SOFC design configurations: tubular and planar, as presented in Figure 4.4. The comparison between these two design

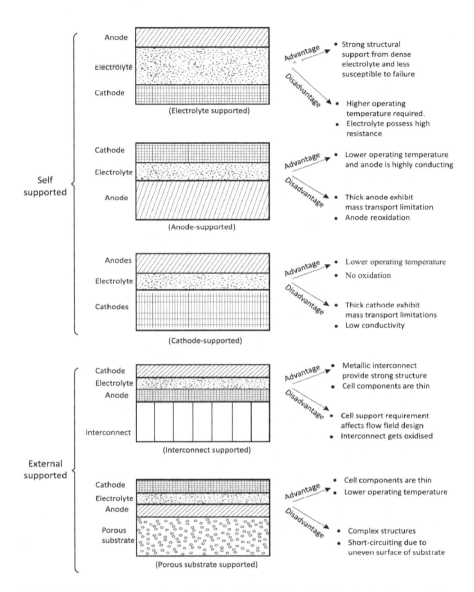

FIGURE 4.2 Different configurations for SOFC single cell. (Reprinted with permission from Ref. [1].)

configurations is summarized in Table 4.1. The tubular design is a well-known design for SOFC, although it is difficult to produce, and it exhibits low power density compared to that of the planar type. However, it has some advantages such as ease of accomplishment, both gas sealing and interconnection of single cells during stack manufacturing. On the other hand, it is not only easier to fabricate the planar SOFCs, but they also can provide much higher power density in comparison with the tubular SOFCs, which they cannot reach due their long current collection path as a result of structural disadvantages. Both the planar and tubular designs are not suitable for

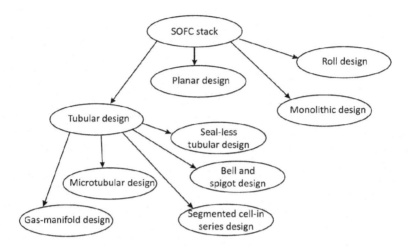

FIGURE 4.3 Different designs proposed for SOFC stacks. (Reprinted with permission from Ref. [1].)

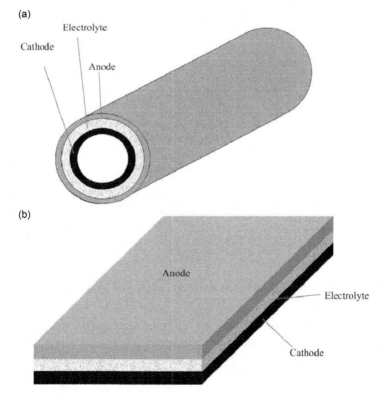

FIGURE 4.4 Schematic of (a) tubular and (b) planar design of SOFC. (Reprinted with permission from Ref. [2].)

TABLE 4.1

Comparison between the Tubular and Planar SOFCs

Property	Tubular	Planar
Power density	Low	High
Volumetric power density	Low	High
High-temperature sealing	Easy	Difficult
Start-up and shutdown	Fast	Slow
Interconnector fabrication	Difficult	High cost
Production cost	High	Low
Thermal cycling stability	High	Low

Source: Data adapted and modified from Ref. [2].

mobile applications where both high voltage/performance in a limited space and fast start-up are needed. When subjected to high heating/cooling rates required for mobile applications, the former suffers from structural instability in spite of the relatively high power density. However, the latter provides thermal stability due to the symmetric circular nature. However, its low power density hampers its application in portable devices [1,2].

Each SOFC design may have pros and cons, but the flat-plate planar, seal-less tubular, and microtubular designs have attracted much attention and have been more studied and developed recently. However, there can be found several different designs reported in the literature as an alternative to these major SOFC cell/stack designs. In general, there are four criteria for designing the SOFC as follows:

- **Current path**: The current path basically depends on the SOFC stack design. For instance, for the seal-less tubular SOFC, the current travels across the electrolyte/interconnect thickness and along the surrounding of the anode and cathode. Also, for the monolithic SOFC designs, which are basically electrolyte-supported, the current follows a multicell monolithic structure. Figure 4.5 demonstrates the current path in monolithic SOFC design;
- **Cell-to-cell connections**: In the stack configuration, the SOFC cells can be connected in either parallel or series to provide a cohesive voltage output of the stack. However, in the series arrangements, the voltage output is higher than that of the parallel connections. Hence, the series arrangement is the most common connection to build voltage;
- **Gas flow configurations**: The fuel and oxidant in the SOFC stack can be arranged to be co-flow, counter-flow, or cross-flow, as shown in Figure 4.6. Different flow patterns such as serpentine-, spiral-, radial-, or Z-flow patterns, as shown in Figure 4.7, can be implied, especially in the planar design with regards to the specific flow configuration. The flow channels are used to promote a uniform gas distribution as well as a uniform mass and heat transport of each cell of the stack. Therefore, the arrangement and shape of

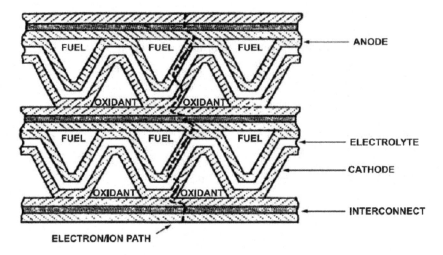

FIGURE 4.5   Current path in monolithic SOFC design. (Reprinted with permission from Ref. [3].)

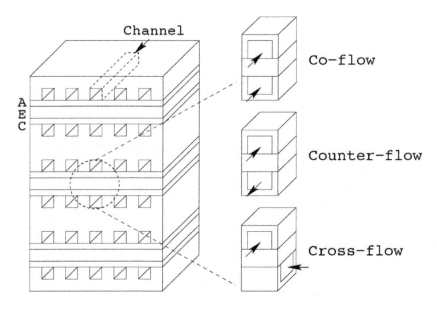

FIGURE 4.6   The principal flow configurations in a planar SOFC stack: co-flow, counter-flow, and cross-flow (A: anode, E: electrolyte, C: cathode). (Reprinted with permission from Ref. [4].)

the flow channels vary to optimize the design of the stack. In some SOFC designs, the flow channels are designed into the electrode, while most of the SOFC designs integrate the flow channels as a part of the interconnect. The contact area between the interconnect and the electrodes, connected by the flow fields, plays a key role in decreasing the contact resistance losses;

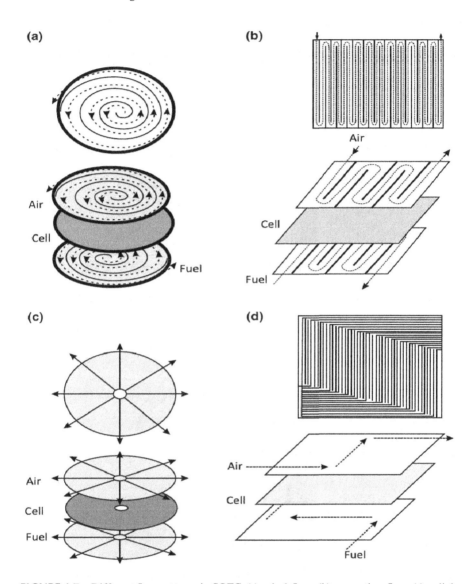

FIGURE 4.7   Different flow patterns in SOFC: (a) spiral-flow; (b) serpentine-flow; (c) radial-flow; (d) Z-flow. (Reprinted with permission from Ref. [1].)

- **Gas manifolds**: The gas manifolds which provide routes for gas supply and product/unreacted gases removal can be either external or integral, as shown in Figure 4.8. The external gas manifolds are designed and constructed separately from the cell and interconnect component of the stack, while the integral gas manifolds are designed as a part of interconnect or cell. The sealing is required for the manifolds to avoid gas leakage and drop in cell performance. The manifolds should also be able to distribute the gas flow uniformly and provide low pressure drop [1,3,5].

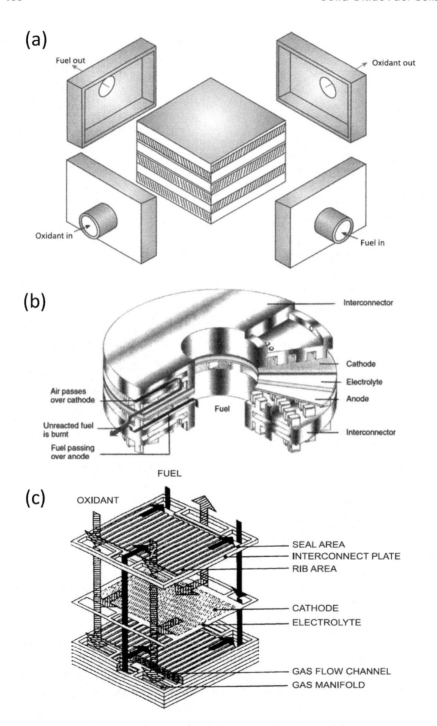

FIGURE 4.8 (a) External and (b, c) integral gas manifolding in SOFC. (Reprinted with permission from Refs. [1,3].)

4.4 PLANAR CELL DESIGN (FLAT PLATE DESIGN)

In planar design, as illustrated in Figure 4.9 in its most generic version, the cell components are configured as thin and flat plates. The electrolyte and the electrodes are sandwiched between the interconnects placed at the top and bottom. The air channels are built between the cathode and the interconnect, while the fuel channels exist between the anode and the interconnect. The interconnects ribbed on both sides also serve as a bipolar gas separator contacting the cathode and the anode of the adjoining cells in the stack. The planar cells are fabricated by low-cost and simple conventional ceramic processing techniques such as slurry sintering, screen printing, tape casting, or plasma spraying. Moreover, the planar design offers high volumetric power densities, simple series electrical connections, low ohmic polarization, and ease of heat removal [1,6–8].

However, the planar geometries have some drawbacks, including difficulty in obtaining stable mechanical structure, failure of cell stack even by moderate stress, high start-up time, and gas-tight high-temperature sealant requirement to avoid gas leaks in a bipolar configuration.

Currently, anode-supported, cathode-supported, and electrolyte-supported planar SOFCs are being developed. In the electrolyte-supported cells, the electrolyte (typically YSZ) thickness is 50–150 μm which makes the ohmic resistance high. Hence, such cells are used in an operating temperature of ~800°C. In the electrode-supported designs, a lower electrolyte thickness can be used (5–20 μm), which decreases their ohmic resistance. Thus, such cells can be used at lower operating temperatures (~600°C). The Ni-YSZ cermet is typically selected as anode and supporting electrode

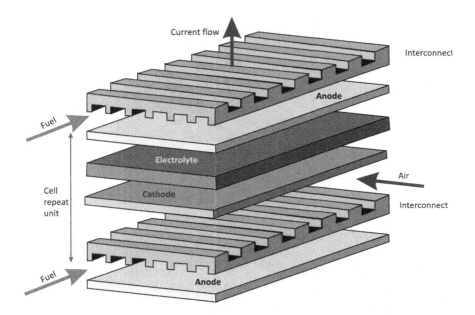

FIGURE 4.9 Planar design for SOFC with ribbed interconnects. (Reprinted with permission from Ref. [1].)

due to its superior mechanical strength, high thermal and electronic conductivities, and minimal chemical interaction with the electrolyte. For such anode-supported SOFCs, Kim et al. [9] reported the power density as high as 1,800 mW cm^{-2} at 800°C. Cathode used in planar SOFCs typically consists of either Sr-doped lanthanum manganite (LSM), Sr-doped lanthanum ferrite, or LSM + YSZ applied to the electrolyte.

4.5   TUBULAR CELL DESIGN (SEAL-LESS)

The first tubular configuration, initially made by Siemens Westinghouse, USA, consists of two tubes, including an outer tube and an inner tube. The outer surface of the outer tube is the anode side of the cell, and the inner surface is the cathode side of the cell tube. The solid oxide electrolyte is placed between the anode and the cathode sides. The inner tube, known as air injection/guidance tube and composed of alumina, guides the injected preheated air toward the bottom of the cell tube. The preheated air then flows over the cathode surface of the cell tube through a gap between the cell tube and the injection tube. One end of the cell tube is closed, which eliminates seal requirement. The fuel gas flows over the anode surface through a gap among the cell tubes. The oxygen ions migrate through the electrolyte from the cathode side to the anode side and react with fuel, creating an electrical current. Each tube in the tubular stack is made of this configuration [10]. The schematic of the tubular SOFC configuration is illustrated in Figure 4.10.

In the tubular cells, the cell components are deposited in the form of thin layers on the LSM cathode tube [11]. The cathode tube is fabricated by the extrusion method, followed by sintering such that a porosity of 30%–35% can be obtained. The YSZ

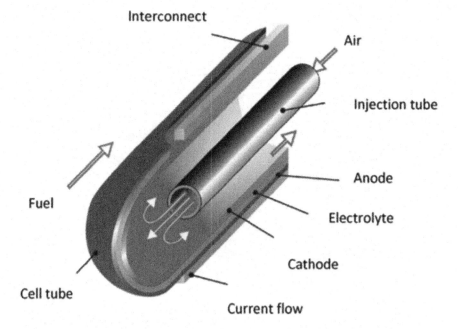

FIGURE 4.10   Schematic of tubular SOFC. (Reprinted with permission from Ref. [10].)

electrolyte is deposited by electrochemical vapor deposition (EVD) technique in the form of a 40 μm thick, dense layer [12]. The Ni-YSZ anode is deposited either by the sintering of Ni-YSZ slurry or by nickel slurry application, followed by EVD of YSZ. Along the length of the cell, the doped lanthanum chromite interconnect strip is deposited by plasma spraying technique [13]. An actual tubular cell currently used may consist of a 2.2 cm diameter air electrode tube with 2.2 mm wall thickness and a length of 180 cm. The length of the tube can increase to improve the cell power and reduce the number of cells required to produce a specific power output. Also, the diameter of the cell can increase to accommodate larger pressure drops [1]. So far, a large number of tubular cells have been tested for over 25,000 h. The cells have performed satisfactorily under different operating conditions with less than 0.1% performance degradation per 1,000 h. Such tubular cells have shown a power density of 250–300 mW cm^{-2} at 1,000°C [6]. These low power densities make the tubular SOFCs attractive only for stationary power generation, not for transportation applications. In contrast, the planar SOFCs can achieve high power densities up to about 2,000 mW cm^{-2} [6,9]. Moreover, the ohmic losses in the tubular SOFCs are much higher in comparison with the tubular SOFCs due to their longer electrical current paths [10]. Thus, when tubular SOFCs are stacked together, the long current paths limit the cell performance, as shown in Figure 4.11.

The major advantages of the tubular SOFCs over other types of SOFCs are:

- A variety of hydrocarbon-based gases or their synthesis derivatives, e.g., natural gas, coal, JP-8, and biomass can be used as fuel sources;

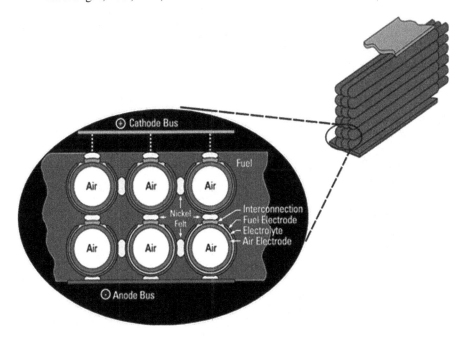

FIGURE 4.11 Demonstration of long current paths because of cell-to-cell connections in the tubular SOFC. (Reprinted with permission from Ref. [14].)

- CO present in the synthesis gas can be oxidized to CO_2 and generate electrical energy within the tubular SOFCs;
- Higher energy efficiencies;
- Suitable for mid- to large-scale applications;
- No sealing requirement to isolate fuel from oxidant due to the closed one end of the cell tube [1,10].

4.6 MICROTUBULAR CELL DESIGN

The tubular SOFCs which have a tube diameter smaller than 3–5 mm are called microtubular SOFCs. Hence, microtubular SOFC can be considered as a geometrically modified tubular SOFC by decreasing the tube diameter to the millimeter scale. Microtubular cells were first invented by Kevin Kendall [15] from extruded 8YSZ tubes with 1–5 mm tube diameter and 100–200 µm wall thickness in the early 1990s, as shown in Figure 4.12a. After more than 20 years of the development period, Meadowcroft et al. [16] have recently tested their 250 W microtubular SOFC supported on the extruded NiO-8YSZ cermet anode with 10 µm YSZ electrolyte and

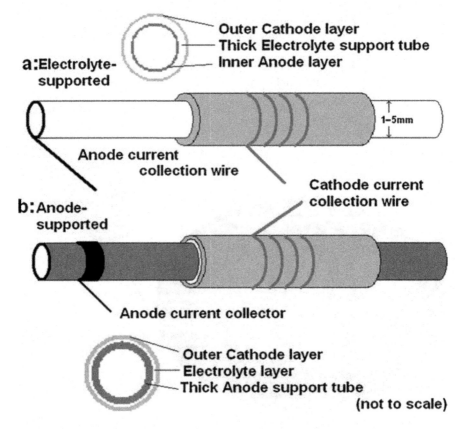

FIGURE 4.12 Typical microtubular SOFC designs. (a) electrolyte-supported design; (b) anode-supported design. (Reprinted with permission from Ref. [17].)

6 μm samaria-doped ceria interlayer coated with a 50 μm $La_{1-x}Sr_xCo_{1-y}Fe_yO_{3-\delta}$ cathode. The results showed 55% fuel utilization under propane fuel with a reformer based on partial oxidation fuel processing.

Despite a simple size modification, microtubular SOFCs have two dramatic benefits in comparison with the classical tubular SOFCs: high power density and thermal shock resistance. Microtubular SOFCs provide relatively high power density since the power density scales with the reciprocal of the tube diameter. Hence, a 1 mm diameter microtubular SOFC provides ten times more power output of stack per stack volume, depending on the tube diameter and a real power density, p, of the stack. If l is the tube length, then the power output of a single tube is πDlp, and the stack volume is πD^2L, giving a power density of $\pi p/4D$. Microtubular SOFCs also possess high thermal shock resistance since the large diameter of tubular SOFC is susceptible to damage due to cracking, but in the microtubular SOFCs no such cracking has been observed. This high thermal shock resistance allows microtubes to be heated in a minute to the operating temperature of ~800°C without any damage. These advantages enable a rapid start-up along with a high power density, which makes microtubular SOFCs promising for mobile applications. However, collecting electrical current generated in the microtubular SOFC is not easy due to its small microtube diameter.

Similar to other cell designs, the anode-supported microtubular structure is the most common design adopted to reduce the electrolyte thickness and improve the cell power output. A schematic of such an anode-supported cell is shown in Figure 4.12b. On the other hand, the conventional extrusion process results in a thick-wall tube, limiting fuel diffusion. Hence, many studies on microtubular SOFCs have focused on the development of microtubular hollow fiber SOFCs fabricated by a combined spinning/phase inversion followed by sintering technique, which provides a thin-wall tube and reduces the polarization resistance. Unlike the extruded tubes, hollow fibers have an asymmetrical dense sponge-like structure with finger-like pores. The pore size and content are determined by the process parameters. Importantly, the pore size can decrease gradually from the outside surfaces to the fiber's center, providing improved gas diffusion. It is also possible to produce hollow fibers with dimensions and lengths not achievable by conventional extrusion technique [18]. The reduction in anode polarization loss may also be achievable for hollow fibers in comparison with the conventional extrusion process. Yang et al. [19] successfully made NiO-3YSZ hollow fibers with a diameter of ~1 mm by phase inversion and sintering as the anode support for microtubular SOFCs. They found that the electronic conductivity and the mechanical performance of the fabricated samples increase with sintering temperature while the porosity decreases. In the other research, Droushiotis et al. [20] showed that the bending strength decreases with the increasing NiO content in NiO-8YSZ hollow fibers while the electronic conductivity tends to increase. They also fabricated dual-layer hollow fibers, including the NiO-gadolinia-doped ceria anode support and gadolinia-doped ceria electrolyte coated with the $La_{1-x}Sr_xCo_{1-y}Fe_yO_{3-\delta}$ -gadolinia-doped ceria cathode to construct a microtubular SOFC in a single step. Although the cell performance was initially low (80 mW cm^{-2} peak power at 550°C) due to the poor current collection caused by the silver spring inside the tube, the highest electrochemical performance of 1,110 mW cm^{-2} peak power was achieved at

600°C from the cell constructed on this fiber after brush painting of multi-layered cathode because of using the thinnest electrolyte layer [21–23]. Chen et al. [24] have also used the phase-inversion process combined with a dip-coating process to fabricate anode-supported microtubular SOFCs. The cells with ~240 μm thick NiO-YSZ support and 10 μm thick YSZ electrolyte demonstrated peak power densities of 752 and 277 mW cm^{-2} at 800°C and 600°C, respectively.

4.7 INTEGRATED PLANAR (SEGMENTED-CELL-IN-SERIES) DESIGN

Integrated planar SOFC, also known as segmented-in-series SOFC is a configuration between planar and tubular pioneered by Rolls-Royce. Figure 4.13 presents a schematic view of an integrated planar SOFC. This design has several advantages over regular planar and tubular designs. In this design, the smaller number of interconnectors is used to lower the weight and cost of the SOFC over other designs. In addition, shorter ionic and electronic current paths of the integrated planar design improve the cell performance, making SOFCs suitable for mobile applications [2,25]. However, its relatively weak sealing and integrity of interconnectors are the main limitations of this design, which can be minimized by a careful cell design and suitable material selection along with a microstructural optimization. The multicell array fabricated on a tubular or planar porous substrate is the primary unit of the stack. Both circular and flattened tubes can also be implemented for this design [2].

Integrated planar SOFCs have extensively been studied in the literature after the first demonstration by Gardner et al. [27] in 2000, who reported the Rolls-Royce efforts on the SOFC development. They fabricated an integrated planar SOFC stack with supported electrolyte membranes with 10–20 μm thickness, having 30 W peak power at 950°C. The fuel (a mixture of H_2-H_2O-N_2:66-17-17 wt%) utilization of the stack was almost 60%. Later, Kim et al. [28] proposed a new fabrication technique for the integrated planar SOFC. They fabricated all cell components except the porous support (produced via uniaxial pressing of 3 mol% Y_2O_3-ZrO_2 with a pore former resin) by the novel direct-writing method called robo-dispensing. The two-cell integrated planar SOFC stack, having 30 μm thick NiO-YSZ anode, 20 μm thick LSM cathode, and 10 μm thick YSZ electrolyte, provided only 35 mW power at 800°C under humidified (3 wt% H_2O) hydrogen fuel. The group later showed that 30% performance improvement is achievable if 20 vol% Al_2O_3 is added to the

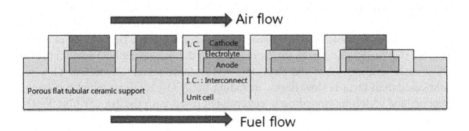

FIGURE 4.13 A schematic view of an integrated planar (segmented-in-series) SOFC. (Reprinted with permission from Ref. [26].)

porous substrate [29]. Lai and Barnett [30] investigated four-cell partially stabilized zirconia-supported integrated planar SOFC stack fabricated by low-cost screen printing technique, except the porous substrate produced via pressing. The results showed that increasing the thickness of the LSM current collecting layer from 11 to 91 μm improves the cell performance by a factor of 2–3 due to area-specific resistance reduction. The highest cell peak power density was 900 mW cm^{-2} based on the active cell area. An et al. [31,32] have recently fabricated a segmented-in-series SOFC by the decalcomania method providing excellent interfacial bonding due to no penetration of cell components into the porous support. They also found the open circuit voltage and maximum power density can increase with decreasing the cell length due to a shorter current path. The highest open circuit voltage and power density were achieved when the cell to cell distance was 2 mm.

Since long service life is also essential for any energy-generating devices like SOFCs, Almutairi et al. [33] studied thermal cycling and long-term durability of the integrated planar SOFC tubes. They reported 1.5% voltage drop per 1,000 h of steady operation at 900°C under 0.17 A cm^{-2} constant load due to an increase in ohmic resistance and interconnect damage. In addition, they observed a large crack on the tube surface after five thermal cycles. Bujalski et al. [34] compared the cyclic performance of an integrated planar SOFC with a tubular and a planar SOFCs under various thermal and current load conditions. They found that the start-up and shut-down procedures of the cell were smaller than the critical size during temperature ramping or falling; thereby, no damage occurred. The small tube was shown to be mechanically stable up to 4,000°C min^{-1}. Therefore, they concluded that the degradation under cycling conditions is size-dependent and suggested that the cell size should be in a millimeter-scale in order to achieve thermal shock resistance.

A reliable integrated planar SOFC with thousands of hours of service life should also possess high redox stability besides long-term durability and thermal cycling. To improve the redox tolerance, nickel is generally added to the porous substrate so that re-oxidation of nickel particles in the anode layer is limited by the oxidized nickel on the porous substrate [35–37]. In order to avoid short-circuiting, the porous substrate should be electronically insulating. Thus, the amount of nickel additive is important. Wang et al. [35] studied the effect of nickel content on the insulating properties of the YSZ porous substrate. For this purpose, NiO-YSZ composites with various nickel contents (0–30 vol%) were made. The results showed that the electronic resistivity of the composite substrate was ≤67 ohm cm at 900°C when the nickel content was <20 vol%. Since the composite substrate with the nickel content of ≤10 vol% also exhibited no obvious cracks during thermal cycling, they concluded that NiO-YSZ with the nickel content of ≤10 vol% is a promising candidate for the support substrate in the segmented-in-series SOFCs. Fujita et al. [36] investigated the redox tolerance of a segmented-in-series SOFC using the Ni-doped MgO porous substrate. The redox was carried out during start and stop operation between 20°C and 775°C. Air or humidified air was used as the oxidation gas during heating and cooling between 20°C and 700°C, while the reduction humidified methane was used to achieve reduction between 700°C and 775°C. The results showed that only 15% of performance loss per cycle occurs after 20 redox cycles due to the presence of nickel.

4.8 CONE-SHAPED DESIGN

Another stack design based on the segmented-in-series concept is the cone-shaped design. In this design, several conically shaped tubular cells are connected in the electrical and gas flow series. Figure 4.14 shows a schematic of a three-cell stack of such designs. This design is different from the integrated planar SOFCs because it has no multicell array constructed on a single support. Here, the stacking is achieved by fitting one cone-shaped cell into the other to form a tubular self-supporting structure. Thus, it can be viewed as a modified tubular SOFC design. In the cone-shaped SOFC design, interconnects serve as both sealant and electrical contact between the anode of one cell and the cathode of the next cell (current collector). Since a high power can be achieved with a small stack size in such designs, it is possible to utilize lightweight and compact cone-shaped stacks for portable applications. However, the main problems that still remain are interconnect integrity and fabrication process due to geometric complexity, similar to the integrated planar SOFC design [2].

The cone-shaped geometry was first introduced by Sui and Liu [38]. They made a three-cell stack with the single unit length of 2 cm and the slip cast cone-shaped 8YSZ electrolyte support. NiO-gadolinia-doped ceria was the anode on the inside of the cone-shaped electrolyte support, while $La_{1-x}Sr_xCo_{1-y}Fe_yO_{3-\delta}$-gadolinia-doped ceria was used as the cathode on the outside of the electrolyte. The manufactured stack provided the peak power density of 33 mW cm^{-2} at 800°C operating temperature. They later improved the stack performance by using samaria-doped ceriaas the electrolyte. This improved design, having 0.85 cm^2 single unit active area, produced the peak power density of 300 mW cm^{-2} at 700°C operating temperature [39]. However, poor mechanical properties of the samaria-doped ceriaelectrolyte as well as the total ohmic loss raised by the thick electrolyte layer, hamper its use in a SOFC stack. Zhang et al. [40] significantly improved the cell performance by fabricating a cone-shaped NiO-YSZ anode-supported structure. They used YSZ and LSM-YSZ as the electrolyte and cathode, respectively. Ag paste was also used as an interconnect and sealant. The single fabricated unit exhibited 800 mW cm^{-2} maximum power density at 800°C operating temperature. However, the cell performance was limited by the interfacial resistance with a low ohmic loss due to the use of a thin electrolyte layer.

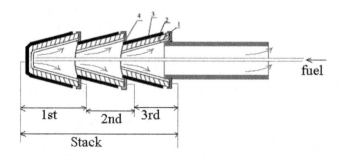

FIGURE 4.14   A schematic view of a cone-shaped three-cell stack. 1. Anode; 2. Electrolyte; 3. Cathode; 4. Interconnect. (Reprinted with permission from Ref. [38].)

It has been shown that the fabrication technique plays a key role in improving the performance of cone-shaped SOFCs. For example, Bai et al. [41], who offered the effective and feasible dip-coating method instead of the traditional slip coating technique, produced a single cone-shaped SOFC composed of the NiO-YSZ anode support, YSZ electrolyte (35.9 µm thick), and LSM-YSZ cathode. The cell provided a maximum power density of 1,080 and 1,350 mW cm^{-2} at 800°C and 850°C operating temperatures, respectively. A two-cell stack with a total active area of 11.6 cm^{-2} fabricated based on the abovementioned design also showed the maximum total power of 3,700 mW at 800°C. In the other study, Ding and Liu [42] produced a single cone-shaped NiO-YSZ anode-supported cell with 7 µm thick YSZ electrolyte using dip-coating technique. The cell provided a peak power density of 1,780 mW cm^{-2} at 800°C, while the two-cell stack with an active area of 10.65 cm^{-2} exhibited maximum power of 2,750 mW at the same operating temperature. The stack also tolerated 12 thermal cycles between the room temperature and 800°C operating temperature. Besides slip casting and dip-coating methods, there are also other fabrication techniques conducted to produce cone-shaped supports such as injection molding [43], phase inversion [44], and gel-casting [45].

4.9   FLAT-TUBE DESIGN

Another common SOFC geometrical configuration is the flat-tube SOFC design, which was developed by Siemens Westinghouse, USA, based on their formerly tubular SOFC design. In 2001, Hassmann [46] first indicated the transition from tubular to flat-tube SOFCs. Due to the low power density of the conventional tubular design, the flat-tube design was introduced to reduce the internal ohmic loss and improve the cell power density. In this design, tubular and planar configurations are combined to provide high cell power density as well as better thermal stability and sealing. However, the flat-tube design has exhibited a relatively high thermal resistance [2,26]. Figure 4.15 illustrates a schematic view of a flat-tube SOFC. In the flat-tube design, the cells are generally constructed on the extruded anode support, having multiple gas channels for fuel passage. The oxidant flows through a conductive porous media located between the unit cells in the stack. In order to electrically connect the unit cells in the stack and separate fuel and oxidant gases, a thin and dense film of interconnect is coated on one side of the unit cells.

Since the interconnect is the most important cell component in the flat-tube SOFCs, the majority of studies have focused on the development of interconnect materials for the flat-tube design. Kim et al. [47] fabricated NiO-8YSZ anode-supported flat-tube SOFCs with $La_{0.75}Ca_{0.27}CrO_3$ ceramic interconnect as well as $(La_{0.85}Sr_{0.15})_{0.9}MnO_3$ dip-coated Fecralloy (Fe-20Cr-5Al) and SUS430 (Fe-16Cr) metallic interconnects. For the ceramic interconnect, the results showed that the dense interconnection layer was achieved when the plasma spray method was employed. The electronic resistance of the coated SUS430 metallic interconnects was measured 148 mohm cm^2 at 750°C. However, it decreased to 43 mohm cm^2 after 450 h of operation due to the phase change in the protective layer. Park et al. [48] developed fully dense La-doped $SrTiO_3$ interconnect materials for the anode-supported flat-tube SOFCs. The results showed that $Sr_{0.8}La_{0.2}TiO_3$ exhibited higher fracture strength and thermal expansion,

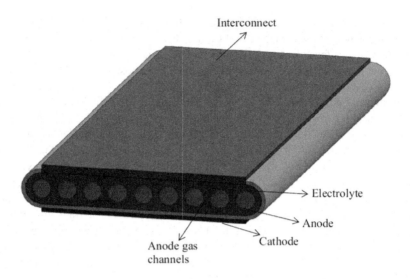

FIGURE 4.15 A schematic view of a flat-tube SOFC design. (Reprinted with permission from Ref. [2].)

while $Sr_{0.7}La_{0.3}TiO_3$ showed higher electronic conductivity within the SOFC operating temperature range. They also developed multi-layer ceramic interconnects on the porous NiO-8YSZ anode support. The unit flat-tube cell was constructed on the NiO-8YSZ anode support made by the extrusion process. The proposed interconnect film showed high electronic conductivity and stability in the dual atmosphere (air and humidified hydrogen). The unit cell also provided maximum power density of 360 mW cm^{-2} at 750°C operating temperature [49]. Pi et al. [50] also developed glass-reinforced Ag composite interconnects coated on the NiO-8YSZ anode support for intermediate-temperature flat-tube SOFC. It was found that all of the composite interconnects possess a dense structure with very low gas leakage. The interconnect, having 10 wt% glass, exhibited the highest electronic conductivity and phase stability. The flat-tube cell, using Ag + 10 wt% glass interconnect, also showed the best performance with the peak power density of 275 mW cm^{-2} at 700°C operating temperature. Hosseini et al. [51] developed a 20 μm thick $La_{0.4}Ca_{0.6}Ti_{0.4}Mn_{0.6}O_{3-\delta}$ composite interconnect coated on the NiO-YSZ flat-tube anode support via the screen printing method which provides 208 mW cm^{-2} cell peak power density at 800°C under 3% humidified hydrogen fuel.

There are also other studies dealing with the different aspects of the flat-tube SOFCs. For example, Kim et al. [52] investigated the electrochemical performance and long-term stability of various materials as a cathode current collector for the flat-tube SOFCs. Among those materials, it was found that wound Ag wire with $La_{0.6}Sr_{0.4}CoO_3$ paste can be a useful cathode current collecting structure for anode-supported flat-tube SOFCs. Lim et al. [53] and Lee et al. [54] fabricated a 1 kW class anode-supported flat-tube SOFC stack with active cell area of 90 cm^2 and three ribs inside the anode-supported flat-tube. The cell utilized NiO-YSZ as the anode support, 8YSZ as the electrolyte, and LSM/YSZ-LSM-$La_{1-x}Sr_xCo_{1-y}Fe_yO_{3-\delta}$ as the three-layered cathodes. The stack was constructed with 30 unit bundles connected in series where each had two flat-tube cells connected in parallel. The stack presented the peak power of 921 W at 750°C operating temperature under 3% humidified hydrogen as a fuel and air as an oxidant.

4.10   HONEYCOMB DESIGN

Honeycomb is another design offering the advantages of using both planar and tubular SOFC designs like the integrated planar, the cone-shaped, and flat-tube designs. The honeycomb design provides high volumetric power density as well as high mechanical strength, which leads to high durability. This design consists of honeycomb channels in an alternating sequence used as the routes for the passing of fuel and oxidant. Figure 4.16 presents a schematic view of a honeycomb SOFC design. The early configuration of this design was electrolyte-supported. However, this configuration showed a high ohmic loss, thus, a low electrochemical performance due to the use of a thick electrolyte layer. In order to solve this problem, the electrode-supported configuration has been suggested where one electrode provides the structural support of the honeycomb, while the other electrode is inside of each channel. Therefore, each channel can act as an independent SOFC. Figure 4.17 illustrates schematic views of both electrolyte-supported and electrode-supported honeycomb SOFCs. The main issues with this design are still interconnect and electrical lead [2,26].

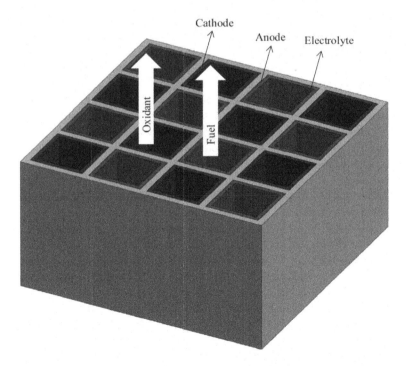

FIGURE 4.16   A schematic view of a honeycomb SOFC design. (Reprinted with permission from Ref. [2].)

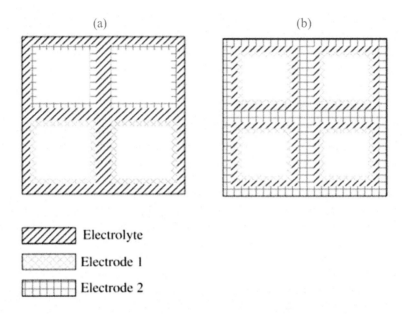

Electrolyte

Electrode 1

Electrode 2

FIGURE 4.17 Schematic view of (a) electrolyte-supported and (b) electrode-supported honeycomb SOFC. (Reprinted with permission from Ref. [26].)

The honeycomb SOFC design was first invented by ABB in 1992 and reported by Wetzko et al. [55], who presented the design as "condensed-tubes" like structure. They fabricated the 8YSZ electrolyte-supported honeycomb SOFC stack via the extrusion process. The stack, composed of five honeycomb unit cells with 4×4 quadratic structure and with a cell pitch of 6 mm and a thickness of 700 µm for the internal cell walls, provided the peak power density of 100 mW cm^{-2} at 800°C operating temperature under hydrogen fuel. However, the open porosities within the electrolyte layer resulted in low open circuit voltage. Later on, Xia et al. [56] and Zha et al. [57] developed the 8YSZ electrolyte-supported honeycomb SOFC with a graded cathode, composed of four different layers of LSM-gadolinia doped ceria, LSM-$La_{1-x}Sr_xCo_{1-y}Fe_yO_{3-\delta}$ -gadolinia doped ceria, $La_{1-x}Sr_xCo_{1-y}Fe_yO_{3-\delta}$ -gadolinia doped ceria, and $La_{1-x}Sr_xCo_{1-y}Fe_yO_{3-\delta}$. Their honeycomb SOFCs showed better electrochemical performance.

Due to the low performance of 8YSZ electrolyte-supported honeycomb SOFCs caused by employing thick electrolyte layers, researchers have mainly focused on enhancing the performance of honeycomb SOFCs by employing alternative electrolyte materials with higher ionic conductivity or fabricating the electrode-supported systems. Yamaguchi et al. [58–61] and Shimizu et al. [62] developed LSM cathode-supported honeycomb SOFCs fabricated by extrusion with quadratic and circular structures. The cell consisted of scandia-stabilized zirconia as the electrolyte and NiO-gadolinia-doped ceria as the anode made via dip-coating and slurry injection methods. Their cell exhibited a high volumetric power density of above 1,000 and 2,000 mW cm^{-3} at 600°C and 650°C, respectively, under wet hydrogen fuel [58]. The cathode-supported honeycomb SOFC also showed a durable performance for the rapid start-up operation even with a 100°C min^{-1} heating rate [59]. They also fabricated a two-cell cathode-supported honeycomb SOFC stack, exhibiting a volumetric power density of 600 mW cm^{-3} at 600°C operating temperature [60]. The cell performance was later improved to 2,000 mW cm^{-3} volumetric power density by utilizing a slurry injection method for the anode and electrolyte coatings [62]. Fukushima et al. [63] also built NiO-8YSZ anode-supported honeycomb SOFCs, showing the peak power density of 220 mW cm^{-3} at 850°C operating temperature with 22% fuel utilization under hydrogen fuel. On the other hand, Zhong et al. [64,65] improved the performance of honeycomb SOFCs by replacing the 8YSZ electrolyte support with the LSGM ($La_{0.9}Sr_{0.1}Ga_{0.8}Mg_{0.2}O_3$) electrolyte support. They used NiFe-LSGM as the anode and $Sm_{0.5}Sr_{0.5}CoO_3$ as the cathode. The cell with one channel exhibited the maximum power density of 500 mW cm^{-2} at 700°C operating temperature, whereas the cell with five channels provided the maximum power of 1,300 mW at 800°C operating temperature [64]. They also built a stack using two single-channel honeycombs in series. The stack provided the peak power output of 1,800 mW at 800°C operating temperature [65].

REFERENCES

1. Kaur, G. 2016. *Solid Oxide Fuel Cell Components*. Springer International Publishing: Cham, Switzerland.
2. Timurkutluk, B., Timurkutluk, C., Mat, M.D., and Kaplan, Y. 2016. A review on cell/stack designs for high performance solid oxide fuel cells. *Renewable and Sustainable Energy Reviews* 56:1101–1121.
3. Minh, N.Q. 2004. Solid oxide fuel cell technology—features and applications. *Solid State Ionics* 174(1–4):271–277.
4. Hernández-Pacheco, E., and Mann, M.D. 2006. A computer model for a high temperature fuel cell. *Revista mexicana de física E* 52(2):119–125.
5. Singhal, S.C., and Kendall, K. 2003. *High-Temperature Solid Oxide Fuel Cells: Fundamentals, Design and Applications*. Elsevier: Oxford, UK.
6. Singhal, S.C. 2002. Solid oxide fuel cells for stationary, mobile, and military applications. *Solid State Ionics* 152:405–410.
7. Singhal, S.C., and Dokiya, M. 1999. Solid oxide fuel cells (SOFC VI). In *Proceedings of the Sixth International Symposium*, The Electrochemical Society, Pennington, NJ, USA.
8. Ghosh, D., Tang, E., Perry, M., Prediger, D., Pastula, M., and Boersma, R. 2001. Status of SOFC developments at Global Thermoelectric. *ECS Proceedings Volumes* 2001:100–110.
9. Kim, J.W., Virkar, A.V., Fung, K.Z., Mehta, K., and Singhal, S.C. 1999. Polarization effects in intermediate temperature, anode-supported solid oxide fuel cells. *Journal of the Electrochemical Society* 146(1):69–78.
10. Hajimolana, S.A., Hussain, M.A., Daud, W.A.W., Soroush, M., and Shamiri, A. 2011. Mathematical modeling of solid oxide fuel cells: a review. *Renewable and Sustainable Energy Reviews* 15(4):1893–1917.
11. Singhal, S.C. 2000. Advances in solid oxide fuel cell technology. *Solid Sate Ionics* 135(1–4):305–313.
12. Pal, U.B., and Singhal, S.C. 1990. Electrochemical vapor deposition of yttria-stabilized zirconia films. *Journal of the Electrochemical Society* 137(9):2937–2941.
13. Kuo, L.J., Vora, S.D., and Singhal, S.C. 1997. Plasma spraying of lanthanum chromite films for solid oxide fuel cell interconnection application. *Journal of the American Ceramic Society* 80(3):589–593.

14. Huang, K., and Singhal, S.C. 2013. Cathode-supported tubular solid oxide fuel cell technology: a critical review. *Journal of Power Sources* 237:84–97.
15. Kendall, K. 2010. Progress in microtubular solid oxide fuel cells. *International Journal of Applied Ceramic Technology* 7(1):1–9.
16. Meadowcroft, A.D., Howroyd, S., Kendall, K., and Kendall, M. 2013. Testing micro-tubular sofcs in unmanned air vehicles (UAVs). *ECS Transactions* 57(1):451–457.
17. Howe, K.S., Thompson, G.J., and Kendall, K. 2011. Micro-tubular solid oxide fuel cells and stacks. *Journal of Power Sources* 196(4):1677–1686.
18. Lawlor, V. 2013. Review of the micro-tubular solid oxide fuel cell (Part II: cell design issues and research activities). *Journal of Power Sources* 240:421–441.
19. Yang, N., Tan, X., and Ma, Z. 2008. A phase inversion/sintering process to fabricate nickel/yttria-stabilized zirconia hollow fibers as the anode support for micro-tubular solid oxide fuel cells. *Journal of Power Sources* 183(1):14–19.
20. Droushiotis, N., Doraswami, U., Kanawka, K., Kelsall, G.H., and Li, K. 2009. Characterization of NiO–yttria stabilised zirconia (YSZ) hollow fibres for use as SOFC anodes. *Solid State Ionics* 180(17–19):1091–1099.
21. Droushiotis, N., Othman, M.H.D., Doraswami, U., Wu, Z., Kelsall, G., and Li, K. 2009. Novel co-extruded electrolyte–anode hollow fibres for solid oxide fuel cells. *Electrochemistry Communications* 11(9):1799–1802.
22. Othman, M.H.D., Wu, Z., Droushiotis, N., Doraswami, U., Kelsall, G., and Li, K. 2010. Single-step fabrication and characterisations of electrolyte/anode dual-layer hollow fibres for micro-tubular solid oxide fuel cells. *Journal of Membrane Science* 351(1–2):196–204.
23. Othman, M.H.D., Droushiotis, N., Wu, Z., Kanawka, K., Kelsall, G., and Li, K. 2010. Electrolyte thickness control and its effect on electrolyte/anode dual-layer hollow fibres for micro-tubular solid oxide fuel cells. *Journal of Membrane Science* 365(1–2):382–388.
24. Chen, C., Liu, M., Yang, L., and Liu, M. 2011. Anode-supported micro-tubular SOFCs fabricated by a phase-inversion and dip-coating process. *International Journal of Hydrogen Energy* 36(9):5604–5610.
25. Kim, D.W., Yun, U.J., Lee, J.W., Lim, T.H., Lee, S.B., Park, S.J., Song, R.H., and Kim, G. 2014. Fabrication and operating characteristics of a flat tubular segmented-in-series solid oxide fuel cell unit bundle. *Energy* 72:215–221.
26. Sharifzadeh, M. 2019. *Design and Operation of Solid Oxide Fuel Cells: The Systems Engineering Vision for Industrial Application*. Academic Press, Elsevier: London, UK.
27. Gardner, F.J., Day, M.J., Brandon, N.P., Pashley, M.N., and Cassidy, M. 2000. SOFC technology development at Rolls-Royce. *Journal of Power Sources* 86(1–2):122–129.
28. Kim, Y.B., Ahn, S.J., Moon, J., Kim, J., and Lee, H.W. 2006. Direct-write fabrication of integrated planar solid oxide fuel cells. *Journal of Electroceramics* 17(2–4):683–687.
29. Oh, K., Kim, J., Choi, S.H., Lee, D., and Moon, J. 2012. Influence of reduced substrate shunting current on cell performance in integrated planar solid oxide fuel cells. *Ceramics International* 38(1):695–700.
30. Lai, T.S., and Barnett, S.A. 2007. Effect of cathode sheet resistance on segmented-in-series SOFC power density. *Journal of Power Sources* 164(2):742–745.
31. An, Y.T., Ji, M.J., Hwang, H.J., Park, S.E., and Choi, B.H. 2015. Effect of cell length on the performance of segmented-in-series solid oxide fuel cells fabricated using decalcomania method. *Journal of the Ceramic Society of Japan* 123(1436):178–181.
32. An, Y.T., Ji, M.J., Hwang, H.J., Park, S.E., and Choi, B.H. 2015. Effect of cell-to-cell distance in segmented-in-series solid oxide fuel cells. *International Journal of Hydrogen Energy* 40(5):2320–2325.
33. Almutairi, G., Kendall, K., and Bujalski, W. 2012. Cycling durability studies of IP-SOFC. *International Journal of Low-Carbon Technologies* 7(1):63–68.

34. Bujalski, W., Dikwal, C.M., and Kendall, K. 2007. Cycling of three solid oxide fuel cell types. *Journal of Power Sources* 171(1):96–100.
35. Wang, Z., Mori, M., and Itoh, T. 2012. Evaluation of porous Ni-YSZ cermets with Ni content of 0–30 vol. % as insulating substrates for segmented-in-series tubular solid oxide fuel cells. *Journal of Fuel Cell Science and Technology* 9(2):021004.
36. Fujita, K., Somekawa, T., Horiuchi, K., and Matsuzaki, Y. 2009. Evaluation of the redox stability of segmented-in-series solid oxide fuel cell stacks. *Journal of Power Sources* 193(1):130–135.
37. Somekawa, T., Horiuchi, K., and Matsuzaki, Y. 2012. A study of electrically insulated oxide substrates for flat-tube segmented-in-series solid oxide fuel cells. *Journal of Power Sources* 202:114–119.
38. Sui, J., and Liu, J. 2007. An electrolyte-supported SOFC stack fabricated by slip casting technique. *ECS Transactions* 7(1):633–637.
39. Sui, J., and Liu, J. 2008. Slip-cast $Ce_{0.8}Sm_{0.2}O_{1.9}$ cone-shaped SOFC. *Journal of the American Ceramic Society* 91(4):1335–1337.
40. Zhang, Y., Liu, J., Yin, J., Yuan, W., and Sui, J. 2008. Fabrication and performance of cone-shaped segmented-in-series solid oxide fuel cells. *International Journal of Applied Ceramic Technology* 5(6):568–573.
41. Bai, Y., Liu, J., and Wang, C. 2009. Performance of cone-shaped tubular anode-supported segmented-in-series solid oxide fuel cell stack fabricated by dip coating technique. *International Journal of Hydrogen Energy* 34(17):7311–7315.
42. Ding, J., and Liu, J. 2009. A novel design and performance of cone-shaped tubular anode-supported segmented-in-series solid oxide fuel cell stack. *Journal of Power Sources* 193(2):769–773.
43. Xiao, J., Liu, J., and Ding, J. 2011. Electrochemical performance of cone-shaped tubular anode supported solid oxide fuel cells fabricated by low-pressure injection moulding technique. *ECS Transactions* 35(1):609–614.
44. Wang, H., and Liu, J. 2012. Effect of anode structure on performance of cone-shaped solid oxide fuel cells fabricated by phase inversion. *International Journal of Hydrogen Energy* 37(5):4339–4345.
45. Liu, Y., Tang, Y., Ding, J., and Liu, J. 2012. Electrochemical performance of cone-shaped anode-supported segmented-in-series SOFCs fabricated by gel-casting technique. *International Journal of Hydrogen Energy* 37(1):921–925.
46. Hassmann, K. 2001. SOFC power plants, the Siemens-Westinghouse approach. *Fuel Cells* 1(1):78–84.
47. Kim, J.H., Song, R.H., Song, K.S., Hyun, S.H., Shin, D.R., and Yokokawa, H. 2003. Fabrication and characteristics of anode-supported flat-tube solid oxide fuel cell. *Journal of Power Sources* 122(2):138–143.
48. Park, B.K., Lee, J.W., Lee, S.B., Lim, T.H., Park, S.J., Song, R.H., Im, W.B., and Shin, D.R. 2012. La-doped $SrTiO_3$ interconnect materials for anode-supported flat-tubular solid oxide fuel cells. *International Journal of Hydrogen Energy* 37(5):4319–4327.
49. Park, B.K., Lee, J.W., Lee, S.B., Lim, T.H., Park, S.J., Song, R.H., and Shin, D.R. 2012. A flat-tubular solid oxide fuel cell with a dense interconnect film coated on the porous anode support. *Journal of Power Sources* 213:218–222.
50. Pi, S.H., Lee, S.B., Song, R.H., Lee, J.W., Lim, T.H., Park, S.J., Shin, D.R., and Park, C.O. 2013. Novel Ag–glass composite interconnect materials for anode-supported flat-tubular solid oxide fuel cells operated at an intermediate temperature. *Fuel Cells* 13(3):392–397.
51. Hosseini, N.R., Sammes, N.M., and Chung, J.S. 2014. Manganese-doped lanthanum calcium titanate as an interconnect for flat-tubular solid oxide fuel cells. *Journal of Power Sources* 245:599–608.

52. Kim, J.H., Song, R.H., Chung, D.Y., Hyun, S.H., and Shin, D.R. 2009. Degradation of cathode current-collecting materials for anode-supported flat-tube solid oxide fuel cell. *Journal of Power Sources* 188(2):447–452.

53. Lim, T.H., Park, J.L., Lee, S.B., Park, S.J., Song, R.H., and Shin, D.R. 2010. Fabrication and operation of a 1 kW class anode-supported flat tubular SOFC stack. *International Journal of Hydrogen Energy* 35(18):9687–9692.

54. Lee, S.B., Lee, J.W., Lim, T.H., Park, S.J., Song, R.H., and Shin, D.R. 2011. Development of anode-supported flat-tube solid oxide fuel cell (SOFC) stack with high power density. *ECS Transactions* 35(1):327–332.

55. Wetzko, M., Belzner, A., Rohr, F.J., and Harbach, F. 1999. Solid oxide fuel cell stacks using extruded honeycomb type elements. *Journal of Power Sources* 83(1–2):148–155.

56. Xia, C., Rauch, W., Wellborn, W., and Liu, M. 2002. Functionally graded cathodes for honeycomb solid oxide fuel cells. *Electrochemical and Solid-State Letters* 5(10):A217–A220.

57. Zha, S., Zhang, Y., and Liu, M. 2005. Functionally graded cathodes fabricated by sol-gel/slurry coating for honeycomb SOFCs. *Solid State Ionics* 176(1–2):25–31.

58. Yamaguchi, T., Shimizu, S., Suzuki, T., Fujishiro, Y., and Awano, M. 2008. Development and evaluation of a cathode-supported SOFC having a honeycomb structure. *Electrochemical and Solid-State Letters* 11(7):B117–B121.

59. Yamaguchi, T., Shimizu, S., Suzuki, T., Fujishiro, Y., and Awano, M. 2009. Evaluation of extruded cathode honeycomb monolith-supported SOFC under rapid start-up operation. *Electrochimica Acta* 54(5):1478–1482.

60. Yamaguchi, T., Shimizu, S., Suzuki, T., Fujishiro, Y., and Awano, M. 2009. Fabrication and evaluation of a novel cathode-supported honeycomb SOFC stack. *Materials Letters* 63(29):2577–2580.

61. Yamaguchi, T., Shimizu, S., Suzuki, T., Fujishiro, Y., and Awano, M. 2009. Design and fabrication of a novel electrode-supported honeycomb SOFC. *Journal of the American Ceramic Society* 92:S107–S111.

62. Shimizu, S., Yamaguchi, T., Suzuki, T., Fujishiro, Y., and Awano, M. 2007. Fabrication and properties of honeycomb-type SOFCs accumulated with multi micro-cells. *ECS Transactions* 7(1):651–656.

63. Fukushima, A., Nakajima, H., and Kitahara, T. 2013. Performance evaluation of an anode-supported honeycomb solid oxide fuel cell. *ECS Transactions* 50(48):71–75.

64. Zhong, H., Matsumoto, H., Ishihara, T., and Toriyama, A. 2008. Self-supported $LaGaO_3$-based honeycomb-type solid oxide fuel cell with high volumetric power density. *Solid State Ionics* 179(27–32):1474–1477.

65. Zhong, H., Matsumoto, H., Toriyama, A., and Ishihara, T. 2009. Honeycomb-type solid oxide fuel cell using $La_{0.9}Sr_{0.1}Ga_{0.8}Mg_{0.2}O_3$ electrolyte for high volumetric power density. *Journal of the Electrochemical Society* 156(1):B74–B79.

5 System Design and Optimization

5.1 SOLID OXIDE FUEL CELL (SOFC) SYSTEM DESIGNS AND PERFORMANCE

SOFCs have been described as one of the most promising energy conversion systems due to their many advantages such as high-energy conversion efficiency, low pollution, and versatile fuel input. In the SOFC system, fuel and air are the stack inputs, while electricity, exhaust gas, and hot water or steam exit the system. Such systems include atmospheric SOFC combined heat and power generation (CHP) systems, pressurized SOFC/turbine hybrid systems, atmospheric SOFC residential and auxiliary power systems, and oxygen separating systems [1]. In order to maximize the use of heat generated from the high-temperature exhaust gas of the typical SOFCs, numerous studies have been focused on atmospheric SOFC-CHP systems [2–4] as well as SOFC tri-generation systems [5–7] and SOFC poly-generation systems [8,9]. The difference between a SOFC system and a SOFC stack is defined as the balance-of-plant (BOP), which can differ for each application with regard to the operating pressure, size of the system, and fuel type [1]. A typical SOFC-CHP uses BOP modules to enable specific electrochemical processes and produce electrical and thermal powers with efficiency as high as 90%, electrical and thermal combined. The SOFC tri-generation systems produce electrical, thermal, and cooling powers.

5.1.1 Atmospheric SOFC-CHP Systems

A simplified schematic of the main BOP equipment for an atmospheric SOFC system built by Siemens Westinghouse and demonstrated in Westervoort, Netherlands is shown in Figure 5.1 [10]. Here, the SOFC stack is integrated with a catalytic burner into a single SOFC generator module and a pre-reformer to reform higher hydrocarbons in the natural gas. Moreover, the fuel is desulfurized before it enters the SOFC stack. Sulfur can poison the nickel-containing anode and deteriorates the stack performance, as discussed in Chapter 2. In general, in order to avoid any performance loss, it is suggested to keep the sulfur level below 0.1 ppm. Natural gas contains far less sulfur compared to the oil-delivered liquid fuels. To remove sulfur compounds, the most common way is to hydrogenate them into hydrogen sulfide (H_2S) using a catalytic reactor and then adsorb the generated H_2S in a separate zinc oxide-containing reactor at an elevated temperature up to 450°C. Using activated carbon to adsorb sulfur compounds is relatively easier since it works at room temperature; however, it is more costly [1].

In order to avoid carbon formation in the SOFC stack, it is needed to pre-reform heavy hydrocarbons to methane, hydrogen, and carbon monoxide. In general, steam

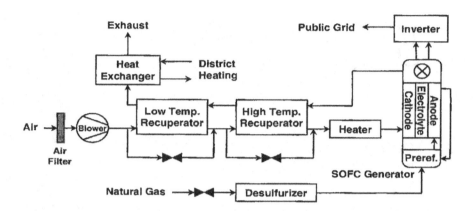

FIGURE 5.1   The main components of a typical atmospheric SOFC system. (Reprinted with permission from Ref. [1].)

reforming is used. To supply steam to the reformer, the recirculating part of the SOFC exhaust stream is used, as shown in Figure 5.1. This also results in increasing the overall fuel utilization.

Since the concentration of fuel decreases toward the exhaust side of the stack, causing the cell voltage drop, only a certain amount of fuel can be electrochemically converted to electricity and heat. The maximum practical amount considered for fuel utilization is 85%–90%. To burn the sulfur air from the cathode side as well as the remaining fuel from the anode side to avoid nickel oxidation, a catalytic burner is utilized.

In order to avoid thermal shock and the subsequent irreversible damage to the stack, the exhaust gas is led through a recuperator to heat the air up to at least 500°C before it enters the stack. An additional exchanger is used to produce hot water. The process steam can be produced instead of hot water by placing the heat exchanger between the two recuperators, as shown in Figure 5.1.

An inverter is used to convert the direct current (DC) produced by the stack to alternating current (AC). Generally, a two-step converter is used to transform the stack voltage to a stable DC voltage in the first step and then convert the DC to grid quality AC in the second step. The efficiency of the inverter is usually between 94% and 98%.

A SOFC system also has additional components including a blower to supply air to the system, air filters to clean the air, an air heater to start the SOFC generator module and to run at low load when the stack temperature is low, control equipment and user interface, purge gas systems to avoid stack damage during start-up and shut-down, and start-up steam generator to provide steam for the pre-reformer [1].

The electrical efficiency for such SOFC-CHP systems is considered about 45%–50% based on lower heating value of the fuel [10,11], while it is feasible to reach a total thermal and electrical efficiency of 85%–90%. In such SOFC systems using natural gas as the fuel, emissions are negligible except for CO_2 due to the clean fuel used and lower operating temperature of the stack compared to a conventional burner. However, the desulfurizer used to remove sulfur from the natural gas fuel produces a low amount of SO_x and particulates [1].

5.1.2 RESIDENTIAL, AUXILIARY POWER, AND OTHER ATMOSPHERIC SOFC SYSTEMS

The general layout of residential and auxiliary power units (APUs) is almost similar to the atmospheric system described above. A 5 kW gasoline-fueled APU along with its main subsystems are shown in Figure 5.2. Electrical efficiency of up to 30% is expected for such an APU system. This system is a close thermal integration where the heat losses are dependent of size and the use of partial oxidation fuel reformers. Partial oxidation reformer is used to preprocess the gasoline and is normally less efficient than a steam reformer. A battery unit is also used in the APU unit, as presented in Figure 5.2, to provide power and level the SOFC stack load. For residential systems operating with natural gas, the electrical efficiency of up to 40% is expected to be achieved.

It is possible to use SOFC technology for the combined production of power and syngas. This syngas, which is a mixture of CO and H_2 generated by the SOFC, can be used as a raw material in different chemical applications [12]. Additionally, it is relatively simple to separate CO_2 from water in the anode exhaust stream of the SOFC system, thus the possibility for carbon sequestration. A CO_2 separating SOFC system based on a pressurized SOFC generator combined with a gas turbine can maintain the electrical efficiency while capturing CO_2 for other applications. By capturing CO_2 from the SOFC, there is no need to use the conventional after-burner section. Instead, the anode exhaust gas is electrochemically oxidized in a separate special after-burner section using a suitable oxygen-selective ceramic membrane. The water vapor is also separated from CO_2 by cooling the exhaust gas, as illustrated in Figure 5.3 [1].

5.1.3 PRESSURIZED SOFC/TURBINE HYBRID SYSTEMS

In order to achieve a very high electrical efficiency of 60%–75%, a pressurized SOFC stack can be combined with a gas turbine. This results in high efficiency even at a very small 1 MW scale with a very low amount of harmful emissions except for CO_2. The simplest design for a gas turbine is shown in Figure 5.4a. The fuel is burned with the compressed air coming from a compressor, and exhaust gas with a temperature of 800°C–1300°C is expanded in a turbine mechanically coupled to the compressor and a generator. The temperature of exhaust gas is reduced to 250°C–600°C due to the expansion in the turbine. The exhaust temperature decreases with increasing the

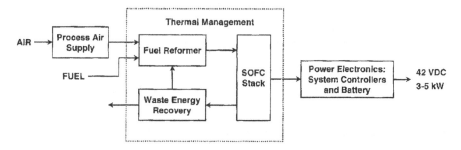

FIGURE 5.2 The main subsystems of a typical SOFC APU. (Reprinted with permission from Ref. [1].)

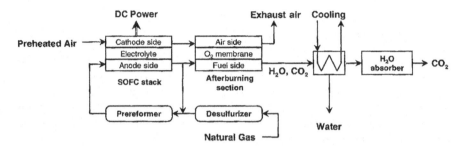

FIGURE 5.3 Simplified configuration for CO_2 capture from an SOFC system. (Reprinted with permission from Ref. [1].)

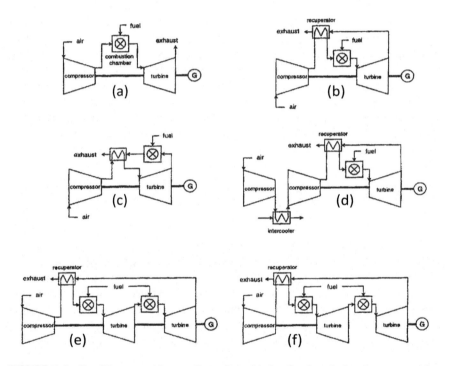

FIGURE 5.4 Possible gas turbine configurations. (a) the simplest design for a gas turbine. By replacing the conventional burner in a turbine with a pressurized SOFC generator, configurations (b–f) can be used for hybrid systems. (Reprinted with permission from Ref. [1].)

pressure ratio between the exhaust and turbine inlet. For small size turbines, the electrical efficiency can be about 20%, while it can increase up to 35% for large industrial turbines. Figure 5.4b–f represents different configurations for conventional gas turbines, which can also be used for hybrid SOFC/turbine systems. In principle, there is a need to replace the conventional combustion chamber in a turbine with a pressurized SOFC generator. In some configurations, a recuperator is used to increase electrical efficiency by decreasing the amount of natural gas required to heat the air [1].

To avoid thermal shock and subsequent damage to the SOFC stack, the air inlet temperature of an SOFC generator should be at least 500°C–650°C. Hence, using heat recuperation seems to be necessary. On the other hand, since the exhaust gas temperature becomes very low to heat the air inlet to the required temperature at high-pressure ratios between the gas turbine inlet and outlet, this pressure ratio is also limited (the ratio needs to be 2–4 unless there is additional heat from conventional burners). The SOFC stack can deliver 65%–80% of the total electrical power output of the hybrid system without additional heat from the conventional burners.

The electrical efficiency of a SOFC/turbine hybrid system is dependent of system size, system configuration (Figure 5.4b–f), the additional heat from the conventional burners, and the performances of the SOFC and the turbine used. This efficiency may vary from 55% to 60% for a small and simple SOFC/turbine hybrid system with the power of 250–1,000 kW to 68% for a 5–10 MW system with intercooler and reheat from a separate SOFC generator. Combining thermal and electrical efficiencies, total efficiency of 85%–90% can be expected [1].

5.1.4 SOFC Tri-Generation Systems

Among the various types of SOFC systems, SOFC tri-generation systems have been the most attractive ones in the market for distributed generation and residential applications [13,14]. Different system configurations, such as coupled designs and decoupled designs, can be utilized for the SOFC tri-generation systems. In the decoupled designs, the waste heat from the SOFC-CHP subsystem is used as a steam generator of absorption refrigeration subsystems, and the heat output of the SOFC tri-generation system is the exhaust heat of the steam generator [15,16]. In the coupled designs, either the SOFC-CHP and refrigeration subsystems can share some components, or the piping of one subsystem goes through the other subsystem components [5]. For instance, Silveira et al. [17] and Shariatzadeh et al. [7] proposed a system in which the exhaust gas from the SOFC stack directly goes to the absorption refrigeration subsystem components and turns back to the BOP components of the SOFC-CHP subsystem.

Residential power is one of the most promising applications for SOFC tri-generation systems. The demand for electricity, cooling power, and heat can frequently vary in this application. Therefore, it is important to investigate the energy outputs of the tri-generation systems and develop strategies to regulate the energy outputs. Recently, Wu and Chen [5] proposed a SOFC tri-generation system that mainly consists of a SOFC-CHP subsystem, an adsorption refrigeration subsystem, and a cooling device between the two subsystems, as shown in Figure 5.5. The SOFC-CHP subsystem consists of a SOFC stack, a reformer, heat exchangers, a burner, and other components. The adsorption refrigeration subsystem consists of two sorption beds, an evaporator, a condenser, and other components. The results indicated that the proposed SOFC tri-generation system provides 4.35 kW electrical power, 2.448 kW exhaust heat power, and 1.348 kW cooling power. Moreover, the energy efficiency of the system was 64.9%. The results also showed that the exhaust heat power varies with the variation of the electrical output power and the cooling power, but the electrical power variations are independent of the cooling power variations.

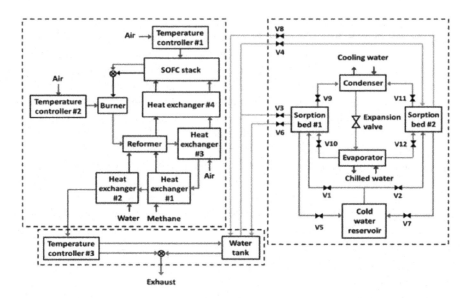

FIGURE 5.5 Proposed SOFC tri-generation system consisting of three subsystems: SOFC-CHP system, adsorption refrigeration system, and water tank equipped with a temperature controller. (Reprinted with permission from Ref. [5].)

In the other study, Moussawi et al. [6] designed a SOFC tri-generation system for domestic applications. Figure 5.6 shows the SOFC-BOP and the recovery system of the tri-generation design. The results indicated that the tri-generation system is energetically and economically superior. The maximum energy and energy efficiency of 65.2% and 45.77%, respectively, and the minimum system cost rate of 22.2 cents kWh[-1] were obtained under on-grid baseload operation.

Shariatzadeh et al. [7] also proposed a new configuration of the SOFC tri-generation system fed by biogas produced from hospital waste. The system consists of a 50 kW tubular SOFC combined with an absorption chiller, a heat recovery steam generator (HRSG), a combustion chamber, a compressor, and other components, as shown in Figure 5.7. The system recovers and uses waste heat as well as generating electricity. This heat results in steam generation in the HRSG, and the steam generated within the absorption chiller produces cooling load. In the proposed system, air enters the compressor, and the flow rate required by the SOFC initially enters the recuperator and finally enters the SOFC cathode. The compressed fuel enters the SOFC anode. However, before the fuel enters, natural gas is preheated and enters the anode side, passing through the SOFC internal reformer. The DC electricity produced by the SOFC stack is converted to AC using a DC/AC inverter. The SOFC outlet flow (superheated water vapor) enters the combustion chamber until its enthalpy becomes higher than before. Thereafter, the flow enters the HRSG, where the required water vapor for the absorption chiller is supplied. The water vapor is sent to the chiller, and then, the cooling load is supplied in the chiller. The results indicated that the proposed system could be economically affordable in the long term.

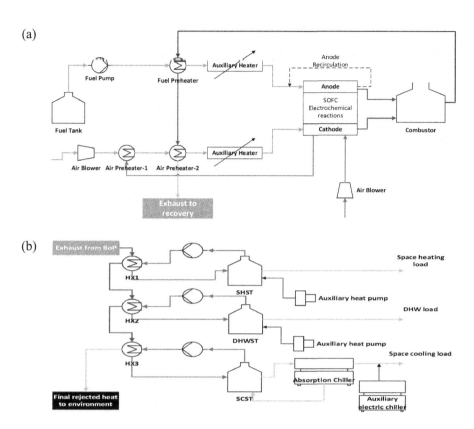

FIGURE 5.6 Schematic of (a) SOFC-BOP system and (b) recovery system of the proposed tri-generation design. (Reprinted with permission from Ref. [6].)

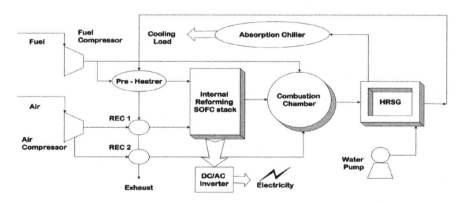

FIGURE 5.7 Schematic diagram of SOFC tri-generation hybrid system. (Reprinted with permission from Ref. [7].)

5.2   OPTIMIZATION STRATEGIES FOR SOFC

SOFC performance improvement has been widely investigated based on modeling and simulation studies in several areas, including advanced identification and estimation [18,19], improved model and observer studies [20–25], the accuracy of management and control [26–28], and optimization of design and operation [29–35]. Although the research on optimization of SOFC is more recent compared to other types of fuel cells such as proton exchange membrane fuel cell and molten carbonate fuel cell, it has been established since 1996 [36,37]. However, the focus on optimization of SOFC is continuously increasing year by year. A review study conducted by Ramadhani et al. [37] revealed an exponential trend of research on the optimization of SOFC systems from 1996 to 2016. The escalation of research interest shows the necessity of having an optimization strategy for the SOFC application.

In general, optimization is defined as finding the best solution among all feasible solutions. Optimization strategies are established to optimize the solutions toward different parameters with specific constraints to minimize or maximize the solution objectives [38,39]. So far, several studies have been carried out to optimize the SOFC performance and implementations. This optimization is a fundamental aspect of the SOFC, which has to be done in different aspects, as illustrated in Figure 5.8.

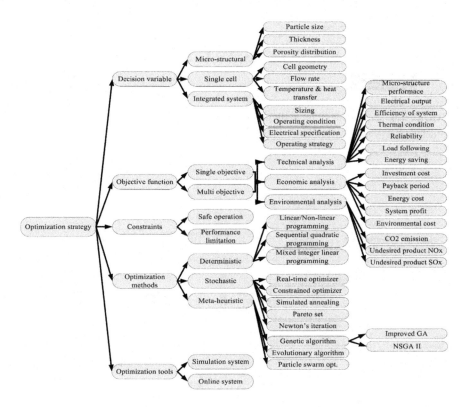

FIGURE 5.8   A scope diagram showing the optimization strategies for SOFC. (Reprinted with permission from Ref. [37].)

The main challenge for the optimization strategy development is to optimize variables, objectives, and constraints, which will be discussed in the following. The optimization methods used for SOFCs are divided into three different categories, i.e., stochastic, deterministic, and meta-heuristic methods. To verify the optimized parameters for maximizing or minimizing the objective functions, simulation, online implementation, or both can be conducted [37].

5.2.1 DECISION VARIABLES

Regarding the geometrical design of the SOFC, it is categorized into two major types: planar and tubular, as discussed in Chapter 4. As the gap between the fuel and air is separately sealed, the tubular design has been more secure than the planar design. Between these two designs, there is also microtubular design, taking advantage of the low resistance of the tubular and planar designs. In addition, microtubular design not only has faster start-up time, but it also possesses higher tolerance to thermal cycling, higher output power density, and better portability feature in comparison with the tubular and planar designs [40].

The different designs of SOFCs, including the planar, tubular, and improved designs, can be implemented into various microstructures, single cells, and integrated systems, affecting the optimized parameters. The important parameters for the microstructure design, which should be optimized, are the thickness of cell components, particularly the electrodes and the electrolytes, grain size, and porosity distribution [40]. Ge et al. [41] also studied the effect of other parameters on the microstructure design such as weight fraction, void fraction, particle size, and density along with the optimization process for the tubular SOFC.

5.2.1.1 Microstructural Parameters

The research on SOFC microstructure optimization has first been conducted by Kleitz and Petitbon in 1996 [42]. The study on the SOFC electrode microstructure and performance revealed that mixing the conducting electrode materials as well as increasing the reaction zone can improve the performance. Since then, there have been many studies carried out on microstructural parameters and optimization. For instance, Li et al. [43] reported that the electrode composition could affect its performance and result in decreasing open polarization resistance and increasing the current exchange density. Ahmed and Sathish [44] optimized the anode material by adding NiO and gadolinia-doped ceria and using graphite as a pore former in the anode substrate. The results revealed that the cell performance increases at three different operating temperatures, 500°C, 550°C, and 600°C, achieving the open circuit voltage and power density of 0.86 V and 48.62 mW cm^{-2}, respectively. Bhattacharyya and Rengaswamy [45] and Bhattacharyya et al. [46] used the isothermal model to optimize the electrode porosity and effective diffusivity of the reacting species. They found that optimizing the cell design as well as operating temperature for an anode-supported tubular cell can significantly improve the cell performance, especially at high current densities. Haanappel et al. [47] optimized the microstructural parameters of an LSM-based anode-supported single cell, including the grain size of the LSM cathode current collector, the thickness of the cathode current collector layer,

LSM to YSZ mass ratio of the cathode functional layer, and calcination effect on the functional cathode layer. The results showed some improvements in the LSM-based SOFC performance by changing the LSM to YSZ mass ratio. In addition, the current density was affected by the thickness and grain size.

5.2.1.2 Single-Cell Parameters

For single-cell applications, researchers have focused on the optimization of the output power and cell efficiency while reducing the cost [48]. There are several parameters that should be taken into consideration for the optimization of a single cell, such as length of triple-phase boundary, cell volume, the thickness of the ribs, and channel height [37]. Bhattacharyya et al. [46] studied the cell design parameters such as cell size, electrode porosity, and species diffusivity and their effects on cell performance. Ji et al. [49] investigated the geometrical features, including channel height, the thickness of ribs and electrodes and their effects on the cell efficiency. In order to minimize thermal stress and improve the output power, Skalar et al. [50] and Feng et al. [51] modeled and optimized the cell thickness and length.

The effects of operating conditions such as fuel and air flow rates, pressure, and temperature of fuel and air, fuel utilization on the single-cell performance have also been investigated and optimized in several studies. For example, Ni et al. [52] found that the activation and ohmic losses are reduced with increasing temperature. In addition, the results revealed that the SOFC output power increases with an increase in the hydrogen content and pressure. Jo et al. [53] have also studied the effect of these parameters on the cell current density as the objective. The results showed that the output current density of the fuel cell is significantly enhanced with temperature, hydrogen molar fraction, and pressure.

5.2.1.3 Integrated System Parameters

As previously discussed, single SOFCs can be arranged into a stack system in parallel or series to increase the output power. In a stack system, the SOFC is attached with BOP components such as heat exchanger, air compressor, reformer, burner, and fuel processor to generate sufficient power at high temperatures. The most important aspects of developing a SOFC stack system are the stack size and efficiency. Additionally, the operating conditions of BOP play a crucial role in improving the overall efficiency of the stack systems. The stack systems have been applied for various applications, e.g., portable power generators, APUs, hybrid vehicles, stand-alone systems, and micro-power generators. In order to maximize the integrated performance for larger applications, the stack system is combined with heat-based generators such as gas turbines, steam turbines, heat engines, and biomass power. The exhaust heat of the stack is also utilized in the CHP system to improve efficiency. To satisfy the cooling demand in some applications, the CHP can also be combined with cooling devices such as the absorption chiller. Such a system is called a combined cooling, heating, and power generation (CCHP) system. The CHP and CCHP systems have been developed for both industrial and residential applications [37].

Designing, controlling, and management are considered as the most important aspects for the optimization of an integrated system. Regarding the designing aspect, optimization of the system design and its size are the key features for minimizing the total cost, depending on the load and application, weather conditions, and government regulations [37]. Wipke et al. [54] optimized the size of the hybrid SOFC-battery system for mobile applications to minimize the cost by considering various parameters such as fuel cell power and battery capacity. Deng et al. [55] have also sized hybrid systems with regard to the size of the wind turbine and CHP system to minimize the environmental emission as well as the cost. Palazzi et al. [31] conducted a thermoeconomic optimization of the design of a 50 kW planar SOFC system for stationary applications. Vijay et al. [56] also optimized an SOFC system with steam heat exchangers and an after-burner to maximize efficiency. They studied the system performance at maximum efficiency or at constant fuel utilization during operation. The results showed a small tradeoff between these two cases.

In addition to design, operating conditions are also required to be optimized to improve the system performance and decrease the cost as well as environmental effects. For example, Shirazi et al. [57] have investigated the optimization of operating conditions, including air compressor pressure ratio, air compressor efficiency, gas turbine efficiency, air, and fuel utilizations, current density, and steam to carbon ratio for an SOFC/turbine hybrid system. In another study, Najafi et al. [58] conducted similar work with optimization of the SOFC/gas turbine hybrid system. They improved the system by adding several parameters such as inlet motive steam temperature, temperature point, brine temperature in the last stage, and the number of desalination stages.

5.2.2 Objective Functions

One of the most important parts in the optimization process is to maximize or minimize the objectives. In SOFC applications, there are several objectives that can be defined based on thermodynamic, economic, and environmental aspects. In a simple system, the optimization is normally performed for one objective function. However, in a complex system such as SOFC, there will be more than one objective, which is called a multi-objective function. One of the objectives may be a constraint that should be satisfied by the others and vice versa [37]. For instance, several constraints, such as cost, cell efficiency, and fuel utilization, can be used to optimize the output power. In the following, the main objectives of the SOFC system optimization based on thermodynamic, economic, and environmental aspects will be elaborated.

5.2.2.1 Thermodynamic Aspects

Regarding the thermodynamic aspects, the objectives, needing to be optimized, can be listed as output energy and power, efficiency, and electrical characteristics. In a single-cell design, the output performance, which should be maximized, includes cell power density, conductivity, current density, cell efficiency, and output voltage. On the other hand, other outputs, such as potential losses, ohmic resistance, and temperature, must be minimized. For example, Tikiz and Taymaz [59] investigated the SOFC performance based on the current density as the objective. They found that the hydrogen flow rate and temperature are the fundamental operating parameters, affecting the cell current density. Shi and Xue [60] maximized the output power of SOFC with regard to the hydrogen and water concentration ratio. Song et al. [61]

optimized the shape of the nanocomposite cathode to minimize the fuel cell resistance using a simplified analysis 2D model. The optimization of microstructural features of SOFC such as pore diameter, porosity and tortuosity, electronic and ionic conductivities, electrode temperature, and diffusivity to achieve the optimum active layer thicknesses for maximizing the net output power and minimizing the total potential losses was studied by Wen et al. [62].

Thermodynamic aspects can also be considered as the objectives in the stack and integrated system. Previous studies showed that electrical output power, overall efficiency, limiting current density, thermal output power, and system temperature affect the system performance. Inui et al. [30] studied the optimization of air utilization and inlet gas temperature in order to minimize the temperature fluctuation of the CHP system and reduce thermal stresses. They proposed an efficient method to lower the temperature that varies significantly for the planar SOFCs without decreasing the single-cell voltage for different flow configurations.

To maximize the system performance, energy usage and saving can also be considered as the optimization objectives. For instance, Wakui and Yokoyama [63] used a simulation approach to design a CHP system with minimal annual primary energy consumption. The optimum design was achieved by variations of operating restrictions and battery utilization. The results revealed that the system performance and the primary energy savings can remarkably be improved by the use of battery.

The summary of mathematical equations for the thermodynamic analysis of objective functions used in the optimization strategy of SOFC application is listed in Table 5.1. Most of the thermodynamic/technical aspects are subject to maximize the output performance of the SOFC system.

5.2.2.2 Economic Aspects

Optimization, the cost of a whole system consisting of the real cost such as electricity cost, investment cost, and fuel consumption cost and the operational cost, including life cycle cost, environmental cost, profitability index, gross operative margins, and payback period, is one of the most important objectives for an economic perspective [37].

During past years, there have been several studies on economic analysis and optimization of the SOFC systems. For example, Calise et al. [64] economically analyzed the synthesis/design of a hybrid SOFC-gas power plant. They considered overall life cycle cost as the objective function. It was determined by sum of the fuel and capital costs. The optimization procedure revealed that the optimized configuration has much lower capital cost than that of the initial configuration. Moreover, the optimal results indicated that the SOFC is not essentially the most expensive component of the plant. Tan et al. [65] developed a thermoeconomic model for the design optimization of a 220 kW SOFC–proton exchange membrane fuel cell hybrid system. In this study, they considered the life cycle cost and the net electrical efficiency as the two objectives to be optimized. The optimization results showed that the life cycle cost of the hybrid system is 3,800–5,600 $ kW^{-1}, and the maximum net electrical efficiency can reach around 63%.

The net value of the unit cost of energy over the lifetime of utilization is expressed by the levelized cost of energy (LCE), calculated by dividing the overall life cycle cost and lifetime energy production. However, for financial and computational analyses,

TABLE 5.1
The List of Thermodynamic Aspects of Objective Functions Used for the SOFC Optimization

Objective Analysis	Objective Function
Maximizing open circuit voltage	$V = V_N - V_{ohm} - V_{act} - V_{conc}$
Maximizing triple-phase boundary length	$l_{TPB} = f_c s N_{j,i}^b \rho_j^{vol}$
Maximizing electronic conductivity	$\sigma\left[S\,cm^{-1}\right] = \dfrac{dI}{\pi r^2 \Delta V}$
Minimizing ohmic drop and diffusion overpotential	Ohmic drop : $U = \dfrac{\rho_e i}{4 th} r^2$ Diffusion overpotential: $\eta_{diff} = \dfrac{RT}{4f} \log \dfrac{P_e l}{P_g}$
Minimizing cathode total resistance	$\eta_{c,\,tot} = \dfrac{R_{LSM}^{eff} R_{YSZ}^{eff}}{R_{LSM}^{eff} + R_{YSZ}^{eff}}\left(\dfrac{n_c(0)}{R_{YSZ}^{eff}} + i_{tot}\delta_c + \dfrac{n_c(\delta_c)}{R_{LSM}^{eff}}\right)$ $R_{c,\,tot} = \dfrac{\eta_{c,\,tot}}{i_{tot}}$
Maximizing cell power	$P_{cell} = M \times \sum_{V_{cell}} j V_{cell}$
Maximizing cell efficiency	$\eta = \sum_{j=1}^{m}\left(rj\,\dfrac{P_{net,\,j}}{P_{system,\,j}}\right)$
Maximizing cell current density	$j_{0,a} = k_a \dfrac{72 \times\left[D_p - (D_p + D_s)n\right]n}{D_s^2 D_p^2\left(1 - \sqrt{1-x^2}\right)} \times \left(\dfrac{P_{H_2}}{P_{ref}}\right)\left(\dfrac{P_{H_2O}}{P_{ref}}\right)\exp\left(-\dfrac{E_{act,\,a}}{RT}\right)$ $j_{0,c} = k_c \dfrac{72 \times\left[D_p - (D_p + D_s)n\right]n}{D_s^2 D_p^2\left(1 - \sqrt{1-x^2}\right)} \times \left(\dfrac{P_{O_2}}{P_{ref}}\right)^{0.25}\exp\left(-\dfrac{E_{act,\,c}}{RT}\right)$
Maximizing electric generation	$P = VI = V \times \left(U_f n F N_{fuel}\right) \times A$
Minimizing temperature of stack	$T_{tube}(z) = \dfrac{\eta_{tube}\left(T_{tube} - T_s\right)d_{tube}\pi z}{f_{gas} C_{gas}}$
Minimizing entropy generation	$g_p = \int\left(g_\mu + g_h + g_m + g_c + g_{ohm}\right)dV + \int g_{rad}dV$
Maximizing system efficiency	$\eta_{elect} = \dfrac{P}{n\Delta H_{fuel} + n\Delta H_{air}} \times 100\%$
Minimizing primary energy consumption	$J_{CGS} = \sum_{m=1}^{M} N(m)\left\{\sum_{k=1}^{K}\left[\varnothing_E(k)E_p(k,m) + \varnothing_G F_p(k,m)\right]\Delta t\right\}$ $+ k\left(\sum_{i=1}^{I}\gamma CGSi + \gamma BT + \sum_{l=1}^{L}\gamma PDI\right)$

(Continued)

TABLE 5.1 (*Continued*)

The List of Thermodynamic Aspects of Objective Functions Used for the SOFC Optimization

Objective Analysis	Objective Function
Maximizing primary energy saving	$PES_{FC} = 1 - \dfrac{E_{fuel}}{\dfrac{E_{el}}{\eta_{el,\,s}} + \dfrac{E_{th}}{\eta_{th,\,s}}}$
Maximizing heat efficiency	$\eta = \dfrac{\Delta H_{react}}{n\Delta H_{fuel} + \dfrac{n}{2}\Delta H_{air}} \times 100\%$
Minimizing stack temperature variation	$dT = \dfrac{1}{3}\displaystyle\sum_{i=1}^{3}\left(\dfrac{\Delta T_{max}}{L_x}\right)^{cell,\,i} dx + \left(\dfrac{\Delta T_{max}}{L_z}\right)dz$

Source: Data adapted and modified from Ref. [37].

other factors such as financing, insurance, and maintenance can be included in the calculation [37]. Amer et al. [66] presented a method for optimization of the power generated by a hybrid renewable energy system to minimize the LCE value. The minimum LCE was found to be 0.0030277 € kW^{-1}.

The summary of mathematical equations for economic analysis of objective functions used in the optimization strategy of the SOFC application is listed in Table 5.2. Most of the economic aspects are subject to minimize the cost.

5.2.2.3 Environmental Aspects

One of the most important environmental aspects of SOFC system optimization is the CO_2 emission rate in the exhaust of the SOFC system. The unit CO_2 emission (in kg MWh^{-1}) for stand-alone electrical power generation systems $\left(\text{Emi}_{CO_2,\,el}\right)$ and tri-generation systems $\left(\text{Emi}_{CO_2,\,tri}\right)$, including electrical power, cooling, and heating are defined as:

$$\text{Emi}_{CO_2,\,el} = \frac{m_{CO_2}}{W_{net}} \times 3600 \tag{5.1}$$

$$\text{Emi}_{CO_2,\,tri} = \frac{m_{CO_2}}{W_{net} + Q_{eva} + Q_c} \times 3600 \tag{5.2}$$

Herein, the m_{CO_2} represents the mass of CO_2 during generation, W_{net} is the electrical output net of the power generation, Q_{eva} is the heat generation, and Q_c defines the cooling power of the system.

Chitsaz et al. [67] studied the environmental performance parameters (i.e., exergy efficiency, exergy destruction rate, and greenhouse gas emissions) for a novel tri-generation system driven by an SOFC. The results indicated that the unit CO_2 emission is significantly higher than that of the tri-generation system. In addition, the tri-generation system achieves about 5% higher exergy efficiency compared to the stand-alone system. In the other research, Kazempoor et al. [68] developed a series

TABLE 5.2

The List of Economic Aspects of Objective Functions Used for the SOFC Optimization

Objective Analysis	Objective Function
Minimizing capital cost	$C_{unit} = C_{Nbase} \left(\dfrac{N_{unit}}{N_{base}} \right)^d$
Minimizing cost of electricity	$\text{CEO} = F_1 \dfrac{R_C C_{cap}}{C_E} + F_2 C_M + F_3 \dfrac{C_F}{\eta_{sys}} - F_3 \dfrac{(\eta_{cog} - \eta_{sys}) C_F}{\eta_{sys} \eta_R}$
Minimizing fuel cost	$\text{CFE} = \dfrac{1}{\left[\left(\dfrac{0.55}{\text{city\_FE}} \right) + \left(\dfrac{0.45}{\text{hwy\_FE}} \right) \right]}$
Minimizing life cycle cost	$C_{LCC \, (\text{SOFC-PEMFC})} = C_{SO} + C_{PEM} + C_{FPS} + c_{fuel} \eta_{fuel} T_{life} t_{oper}$
Maximizing profitability index	$\text{PI} = \dfrac{PV_{in} - PV_{out}}{INV_F + I_s P_e + C_L L_P} - 1$
Maximizing net present value	$\text{NPV} = \displaystyle\sum_{i=0}^{y} \text{CF}_0(i)$
Maximizing gross operative margin	$\text{GOM} = \displaystyle\sum_{i, j, k} \{ R_{i,j}(k) - C_{i,j}(k) \} \cdot \Delta t \cdot g_{i,j}(k)$
Minimizing payback period	$\text{PBP} = (y_{cash} - 1) + \dfrac{\text{cost} - \text{CCF}(y_{cash} - 1)}{\text{CF}y_{cash}}$
Minimizing cost of energy	$\displaystyle\sum (C_{out} E_{out}) K + C_{w,k} W_k = C_{Q,k} E_{Q,k} + \sum (C_{in} E_{in}) + Z_k$
Minimizing levelized electricity cost	$\text{LEC}_i(\text{RC}) = \dfrac{\text{LEC}_i(\text{ESI}) - \text{LEC}_i(\text{FIX})}{\text{LEC}_i(\text{FIX})}$

Source: Data adapted and modified from Ref. [37].

of modeling steps for the performance assessment of the building co-generation and poly-generation systems using SOFC. At the final step, they evaluated the primary energy demand and CO_2 emissions of the building integrated co- and poly-generation SOFC systems. The results indicated that the SOFC-based CHP system can save 66% of energy and reduce emissions for 14%.

The summary of mathematical equations for environmental analysis of objective functions used in the optimization strategy of SOFC application is listed in Table 5.3. Most of the environmental aspects are subject to minimize the unwanted emissions for the SOFC system.

5.2.3 Constraints

Most of the thermodynamic, economic, and environmental objective problems are optimized using one or several constraints. For SOFC applications, two types of

TABLE 5.3

The List of Environmental Aspects of Objective Functions Used for the SOFC Optimization

Objective Analysis	Objective Function
Minimizing CO_2 emission	$z = g_g \sum_1^t \left(G_{t,\,FC} + G_{t,\,aux} \right) + g_e \sum_1^t \left(Qreq_{t,\,ele} + Q_{t,\,f,\,c,\,ele} \right)$
Maximizing C_2 production	$S_{C_2} = \dfrac{2 \left(F_{C_2H_6} + F_{C_2H_4} \right)}{F_{CH_4}^0 - F_{CH_4}}$
Minimizing CO_2 equivalent	$CHG = 0.0001 \cdot FU \cdot HLV \cdot E_{fct}$

Source: Data adapted and modified from Ref. [37].

constraints are used to solve optimization problems: safe operation and specification range. For safe operation, the objectives must be obtained without harming the system because of overpressure, overheat, and overflow. The specification range refers to the limited performance of the components for the system, such as voltage, current density, output power, and heat power [37].

5.2.3.1 Safe Operation

One of the most significant parameters influencing the performance of SOFC-based systems is the working temperature. Since SOFC systems operate at high temperatures, material selection and electrode efficiency for SOFCs have always been major problems. Hence, many studies have been conducted on understanding the range of safe operation for fuel and air temperatures [37]. For instance, Milewski [69] found that the optimal temperature for the cell operation and maximum temperature for the gas turbine inlet is 800°C and 1100°C, respectively, in a SOFC hybrid system. In addition, the system reaches its maximum efficiency by changing SOFC fuel utilization, turbine pressure ratio, heat exchanger effectiveness, and current density with constant control of temperature.

5.2.3.2 Specification Ranges

The range of characteristics of the SOFC systems is related to the material parameters mainly including the compositions and properties of the electrode and electrolyte materials, cell geometry parameters consisting of the electrode and rib thicknesses as well as the length of cell and channel, and electrical parameters including voltage, current density, output power, and heat. For example, Wen et al. [70] considered the cell volume as a constraint and the thickness of interconnect as a constant to optimize a single planar SOFC. The thicknesses of cathode, anode, and electrolyte components were optimized to reach the maximum power output. In another study, Chen and Ni [71] took the social and governmental rules as the legal constraints to optimize a SOFC power generation in a CHP-CCHP system for a hotel building application. They also considered load requirements as the other important constraint for designing the capacity of the co-/tri-generation system.

REFERENCES

1. Singhal, S.C., and Kendall, K. 2003. *High-Temperature Solid Oxide Fuel Cells: Fundamentals, Design and Applications.* Elsevier: Oxford, UK.
2. Kupecki, J. 2015. Off-design analysis of a micro-CHP unit with solid oxide fuel cells fed by DME. *International Journal of Hydrogen Energy* 40(35):12009–12022.
3. Lee, K., Kang, S., and Ahn, K.Y. 2017. Development of a highly efficient solid oxide fuel cell system. *Applied Energy* 205:822–833.
4. Zhang, L., Xing, Y., Xu, H., Wang, H., Zhong, J., and Xuan, J. 2017. Comparative study of solid oxide fuel cell combined heat and power system with multi-stage exhaust chemical energy recycling: modeling, experiment and optimization. *Energy Conversion and Management* 139:79–88.
5. Wu, C.C., and Chen, T.L. 2019. Design and dynamics simulations of small scale solid oxide fuel cell tri-generation system. *Energy Conversion and Management: X* 1:100001.
6. Al Moussawi, H., Fardoun, F., and Louahlia, H. 2017. 4-E based optimal management of a SOFC-CCHP system model for residential applications. *Energy Conversion and Management* 151:607–629.
7. Shariatzadeh, O.J., Refahi, A.H., Rahmani, M., and Abolhassani, S.S. 2015. Economic optimisation and thermodynamic modelling of SOFC tri-generation system fed by biogas. *Energy Conversion and Management* 105:772–781.
8. Worall, M., Elmer, T., Riffat, S., Wu, S., and Du, S. 2017. An experimental investigation of a micro-tubular SOFC membrane-separated liquid desiccant dehumidification and cooling tri-generation system. *Applied Thermal Engineering* 120:64–73.
9. Mastropasqua, L., Campanari, S., and Brouwer, J. 2018. Electrochemical carbon separation in a SOFC–MCFC polygeneration plant with near-zero emissions. *Journal of Engineering for Gas Turbines and Power* 140(1):013001.
10. Modern Power Systems. 1998. Westervoort SOFC: the road to commercialization, 29–32. https://www.modernpowersystems.com/features/featurewestervoort-sofc-the-road-to-commercialization/.
11. Chau, K.T., Wong, Y.S., and Chan, C.C. 1999. An overview of energy sources for electric vehicles. *Energy Conversion and Management* 40(10):1021–1039.
12. Achenbach, E., and Riensche, E. 1994. Methane/steam reforming kinetics for solid oxide fuel cells. *Journal of Power Sources* 52(2):283–288.
13. Braun, R.J., and Kazempoor, P. 2013. Application of SOFCs in combined heat, cooling and power systems. In *Solid Oxide Fuel Cells*: From *Materials* to *System Modeling*, Edited by Zhao, T.S., and Ni, M., The Royal Society of Chemistry: Cambridge, UK, 327–382.
14. Elmer, T. 2016. *A Novel SOFC Tri-Generation System for Building Applications.* Springer: Switzerland.
15. Chitsaz, A., Mehr, A.S., and Mahmoudi, S.M.S. 2015. Exergoeconomic analysis of a trigeneration system driven by a solid oxide fuel cell. *Energy Conversion and Management* 106:921–931.
16. Asghari, M., McVay, D., and Brouwer, J. 2017. Integration of a solid oxide fuel cell with an absorption chiller for dynamic generation of combined cooling and power for a residential application. *ECS Transactions* 78(1):243–255.
17. Silveira, J.L., Leal, E.M., and Ragonha Jr, L.F. 2001. Analysis of a molten carbonate fuel cell: cogeneration to produce electricity and cold water. *Energy* 26(10):891–904.
18. Jayasankar, B.R., Ben-Zvi, A., and Huang, B. 2009. Identifiability and estimability study for a dynamic solid oxide fuel cell model. *Computers & Chemical Engineering* 33(2):484–492.
19. Lockett, M., Simmons, M.J.H., and Kendall, K. 2004. CFD to predict temperature profile for scale up of micro-tubular SOFC stacks. *Journal of Power Sources* 131(1–2):243–246.

20. Xue, X., Tang, J., Sammes, N., and Du, Y. 2005. Dynamic modeling of single tubular SOFC combining heat/mass transfer and electrochemical reaction effects. *Journal of Power Sources* 142(1–2):211–222.
21. Laurencin, J., Delette, G., Lefebvre-Joud, F., and Dupeux, M. 2008. A numerical tool to estimate SOFC mechanical degradation: case of the planar cell configuration. *Journal of the European Ceramic Society* 28(9):1857–1869.
22. Hashimoto, S., Nishino, H., Liu, Y., Asano, K., Mori, M., Funahashi, Y., and Fujishiro, Y. 2008. The electrochemical cell temperature estimation of micro-tubular SOFCs during the power generation. *Journal of Power Sources* 181(2):244–250.
23. Ali, J.M., Hussain, M.A., Tade, M.O., and Zhang, J. 2015. Artificial Intelligence techniques applied as estimator in chemical process systems–a literature survey. *Expert Systems with Applications* 42(14):5915–5931.
24. Ali, J.M., Hoang, N.H., Hussain, M.A., and Dochain, D. 2015. Review and classification of recent observers applied in chemical process systems. *Computers & Chemical Engineering* 76:27–41.
25. Ali, J.M., Hoang, N.H., Hussain, M.A., and Dochain, D. 2016. Hybrid observer for parameters estimation in ethylene polymerization reactor: a simulation study. *Applied Soft Computing* 49:687–698.
26. Hajimolana, S., Hussain, M.A., Soroush, M., Wan Daud, W.A., and Chakrabarti, M.H. 2013. Multilinear-model predictive control of a tubular solid oxide fuel cell system. *Industrial & Engineering Chemistry Research* 52(1):430–441.
27. Hajimolana, S.A., Tonekabonimoghadam, S.M., Hussain, M.A., Chakrabarti, M.H., Jayakumar, N.S., and Hashim, M.A. 2013. Thermal stress management of a solid oxide fuel cell using neural network predictive control. *Energy* 62:320–329.
28. Moghaddam, A.A., Seifi, A., Niknam, T., and Pahlavani, M.R.A. 2011. Multi-objective operation management of a renewable MG (micro-grid) with back-up micro-turbine/fuel cell/battery hybrid power source. *Energy* 36(11):6490–6507.
29. Weber, C., Maréchal, F., Favrat, D., and Kraines, S. 2006. Optimization of an SOFC-based decentralized polygeneration system for providing energy services in an office-building in Tōkyō. *Applied Thermal Engineering* 26(13):1409–1419.
30. Inui, Y., Ito, N., Nakajima, T., and Urata, A. 2006. Analytical investigation on cell temperature control method of planar solid oxide fuel cell. *Energy Conversion and Management* 47(15–16):2319–2328.
31. Palazzi, F., Autissier, N., Marechal, F.M., and Favrat, D. 2007. A methodology for thermo-economic modeling and optimization of solid oxide fuel cell systems. *Applied Thermal Engineering* 27(16):2703–2712.
32. Autissier, N., Palazzi, F., Maréchal, F., Van Herle, J., and Favrat, D. 2007. Thermo-economic optimization of a solid oxide fuel cell, gas turbine hybrid system. *Journal of Electrochemical Energy Conversion Storage* 4(2):123–129.
33. López, P.R., González, M.G., Reyes, N.R., and Jurado, F. 2008. Optimization of bio-mass fuelled systems for distributed power generation using particle swarm optimization. *Electric Power Systems Research* 78(8):1448–1455.
34. Sciacovelli, A. 2010. Thermodynamic optimization of a monolithic-type solid oxide fuel cell. *International Journal of Thermodynamics* 13(3):95–103.
35. Zhang, X., Su, S., Chen, J., Zhao, Y., and Brandon, N. 2011. A new analytical approach to evaluate and optimize the performance of an irreversible solid oxide fuel cell-gas turbine hybrid system. *International Journal of Hydrogen Energy* 36(23):15304–15312.
36. Bozorgmehri, S., and Hamedi, M. 2012. Modeling and optimization of anode-supported solid oxide fuel cells on cell parameters via artificial neural network and genetic algorithm. *Fuel Cells* 12(1):11–23.

37. Ramadhani, F., Hussain, M.A., Mokhlis, H., and Hajimolana, S. 2017. Optimization strategies for Solid Oxide Fuel Cell (SOFC) application: a literature survey. *Renewable and Sustainable Energy Reviews* 76:460–484.

38. Grefenstette, J.J. 1986. Optimization of control parameters for genetic algorithms. *IEEE Transactions on Systems, Man, and Cybernetics* 16(1):122–128.

39. Zavattieri, P.D., Dari, E.A., and Buscaglia, G.C. 1996. Optimization strategies in unstructured mesh generation. *International Journal for Numerical Methods in Engineering* 39(12):2055–2071.

40. Buonomano, A., Calise, F., d'Accadia, M.D., Palombo, A., and Vicidomini, M. 2015. Hybrid solid oxide fuel cells–gas turbine systems for combined heat and power: a review. *Applied Energy* 156:32–85.

41. Ge, X.M., Fang, Y.N., and Chan, S.H. 2012. Design and optimization of composite electrodes in solid oxide cells. *Fuel Cells* 12(1):61–76.

42. Kleitz, M., and Petitbon, F. 1996. Optimized SOFC electrode microstructure. *Solid State Ionics* 92(1–2):65–74.

43. Li, W., Xiong, C., Zhang, Q., Jia, L., Chi, B., Pu, J., and Jian, L. 2016. Composition optimization of samarium strontium manganite-yttria stabilized zirconia cathode for high performance intermediate temperature solid oxide fuel cells. *Electrochimica Acta* 190:531–537.

44. Ahmed, A.J., and Sathish, P. 2014. Optimizing anode microstructure of intermediate temperature solid oxide fuel cell based on gadolinium doped ceria. In *3rd International Conference on the Developments in Renewable Energy Technology (ICDRET)*, IEEE, 1–4, Dhaka, Bangladesh.

45. Bhattacharyya, D., and Rengaswamy, R. 2009. Transport, sensitivity, and dimensional optimization studies of a tubular solid oxide fuel cell. *Journal of Power Sources* 190(2):499–510.

46. Bhattacharyya, D., Rengaswamy, R., and Finnerty, C. 2007. Isothermal models for anode-supported tubular solid oxide fuel cells. *Chemical Engineering Science* 62(16):4250–4267.

47. Haanappel, V.A.C., Mertens, J., Rutenbeck, D., Tropartz, C., Herzhof, W., Sebold, D., and Tietz, F. 2005. Optimisation of processing and microstructural parameters of LSM cathodes to improve the electrochemical performance of anode-supported SOFCs. *Journal of Power Sources* 141(2):216–226.

48. Secanell, M., Wishart, J., and Dobson, P. 2011. Computational design and optimization of fuel cells and fuel cell systems: a review. *Journal of Power Sources* 196(8):3690–3704.

49. Ji, Y., Yuan, K., Chung, J.N., and Chen, Y.C. 2006. Effects of transport scale on heat/mass transfer and performance optimization for solid oxide fuel cells. *Journal of Power Sources* 161(1):380–391.

50. Skalar, T., Lubej, M., and Marinšek, M. 2015. Optimization of operating conditions in a laboratory SOFC testing device. *Materials and Technology* 49:731–738.

51. Feng, H., Chen, L., Xie, Z., and Sun, F. 2015. Constructal optimization for a single tubular solid oxide fuel cell. *Journal of Power Sources* 286:406–413.

52. Ni, M., Leung, M.K., and Leung, D.Y. 2007. Parametric study of solid oxide fuel cell performance. *Energy Conversion and Management* 48(5):1525–1535.

53. Jo, D.H., Chun, J.H., Park, K.T., Hwang, J.W., Lee, J.Y., Jung, H.W., and Kim, S.H. 2011. Optimization of physical parameters of solid oxide fuel cell electrode using electrochemical model. *Korean Journal of Chemical Engineering* 28(9):1844.

54. Wipke, K., Markel, T., and Nelson, D. 2001. Optimizing energy management strategy and degree of hybridization for a hydrogen fuel cell SUV. In *Proceedings of 18th Electric Vehicle Symposium*, 1–12, Berlin, Germany.

55. Deng, Q., Gao, X., Zhou, H., and Hu, W. 2011. System modeling and optimization of microgrid using genetic algorithm. In *2nd International Conference on Intelligent Control and Information Processing*, IEEE, 1:540–544, Harbin, China.

56. Vijay, P., Samantaray, A.K., and Mukherjee, A. 2010. Constant fuel utilization operation of a SOFC system: an efficiency viewpoint. *Journal of Fuel Cell Science and Technology* 7(4):0410111–0410117.

57. Shirazi, A., Aminyavari, M., Najafi, B., Rinaldi, F., and Razaghi, M. 2012. Thermal–economic–environmental analysis and multi-objective optimization of an internal-reforming solid oxide fuel cell–gas turbine hybrid system. *International Journal of Hydrogen Energy* 37(24):19111–19124.

58. Najafi, B., Shirazi, A., Aminyavari, M., Rinaldi, F., and Taylor, R.A. 2014. Exergetic, economic and environmental analyses and multi-objective optimization of an SOFC-gas turbine hybrid cycle coupled with an MSF desalination system. *Desalination* 334(1):46–59.

59. Tikiz, I., and Taymaz, I. 2016. An experimental investigation of solid oxide fuel cell performance at variable operating conditions. *Thermal Science* 20(5):1421–1433.

60. Shi, J., and Xue, X. 2011. Optimization design of electrodes for anode-supported solid oxide fuel cells via genetic algorithm. *Journal of the Electrochemical Society* 158(2):B143–B151.

61. Song, X., Diaz, A.R., Benard, A., and Nicholas, J.D. 2013. A 2D model for shape optimization of solid oxide fuel cell cathodes. *Structural and Multidisciplinary Optimization* 47(3):453–464.

62. Wen, H., Ordonez, J.C., and Vargas, J.V.C. 2013. Composite electrode modelling and optimization for solid oxide fuel cells. *International Journal of Energy Research* 37(2):95–104.

63. Wakui, T., and Yokoyama, R. 2015. Optimal structural design of residential cogeneration systems with battery based on improved solution method for mixed-integer linear programming. *Energy* 84:106–120.

64. Calise, F., d'Accadia, M.D., Vanoli, L., and Von Spakovsky, M.R. 2006. Single-level optimization of a hybrid SOFC–GT power plant. *Journal of Power Sources* 159(2):1169–1185.

65. Jun Tan, L., Yang, C., and Zhou, N. 2014. Thermoeconomic optimization of a solid oxide fuel cell and proton exchange membrane fuel cell hybrid power system. *Journal of Fuel Cell Science and Technology* 11(1):011005.

66. Amer, M., Namaane, A., and M'sirdi, N.K. 2013. Optimization of hybrid renewable energy systems (HRES) using PSO for cost reduction. *Energy Procedia* 42:318–327.

67. Chitsaz, A., Mahmoudi, S.M.S., and Rosen, M.A. 2015. Greenhouse gas emission and exergy analyses of an integrated trigeneration system driven by a solid oxide fuel cell. *Applied Thermal Engineering* 86:81–90.

68. Kazempoor, P., Dorer, V., and Weber, A. 2011. Modelling and evaluation of building integrated SOFC systems. *International Journal of Hydrogen Energy* 36(20):13241–13249.

69. Milewski, J. 2011. SOFC hybrid system optimization using an advanced model of fuel cell. *Sustainable Research and Innovation Proceedings* 3:1–9.

70. Wen, H., Ordonez, J.C., and Vargas, J.V.C. 2011. Single solid oxide fuel cell modeling and optimization. *Journal of Power Sources* 196(18):7519–7532.

71. Chen, J.M.P., and Ni, M. 2014. Economic analysis of a solid oxide fuel cell cogeneration/trigeneration system for hotels in Hong Kong. *Energy and Buildings* 75:160–169.

6 Fuel Cell Technology Commercialization

6.1 POTENTIAL ROLE OF FUEL CELLS IN GREEN ENERGY ECONOMY

For many years, fossil fuels, internal combustion (IC) engines, and boilers have been the main components of the traditional energy business chain. For instance, the U.S. Energy Information Administration (EIA) reported that the consumption of fossil fuels, including petroleum, natural gas, and coal, grew by 4% in 2018 from 2017 and accounted for 80% of US total energy consumption, as shown in Figure 6.1. The primary energy consumption in the US reached a record high of 101.3 quadrillion British thermal units (Btu) in 2018. The increase in 2018 has been the largest increase in energy consumption, in both absolute and percentage terms, since 2010 [1]. On the other hand, China is presently facing its energy crisis because of the industrial needs and population growth. For example, coal consumption in China accounted for 67.7% and 66% of the world's coal consumption in 2004 and 2013, respectively [2]. However, as the world population and industrial requirements grow, traditional energy sources cannot be able to meet the increasing demands or sustainable energy targets. Hence, many countries have invested in green industries and the low-carbon economy as a long-term objective to ensure energy supply can align with their sustainable environment targets and economic development [3–5].

The integration of renewable energy sources such as solar energy, tidal energy, wind energy, and bioenergy has provided new business opportunities for the energy

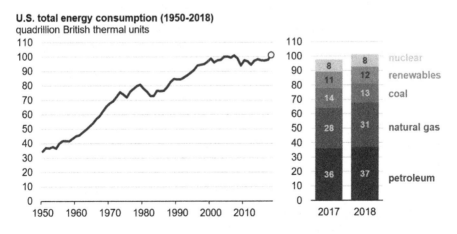

FIGURE 6.1 US total energy consumption. (Reprinted with permission from Ref. [1].)

industry and lead to a low-carbon green economy. For example, EIA reported that renewable energy consumption in the US reached a record high 11.5 quadrillion Btu in 2018, rising 3% from 2017, largely driven by adding new wind and solar power plants [1]. Although various types of renewable energy sources have been developed, every type of renewable energy needs to possess the same function and reliability of the traditional current fossil fuel model. However, the independent statistics and analyses revealed that there is still a great deal of room to invest in the development of renewable energy sources for energy saving and reducing emissions, which can be obtained by improving the efficiency of existing energy systems. Due to high efficiency in energy conversion, fuel cells have been considered as an interesting alternative to the existing energy systems and traditional engines [5]. Figure 6.2 illustrates the important role of fuel cell technology both in supplying renewable energies and in reducing emission and saving energy. Figure 6.2a shows a traditional chain of fossil fuels economy, while Figure 6.2b represents a new economic model with alternative green energy sources. There are two approaches through which fuel cells can play a key role in the renewable energy technology and economy: (1) new energy storage technology, as shown in Figure 6.2b, and (2) highly efficient engine technology for energy conversion and emission reduction, as shown in Figure 6.2a.

In general, solar, wind, and tidal energies are considered as the primary electricity generators in the business chain of renewable energy, which can be used for power supply ranging from driving cars to heating buildings. However, these renewable energy resources mainly depend on geological conditions. Their temporal and spatial

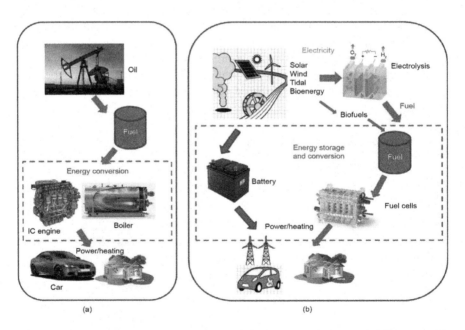

FIGURE 6.2 The energy economy. (a) Fossil fuels energy economy and (b) renewable energy economy. (Reprinted with permission from Ref. [5].)

distributions lead to a demand for the storage of energy, which has been a long-standing challenge, especially for large-scale applications. So far, various technologies such as researchable batteries [6,7], pumped-storage hydroelectricity [8], fuel cells [5,9,10], and phase-change materials [11] have been developed to address the energy storage issue for large-scale applications to be viable as an alternative to fossil fuels. Among the aforementioned technologies, batteries and fuel from electrolysis associated with fuel cells have been the greatest potential alternatives.

Rechargeable batteries such as Li-ion and Ni–Mn batteries have been widely used for various light-duty power sources, including mobile phones, laptops, uninterruptible power supply, etc. due to their convenience in storing energy. However, their remarkable disadvantages, such as long recharge times, aging, lower energy density than other fuels, e.g., hydrogen and methane, higher manufacturing cost, and environmental impacts, have hampered their usage in large-scale or heavy-duty applications [12]. These disadvantages of rechargeable batteries can also negatively affect their usage in the automotive industry. Charging batteries takes many hours at 240 V and various amperage depending on the size of the battery, while fuel cells bypass the charging issue by using energy-dense fuels such as hydrogen and methanol. A fuel cell vehicle (FCV) with sufficient hydrogen fuel can drive for 200–300 mi (322–482 km) and can refill in 3–5 min. On the other hand, conventional batteries have an average normal lifespan of four years and need to be replaced every few years, even if they are not used due to leaching and aging. In contrast, fuel cells degradation depends on the operation hours with the goal of 8,000 h or more operation (~150,000 mi). In addition, the cost of a battery pack is much higher than the cost of a fuel cell with a similar capacity [5,13]. Hence, battery-powered vehicles are mainly used for light-duty operations and short distances.

Hybrid vehicles (HVs) have also been developed to utilize a hybrid powertrain system consisting of an engine (i.e., IC engine or fuel cell) with a battery pack in one vehicle. The concept of hybrid fuel cell electric vehicles has been proposed to combine the features of an electric vehicle and an FCV, which can facilitate a long-range drive and a short refueling time using hybrid power [14–17]. In addition, in hybrid fuel cell electric vehicles, the battery can continue to drive if the fuel cell breaks down, which results in higher reliability of the HV. Although hybrid fuel cell electric vehicles with two sets of power systems in one vehicle have many advantages, the complexity of their control system, along with their higher cost compared to a single powertrain system, hampers their commercialization.

Although it is not easy to precisely predict the future of fuel cell technology in the green energy economy, this exciting technology is highly potential to bridge the gap between reduction in usage of fossil fuels and the rise of renewable energy due to its high efficiency and low emissions [18]. Over the past few years, the US, Japan, and Europe have invested more than 22 billion USD in research and development of fuel cell technology [19]. For instance, the number of patents granted to companies in Japan was 23 times higher in 2010 compared to those granted in 2000 [20]. Hence, the fuel cell technology can lead to environmentally friendly and low-carbon industry/economy if it can successfully be utilized as an alternative to fossil fuel energy sources and IC engines.

6.2 MAJOR BARRIERS FOR FUEL CELL TECHNOLOGY COMMERCIALIZATION

In general, there are three main challenges for the commercialization of any new product, they are quality, cost, and acceptance by end-users. Given the quality and the price of fuel cells being comparable to IC engines and gas turbines, fuel cells would be accepted by end-users and have no natural opposition as an environmentally friendly alternative technology to conventional energy sources. This has been reflected by an increase in public acceptance and large investments in fuel cell technology over the past few years [18]. However, the biggest barriers are still economic issues (e.g., hydrogen production and infrastructure economy as well as fuel cell costs) and technical issues (e.g., low reliability and durability) compared to the turbine and IC engines. Figure 6.3 shows a chart describing these two barriers to achieve the commercialization of fuel cells. One is a path for hydrogen production and infrastructure economy, and the other is the commercialization of fuel cell itself.

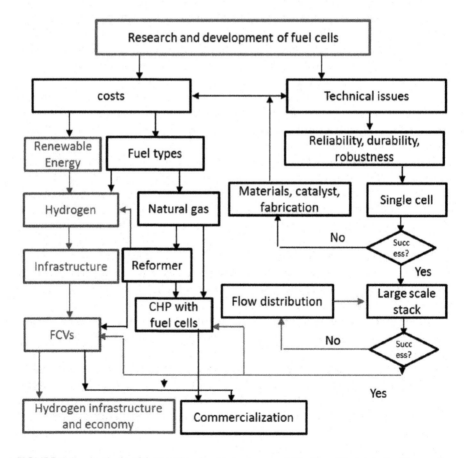

FIGURE 6.3 Analysis of fuel cell technology commercialization. (Reprinted with permission from Ref. [18].)

6.2.1 ECONOMIC CHALLENGES

6.2.1.1 Hydrogen Economy

Hydrogen economy includes hydrogen fueling infrastructure and fuel cell technology, which have a close relationship to each other. However, there are major challenges with hydrogen fueling infrastructure and fuel cell technology, turning them to post 2030 technologies [21–24]. The fuel cell commercialization initially created a chicken–egg relationship between fuel cell technology in vehicles and hydrogen fueling infrastructure, described as no hydrogen fueling infrastructure; thus, no hydrogen vehicles and vice versa. The automotive industries and research communities suggested that the cost of FCVs and hydrogen can be comparable to the current IC engine vehicles over time if governments support mass vehicle purchases and hydrogen fueling infrastructure deployment [25–33]. Although this would help to reduce the cost of massive manufacturing of hydrogen fuel cells to some extent, the technical issues and cost of the fuel cell itself are still needed to be addressed and could not be solved by a hydrogen fueling infrastructure. Over the past few decades, many studies have been conducted on economic analysis of a hydrogen fueling infrastructure and FCVs deployment, but none of them has technically analyzed the effect of hydrogen fueling infrastructure on fuel cell cost reduction and technical issues of fuel cells (i.e., reliability and durability) [18,31,34–38]. Similar to a petroleum infrastructure, a hydrogen fueling infrastructure is established to ensure reliable fuel supply to FCVs, not to reduce the fuel cell's costs and technical issues. For the market acceptance of fuel cells, it seems that the importance of the hydrogen fueling infrastructure on the costs of fuel cells has been overestimated since hydrogen is not the only energy source for fuel cells [e.g., natural gas can be used for solid oxide fuel cells (SOFCs)]. In addition, hydrogen does not affect the manufacturing costs and technical problems of fuel cells (i.e., it only relates to operational costs), and hydrogen fueling infrastructure can only ensure a reliable hydrogen supply to vehicles. Therefore, it can be concluded that the investments in the hydrogen fueling infrastructure cannot reduce the costs of the fuel cell itself and address its technical problems, including robustness, reliability, and durability [18,33]. In fact, it is needed to analyze the fuel cell costs independent of a hydrogen economy.

6.2.1.2 Fuel Cell Costs

The costs of a fuel cell-stack can be divided into three categories: (1) material and component costs; (2) labor costs including design, fabrication, and transport; and (3) capital costs of the manufacturing equipment [18,39]. It should be noted that the materials and component costs only depend on the market and technological breakthrough, while mass manufacturing can reduce labor and capital costs. To date, although fuel cells seem to be less expensive than IC engines due to the lack of moving parts, they are more expensive and less durable and reliable to some extent than the IC engines [40,41]. Elnozahy et al. [42] have carried out a comparative study on the costs of FCVs and IC engine vehicles. They found that the total cost of a proposed FCV is around 24,355 USD. In comparison, the total cost of an IC engine vehicle (Honda Civic Sedan) is around 15,805 USD, considering the materials, fuel, and manufacturing costs. However, the proposed FCV represents the efficiency of

60%–70%, which is remarkably higher than that of the IC engine vehicle (10%–16%). This causes a significant reduction in fuel cost for FCVs and total cost over a long period of operation due to higher thermal efficiency. It should be noted that U.S. Department of Energy (DOE) has currently reported 40%–60% thermal efficiency for fuel cell operations [43]. However, fuel cell efficiency can be even greater that 90% when configured for CHP [5].

Although the total cost of fuel cells, mostly resulted from the manufacturing costs, is still higher than that of the IC engines, the repair and maintenance costs are necessary for stack services, which should also be considered as a primary factor for end-user acceptance. The failure of a component in a fuel cell-stack can lead to failure of the whole stack, which can add 100% of the cost of assembly, conditioning, and balancing of the stack system. Yang [44] showed that the cost of assembly, conditioning, and balancing of the stack for an 80 kW proton exchange membrane fuel cell stack system could be around 22% of the whole stack system cost. Therefore, it is clear that the reliability and durability of a fuel cell system play a pivotal role in reducing the cost of the whole system and increasing end-user acceptance. In order to outperform conventional competitors such as IC engines, a comprehensive strategy is necessary to improve reliability and durability of fuel cells, which will be discussed in the following sections.

Regarding the SOFC systems, the U.S. DOE specific goal is to meet a stack cost target of 225 USD kW^{-1} and a system cost target of 900 USD kW^{-1} [45]. The cost breakdown for a 5 kW SOFC power plant is depicted in Figure 6.4. In addition, Figure 6.5 shows the stack cost volume trends for the 100 and 250 kW SOFC stacks. As can be seen in Figure 6.4, the stack represents 32% of the whole plant cost at production quantities of ≥50,000 units year^{-1}. The stack, along with fuel and air supply, is the most expensive part of the SOFC power plant. Moreover, the biggest portion of the SOFC stack cost is related to the materials and component costs, as shown in Figure 6.5. Therefore, a major path to lower the fuel cell cost is to develop cheaper materials with higher stability within the fuel cell operating temperature range.

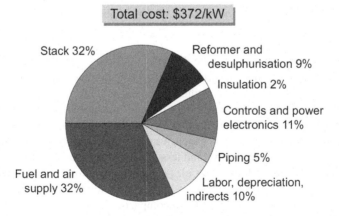

FIGURE 6.4   Cost breakdown for a 5 kW SOFC power plant. (Reprinted with permission from Ref. [46].)

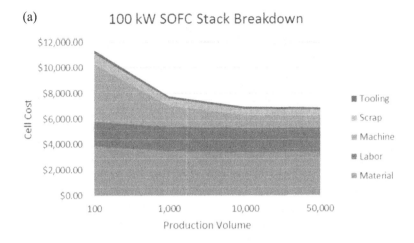

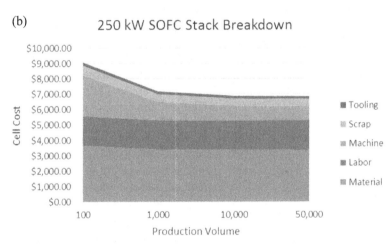

FIGURE 6.5 The stack cost volume trends for (a) 100 kW SOFC stack and (b) 250 kW SOFC stack. (Reprinted with permission from Ref. [47].)

Hence, numerous studies have been directed to solve material issues. However, there has been not sufficient progress toward the ultimate goal of marketability, which results in incremental costs of research and development of fuel cells [18]. For an SOFC system, less expensive materials, high power density, mass-reproducible system design, and high-volume markets are the key criteria that should be met to compete with today's competitive power market.

6.2.2 Technical Challenges: Durability and Reliability

The U.S. DOE has established specific technical targets for fuel cell durability. The 2020 technical targets of proton exchange membrane fuel cells for transportation application is to meet 5,000 h durability with cycling. The U.S. DOE has also developed the technical targets of SOF-CHP systems for stationary applications with an operating lifetime of 60,000 and 80,000 h for 1–25 kW residential and light commercial systems and 100 kW–3 MW systems, respectively, by 2020 [48,49]. Durability target, which is a lifetime within the repaired rate and cost in the planned maintenance, is not sufficient as the only target for end-users acceptance. Reliability of fuel cells representing fuel cell-stack performance without maintenance along with their availability may play a more important role for end-users acceptance since an unexpected repair and maintenance can cause delays in the fuel cell operation for different applications. Durability and reliability indicate a higher availability, meaning fewer repairs and maintenance as well as a warranty of use for end-users. Hence, the major technical challenge for fuel cell marketing and scaling-up is not only low durability but also low reliability, which should be taken into consideration as a technical target.

Addressing the reliability issue of fuel cells is essential to lower the cost and improve the availability of fuel cells. The repair and maintenance costs due to unexpected defects and failure could be high [10]. The failure is mainly caused by accumulated degradation of materials and cell components due to poor water and thermal management, corrosion and chemical reactions, fuel and oxidant starvation, etc. It has been shown that optimization of flow channel design and controlling the flow conditions can significantly improve the water and thermal managements and extend the fuel cells lifetime without any other alterations (i.e., materials or catalysts) [18,50–52]. The channels in a cell, as well as cells in a stack, need to operate under the same conditions.

A general procedure for the development of fuel cells, proposed by Wang [18] and shown in Figure 6.6, is to first solve any problems associated with materials, catalysts, or sealing on a small scale. It is clear that there are no significant problems of durability and reliability such as degradation and water flooding for a successful

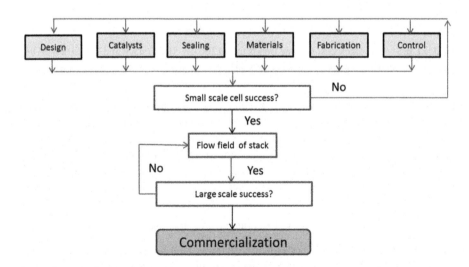

FIGURE 6.6   Flow chart of fuel cell scaling-up. (Reprinted with permission from Ref. [18].)

small-scale fuel cell. Thereafter, the fuel cell can enter the next stage, which is scaling up where the identical individual cells are stacked together through the designed flow fields. The challenge here is to keep flow rate and pressure drop uniform for all the cells in stack [53,54]. Although outward appearances of failures are materials, catalysts, thermal, and water issues, addressing the issue of uneven flow distribution within the stack can significantly solve the technical problems of scaling-up (i.e., durability, reliability, and robustness) toward the commercialization of fuel cells. On the other hand, a uniform flow distribution can also improve fuel cell efficiency in scaling-up to some extent. However, the improvement of fuel cell efficiency may not be a primary goal since fuel cells are already much more efficient than their conventional competitors such as IC engines. Therefore, the effective design of flow fields has been the greatest challenge for fuel cell performance improvement and cost reduction. In order to assess the deployment of the fuel cell technologies, it is required to carefully examine the framework and scaling-up approach with respect to the operation and potential risks associated with the fuel cell systems. The durability and reliability of technical barriers are the critical barriers for fuel cell commercialization.

6.3   FUTURE STRATEGIES FOR FUEL CELL TECHNOLOGY COMMERCIALIZATION

It is clear that fuel cells have a significant market potential as a substitute for their conventional counterparts due to their higher efficiency in comparison with IC engines, gas turbines, and boilers. According to the U.S. DOE and other reports, many fuel cell systems have been installed and operated so far. For example, since 2009, a significant increase in the deployment of fuel cell backup systems has been reported in the US [55]. Over 7000 fuel cell systems with the total power of 16.3 MW and 2000 backup power systems used for telecommunication systems have been installed. In Connecticut, a 400 kW fuel cell has been installed at CTtransit to supply power for the transit agency's maintenance and storage facility [56]. In California, a 2.8 MW fuel cell system was installed for a wastewater treatment plant. However, building commercial markets for the deployment of fuel cells still strongly depends on government subsidies. The prerequisite for commercialization and massive production of fuel cells would still be their cost, durability, and reliability, which are not comparable to those of IC engines, boiling heating systems, and gas turbines. Therefore, the first step on the way of fuel cell commercialization and widespread use of fuel cell technology is to solve their cost, durability, and reliability issues. In addition, the second step is to develop a new business model and define standards for the industry [5,18].

Multidisciplinary collaboration is the best way to solve the cost, durability, and reliability issues under a proper theoretical guideline since the conventional trial and error methods without a theoretical guide cannot be sufficient to improve the fuel cell performance and reduce the cost. The massive production of fuel cells is less related to this step. In this first step, a systematic approach is to design an efficient flow field to properly improve the performance and link materials, shape, configuration, and flow conditions. This theory can direct us to more informative experiments and modeling to find the combinations of flow field design with materials and fabrication techniques to achieve the targeted fuel cell with higher performance and lower cost [18].

As the second step, it is necessary to explore larger-scale production and distribution and develop a new business model and standardization after a successful scaling-up for further cost reduction and marketing. Once cost, durability, and reliability targets are achieved, the fuel cell technologies can be adopted by the public and become economically viable for common domestic applications such as house heating with CHP and backup power [57,58]. Moreover, fuel cells possess higher efficiency and low noise in comparison with conventional engines such as gas turbines and IC engines and do not depend on weather conditions unlike solar and wind energy systems. These advantages can result in lower fuel cost and carbon footprints, which reduce any public, government, and marketplace resistances and accelerate the adoption of fuel cell products. A new energy revolution with the massive adoption of fuel cell systems in the markets will be remarkably effective in reducing global warming and energy security. The maturation of fuel cell technology can strongly impact the hydrogen economy development and further increase the confidence in the use of hydrogen [18]. It can be expected that new advancements in fuel cell technology will accelerate fuel cell commercialization in the near future.

6.4 RECENT SOFC TECHNOLOGY STATUS IN THE US, JAPAN, CHINA, AND EUROPE

The SOFC technology has progressed to its highest level of demonstration and is actively being developed by the number of developers, research institutes, and stakeholders in the US, Japan, China, India, the European Union (Great Britain, Italy, Germany, Switzerland, Denmark, France, etc.), Australia, Singapore, South Korea, Russia, and other countries [59]. The SOFC technology has shown its longevity, applicability, and versatility in different applications and entered the various market segments in the US, Europe, and Asia. Although the market capacity is fairly large, the commercialization of SOFC technology still requires a magnitude of investments and further cost reduction and performance improvement. More information will be provided in Chapter 7.

In the US, DOE Office of Fossil Energy and FE's National Energy Technology have been developing low-cost and efficient SOFC-based power systems under their SOFC program that produce electric power from coal or natural gas with intrinsic carbon capture capabilities for power generation applications ranging from small and medium distributed generation to large central-station generation. The current SOFC program includes cell development as well as core technology research to improve durability, reliability, and robustness of cell, stack, and system technology. To date, small natural gas-fueled systems for distributed generation applications are commercially available. These SOFC-based distributed generation systems using conventional heat-power technologies, principally engines and gas turbines, are able to run at significantly high efficiency levels, ranging from 45% to 60% with regards to the system configuration and can serve a range of electric capacity requirements. The current voltage degradation rates for cell and stack are less than 1% per 1,000 h [60,61].

Fuel Cell Energy and LG Fuel Cell Systems have developed two natural gas-fueled SOFC power systems, 50 and 200 kW, respectively, as shown in Figure 6.7. The atmospheric pressure 50 kW SOFC system built and tested by Fuel Cell Energy has possessed 55% electrical efficiency [net AC/higher heating value (HHV)] with a voltage degradation rate of 0.9% per 1,000 h over 1,500 h of operation. This system was configured for all-electric duty in the proof-of-concept demonstration but can be equipped for the SOFC exhaust heat recovery and work in the CHP mode. The 200 kW pressurized SOFC system built and operated by LG and integrated into the grid has delivered 57% electrical efficiency (net AC/HHV). In addition, LG built and tested a 250 kW pressurized SOFC system at Stark College campus in North Canton, OH, as shown in Figure 6.8, which operated for 1,300 h with 55% electrical efficiency (net AC/HHV). In 2019, a 200 kW SOFC system built by Fuel Cell Energy was successfully installed and operated at NRG Energy Center in Pittsburgh, PA, as shown in Figure 6.9. So far, the system has operated for over 1,500 h. Fuel Cell Energy is also developing a kW-scale solid oxide electrolyzer cell (SOEC) system with an overall efficiency of over 75% and hydrogen production capability of 4 kg day^{-1} [59–61].

As of today, Bloom Energy's SOFC technology has provided over 350 MW of power plants to hundreds of customers across the world such as Home Depot, IKEA,

FuelCell Energy 50 kWe POC
- Atmospheric-pressure
- ~50 kWe AC to grid
- Efficiency = 55% (net AC/HHV)
- Degradation rate = 0.9%/1000 hrs
- 1,500 hrs operation
- Overall dimensions:
  4.4m(l) x 2.1m(w) x 3.1m(h)
- TRL 6

Photo Courtesy FuelCell Energy

LG 200 kWe POC
- Pressure = 5 bara
- ~200 kWe AC to grid
- Efficiency = ~57% (net AC/HHV)
- 2,000 hrs operation
- TRL 6

Photo Courtesy LG Fuel Cell Systems

FIGURE 6.7 Proof-of-concept SOFC power systems. (Reprinted with permission from Ref. [61].)

FIGURE 6.8 The 250 kW SOFC power system built by LG Fuel Cell Systems. The system was shut down after 1,300 h operation as LG decided to exit fuel cell business. (Reprinted with permission from Ref. [59].)

FIGURE 6.9 The 200 kW SOFC power system built by Fuel Cell Energy. The system is operating in NRG Energy Center in Pittsburgh, PA. (Reprinted with permission from Ref. [59].)

Macy's, Google, Nokia, eBay, Walmart, etc., as a result of an increase in demand for reliable and clean energy/electricity due to increasing electricity costs and aging grid infrastructure [59].

As an SOFC stacks and systems developer and manufacturer, WATT Fuel Cell Corporation has built a ceramic-based microtubular SOFC that operates on common, readily available commercial hydrocarbon fuels such as propane, natural gas, JP-8, and diesel at high temperatures. In order to fabricate the fuel cell tube, WATT has developed an additive manufacturing technique using automated printing processes.

In 2015, WATT Fuel Cell Corporation developed a 1 kW hybrid SOFC system which is portable, efficient, quiet, and lightweight (<13.6 kg) and has the capability to use hydrocarbon fuels in conjunction with wind or solar power technologies to optimize power output, energy storage, and fuel consumption [62].

Atrex Energy, as a commercial supplier, has been developing highly efficient, high power density, and reliable SOFCs for defense applications for more than eight years. So far, the company has delivered over 650 fuel cell systems in various remote power markets using natural gas or LPG as fuels, collectively accumulating over 6 million hours of lifetime operation. Recently, Atrex Energy successfully demonstrated an advanced 1,500 W SOFC power system for U.S. Department of Defense (DOD), operating on desulfurized JP-8 fuel with an electrical efficiency of ~40% [net DC/lower heating value (LHV)] [63].

In Japan, a demonstrative research project was carried out by the New Energy Foundation of Japan and New Energy and Industrial Technology Development Organization (NEDO) from 2007 to 2011 [64]. In this project, 233 units of SOFC systems were installed, operated, and evaluated in various sites. The results revealed that residential SOFC systems could significantly lower energy consumption and CO_2 emissions in households. JX Nippon Oil & Energy Corporation and Aisin also released a residential SOFC CHP system in 2011 and 2012. Kyocera Corporation manufactured tubular cell-stacks, operating at 700°C–750°C with 46.5% power generation efficiency (LHV) and 90% overall energy efficiency (LHV) [65]. Simulated by Kyocera Corporation's success, a number of ceramic manufactures in Japan started to develop SOFC cell-stacks for residential applications.

Fuel cell technologies are strongly supported by the government and already put into practical use in both mobile and residential applications in Japan. NEDO SOFC project "Technology Development for Promoting SOFC Commercialization" started since 2013 is focusing on developing high efficient ad tough SOFC stacks which increase the SOFC usage in the market. In a new project started in 2018, the durability of cell-stacks high target has been being developed. A new 2020 SOFC project is also expected to focus on metal-supported SOFCs [59,66]. Based on the results of NEDO project, Miura Co., Ltd., one of the major commercial and industrial boiler companies, has developed a 4.2 kW SOFC cogeneration system for commercial use, which achieved 48% electrical efficiency and 90% overall energy efficiency [67]. In addition, Hitachi Zosen Corporation has installed a SOFC for a demonstration test at the Sakuya Konohana Kan located inside the Flower Expo Memorial Park Tsurumi Ryokuchi in Osaka. The test conducted in 2018 achieved a 52% power generation efficiency, which is one of the highest in industry. To evaluate safety, durability, and reliability, the SOFC system will operate continuously for 4,000 h under actual operating conditions [68].

Mitsubishi Hitachi Power Systems (MHPS) is also developing a 250 kW hybrid-SOFC cogeneration system consisting of efficient pressurized SOFC cell-stacks and a micro gas turbine driven by residual reformed gas from the SOFC anode. The system enables to provide high power generation efficiency of 55% (LHV). TOTO has developed tubular cell-stacks with a thin lanthanum gallate electrolyte, operating at low temperatures. NGK Spark Plug has developed planar type cell-stacks with high power density. The company also supplies cell-stacks to MHPS [59,66].

In China, there are hundreds of universities and research and development (R&D) organizations focusing on energy and fuel cell research. In addition, the government has established many plans to promote new energy development [2]. In particular, in 2016, the Chinese government has issued seven key targets within the *13th Renewable Energy Development Five-Year Plan* toward renewable energy deployment until 2020 [69]. China has also dedicated significant resources for SOFC technology development. For example, CNCF, Suzhou Huatsing Power, and SOFCMAN Energy have significantly developed the SOFC and SOEC technologies during the last decade. CNCF has scaled up the technologies in Shanghai and Wuxi to build CNCF Mars 30 kW SOFC and CNCF Jupiter 80 kW SOEC, respectively. Suzhou Huatsing Power, one of the biggest and leading SOFC developers in China, has cooperated with universities on 973 national projects since 2010. Suzhou Huatsing Power is currently supplying 5 kW SOFC stacks to packagers for integration throughout China and testing a 10 kW SOFC stack in Dongguan. The company also plans for 2020 to build and test a 20 kW SOFC stack as a part of the renewable energy project in Xuzhou. SOFCMAN, located in Ningbo City, is manufacturing a 1,300–2,200 W anode-supported SOFC, operating at 750°C with 94.3% fuel utilization and 72.5% electrical efficiency (LHV) [59].

In Europe, many manufacturers are also developing their original SOFCs for mobile and residential applications. Ceres Power in the UK has developed a low-cost metal-supported cell operating at 500°C–600°C. The company has signed a Joint Development Agreement with many worldwide partnerships to expand its business and develop stacks using Ceres Power's metal-supported Steel Cell technology for different power equipment applications. The partners are Weichai (30 kW, battery EV range extender for electric bus), Doosan (5–20 kW, stationary power), Honda (stationary power), Bosch (10 kW, data centers), Cummins (10 kW, data centers), and Miura (4.2 kW, CHP 50% LHV). Ceres Power has recently announced its latest 5 kW SOFC stack and V5 Steel Cell® technology with higher power density, shorter start-up times, reduced vibration, higher efficiency, and lower degradation. Ceres Power has a 2 MW year^{-1} manufacturing facility in the UK [59,70,71].

SOLIDpower in Italy is the manufacturer of BlueGEN, the most efficient micro-CHP appliance in the world with a 1.5 kW power unit and 60% electrical efficiency. It has already tested for data centers application and development of new commercial SOFC technology. So far, several BlueGEN units tested at Microsoft Stark Data Center, and one unit with a net electrical efficiency of 59.6% tested at Brookhaven National Laboratory's Sustainable Energy Tech Laboratory. SOLIDpower has also developed the anode-supported cell-stacks, called EnGEN, with a 2.5 kW power unit and 90% overall efficiency that supply clean electricity and heat in the form of hot water. They are operating the system under a project funded by the European Commission's Fuel Cells and Hydrogen Joint Undertaking [59,66].

Sunfire has developed the PowerCore technology, an electrolyte solid oxide cell-stack, to manufacture high-temperature fuel cells and steam SOECs. Their system work in either a 150 kW SOEC mode with an efficiency of 82% or a 30 kW SOFC mode with 625 mohm cm^2 ASR. Elcogen has been focusing on developing the 1,000 and 3,000 W anode-supported SOFC cell-stacks, operating at 650°C with an efficiency of about 70%. Elcogen recently received a $12 million loan to manufacture a facility in Estonia. HEXIS/Viessmann, as a center of expertise for research, is developing and producing the next-generation 1.5 kW SOFC systems with 40% electrical efficiency and 93% overall efficiency. Their ~40,000 h testing is probably the longest test ever operated [59].

REFERENCES

1. U.S. Energy Information Administration. 2020. Monthly energy review. Report. Washington, DC: Office of Energy Statics, U.S. Department of Energy, Report No.: DOE/EIA-0035 (2020/2). https://www.eia.gov/todayinenergy/detail.php?id=39092.
2. Lu, Y., Cai, Y., Souamy, L., Song, X., Zhang, L., and Wang, J. 2018. Solid oxide fuel cell technology for sustainable development in China: an over-view. *International Journal of Hydrogen Energy* 43(28):12870–12891.
3. Jacobsson, S., and Lauber, V. 2006. The politics and policy of energy system transformation-explaining the German diffusion of renewable energy technology. *Energy Policy* 34(3):256–276.
4. Peidong, Z., Yanli, Y., Yonghong, Z., Lisheng, W., and Xinrong, L. 2009. Opportunities and challenges for renewable energy policy in China. *Renewable and Sustainable Energy Reviews* 13(2):439–449.
5. Wang, J., Wang, H., and Fan, Y. 2018. Techno-economic challenges of fuel cell commercialization. *Engineering* 4(3):352–360.
6. Goodenough, J.B., and Park, K.S. 2013. The Li-ion rechargeable battery: a perspective. *Journal of the American Chemical Society* 135(4):1167–1176.
7. Li, Y., Yang, J., and Song, J. 2017. Design structure model and renewable energy technology for rechargeable battery towards greener and more sustainable electric vehicle. *Renewable and Sustainable Energy Reviews* 74:19–25.
8. Jiang, R., Wang, J., and Guan, Y. 2011. Robust unit commitment with wind power and pumped storage hydro. *IEEE Transactions on Power Systems* 27(2):800–810.
9. Wang, J. 2015. Theory and practice of flow field designs for fuel cell scaling-up: a critical review. *Applied Energy* 157:640–663.
10. Wang, J. 2017. System integration, durability and reliability of fuel cells: challenges and solutions. *Applied Energy* 189:460–479.
11. Ramakrishnan, S., Wang, X., Sanjayan, J., and Wilson, J. 2017. Thermal performance of buildings integrated with phase change materials to reduce heat stress risks during extreme heatwave events. *Applied Energy* 194:410–421.
12. Peters, J.F., Baumann, M., Zimmermann, B., Braun, J., and Weil, M. 2017. The environmental impact of Li-Ion batteries and the role of key parameters–a review. *Renewable and Sustainable Energy Reviews* 67:491–506.
13. Drive US. 2017. *Fuel Cell Technical Team Roadmap.* US DRIVE: Southfield, MI.
14. Emadi, A., Rajashekara, K., Williamson, S.S., and Lukic, S.M. 2005. Topological overview of hybrid electric and fuel cell vehicular power system architectures and configurations. *IEEE Transactions on Vehicular Technology* 54(3):763–770.
15. Rahman, K.M., Patel, N.R., Ward, T.G., Nagashima, J.M., Caricchi, F., and Crescimbini, F. 2006. Application of direct-drive wheel motor for fuel cell electric and hybrid electric vehicle propulsion system. *IEEE Transactions on Industry Applications* 42(5):1185–1192.
16. Lane, B., Shaffer, B., and Samuelsen, G.S. 2017. Plug-in fuel cell electric vehicles: a California case study. *International Journal of Hydrogen Energy* 42(20):14294–14300.
17. Brunet, J., and Ponssard, J.P. 2017. Policies and deployment for fuel cell electric vehicles an assessment of the Normandy project. *International Journal of Hydrogen Energy* 42(7):4276–4284.

18. Wang, J. 2015. Barriers of scaling-up fuel cells: cost, durability and reliability. *Energy* 80:509–521.
19. Behling, N. 2012. Solving the fuel cell dilemma. *Fuel Cells Bulletin* 2012(11):12–14.
20. Blanchette Jr, S. 2008. A hydrogen economy and its impact on the world as we know it. *Energy Policy* 36(2):522–530.
21. Haeseldonckx, D., and D'haeseleer, W. 2011. Concrete transition issues towards a fully-fledged use of hydrogen as an energy carrier: methodology and modelling. *International Journal of Hydrogen Energy* 36(8):4636–4652.
22. Bakker, S. 2010. The car industry and the blow-out of the hydrogen hype. *Energy Policy* 38(11):6540–6544.
23. Bossel, U. 2006. Does a hydrogen economy make sense? *Proceedings of the IEEE* 94(10):1826–1837.
24. Romm, J.J. 2004. *The Hype about Hydrogen: Fact and Fiction in the Race to Save the Climate.* Island Press: Washington, DC.
25. Agnolucci, P. 2007. Hydrogen infrastructure for the transport sector. *International Journal of Hydrogen Energy* 32(15):3526–3544.
26. Mercuri, R., Bauen, A., and Hart, D. 2002. Options for refuelling hydrogen fuel cell vehicles in Italy. *Journal of Power Sources* 106(1–2):353–363.
27. Oi, T., and Wada, K. 2004. Feasibility study on hydrogen refueling infrastructure for fuel cell vehicles using the off-peak power in Japan. *International Journal of Hydrogen Energy* 29(4):347–354.
28. Baufume, S., Grüger, F., Grube, T., Krieg, D., Linssen, J., Weber, M., Hake, J.F., and Stolten, D. 2013. GIS-based scenario calculations for a nationwide German hydrogen pipeline infrastructure. *International Journal of Hydrogen Energy* 38(10):3813–3829.
30. Stephens-Romero, S.D., Brown, T.M., Kang, J.E., Recker, W.W., and Samuelsen, G.S. 2010. Systematic planning to optimize investments in hydrogen infrastructure deployment. *International Journal of Hydrogen Energy* 35(10):4652–4667.
31. Yang, C., and Ogden, J.M. 2013. Renewable and low carbon hydrogen for California–modeling the long term evolution of fuel infrastructure using a quasi-spatial TIMES model. *International Journal of Hydrogen Energy* 38(11):4250–4265.
32. Hardman, S., and Steinberger-Wilckens, R. 2014. Mobile phone infrastructure development: lessons for the development of a hydrogen infrastructure. *International Journal of Hydrogen Energy* 39(16):8185–8193.
33. Stolzenburg, K., Tsatsami, V., and Grubel, H. 2009. Lessons learned from infrastructure operation in the CUTE project. *International Journal of Hydrogen Energy* 34(16):7114–7124.
34. Chiesa, P., Consonni, S., Kreutz, T., and Williams, R. 2005. Co-production of hydrogen, electricity and CO_2 from coal with commercially ready technology. Part A: performance and emissions. *International Journal of Hydrogen Energy* 30(7):747–767.
35. Feng, W., Wang, S., Ni, W., and Chen, C. 2004. The future of hydrogen infrastructure for fuel cell vehicles in China and a case of application in Beijing. *International Journal of Hydrogen Energy* 29(4):355–367.
36. Hua, T., Ahluwalia, R., Eudy, L., Singer, G., Jermer, B., Asselin-Miller, N., Wessel, S., Patterson, T., and Marcinkoski, J. 2014. Status of hydrogen fuel cell electric buses worldwide. *Journal of Power Sources* 269:975–993.
37. Mayer, T., Kreyenberg, D., Wind, J., and Braun, F. 2012. Feasibility study of 2020 target costs for PEM fuel cells and lithium-ion batteries: a two-factor experience curve approach. *International Journal of Hydrogen Energy* 37(19):14463–14474.
38. Ogden, J.M. 1999. Developing an infrastructure for hydrogen vehicles: a Southern California case study. *International Journal of Hydrogen Energy* 24(8):709–730.

39. Marcinkoski, J., James, B.D., Kalinoski, J.A., Podolski, W., Benjamin, T., and Kopasz, J. 2011. Manufacturing process assumptions used in fuel cell system cost analyses. *Journal of Power Sources* 196(12):5282–5292.

40 Odeh, A.O., Osifo, P., and Noemagus, H. 2013. Chitosan: a low cost material for the production of membrane for use in PEMFC-A review. *Energy sources, Part A: Recovery, Utilization, and Environmental Effects* 35(2):152–163.

41. Sun, Y., Delucchi, M., and Ogden, J. 2011. The impact of widespread deployment of fuel cell vehicles on platinum demand and price. *International Journal of Hydrogen Energy* 36(17):11116–11127.

42. Elnozahy, A., Rahman, A.K.A., Ali, A.H.H., and Abdel-Salam, M. 2014. A cost comparison between fuel cell, hybrid and conventional vehicles. In *Proceedings of the 16th International Middle-East Power Systems Conference—MEPCON* 2014:23–25, Cairo, Egypt.

43. Curtin, S., and Gangi, J. 2016. State of the states: fuel cells in America. Report. Washington, DC: Energy Efficiency, Renewable Energy's Fuel Cell Technologies Office, U.S. Department of Energy, 7th edition, Report No.: DOE/EE–1493.

44. Yang, Y. 2015. *PEM Fuel Cell System Manufacturing Cost Analysis for Automotive Applications*. Austin Power Engineering LLC: Wellesley, MA.

45. Office of Fossil Energy. Solid oxide fuel cells. https://www.energy.gov/fe/science-innovation/clean-coal-research/solid-oxide-fuel-cells.

46. Collins, D., Williams, M.C., and Surdoval, W. 2007. Application of power electronics with the US DOE distributed generation programme. *International Journal of Energy Technology and Policy* 5(2):163.

47. Battelle Memorial Institute. 2016. Manufacturing cost analysis of 100 and 250 kW fuel cell systems for primary power and combined heat and power applications. Prepared for U.S. Department of Energy.

48. Energy Efficiency & Renewable Energy. DOE technical targets for polymer electrolyte membrane fuel cell components. https://www.energy.gov/eere/fuelcells/doe-technical-targets-polymer-electrolyte-membrane-fuel-cell-components.

49. Energy Efficiency & Renewable Energy. DOE technical targets for fuel cell systems for stationary (combined heat and power) applications. https://www.energy.gov/eere/fuelcells/doe-technical-targets-fuel-cell-systems-stationary-combined-heat-and-power.

50. Jang, J.H., Chiu, H.C., Yan, W.M., and Sun, W.L. 2008. Effects of operating conditions on the performances of individual cell and stack of PEM fuel cell. *Journal of Power Sources* 180(1):476–483.

51. Knights, S.D., Colbow, K.M., St-Pierre, J., and Wilkinson, D.P. 2004. Aging mechanisms and lifetime of PEFC and DMFC. *Journal of Power Sources* 127(1–2):127–134.

52. Pei, P., Yuan, X., Gou, J., and Li, P. 2009. Dynamic response during PEM fuel cell loading-up. *Materials* 2(3):734–748.

53. Wang, J., and Wang, H. 2012. Flow-field designs of bipolar plates in PEM fuel cells: theory and applications. *Fuel Cells* 12(6):989–1003.

54. Wang, J., and Wang, H. 2012. Discrete approach for flow field designs of parallel channel configurations in fuel cells. *International Journal of Hydrogen Energy* 37(14):10881–10897.

55. Wei, M., Chan, S.H., Mayyas, A., and Lipman, T. 2016. Deployment and capacity trends for stationary fuel cell systems in the USA. In *Fuel Cells: Data, Facts, and Figures*, Edited by Stolten, D., Samsun, R.C., Garland, N., Wiley: Weinheim, Germany, 257–269.

56. Jenn, A., Azevedo, I.M., and Michalek, J.J. 2016. Alternative fuel vehicle adoption increases fleet gasoline consumption and greenhouse gas emissions under United States corporate average fuel economy policy and greenhouse gas emissions standards. *Environmental Science & Technology* 50(5):2165–2174.

57. Staffell, I., Green, R., and Kendall, K. 2008. Cost targets for domestic fuel cell CHP. *Journal of Power Sources* 181(2):339–349.
58. Staffell, I., and Green, R. 2013. The cost of domestic fuel cell micro-CHP systems. *International Journal of Hydrogen Energy* 38(2):1088–1102.
59. Williams, M.C., Vora, S.D., and Jesionowski, G. 2020. Worldwide status of solid oxide fuel cell technology. *ECS Transactions* 96(1):1–10.
60. Vora, S.D., Lundberg, W.L., and Pierre, J.F. 2017. Overview of US department of energy office of fossil energy's solid oxide fuel cell program. *ECS Transactions* 78(1):3–19.
61. Vora, S.D., Jesionowski, G., and Williams, M.C. 2019. Overview of US department of energy office of fossil energy's solid oxide fuel cell program for FY2019. *ECS Transactions* 91(1):27–39.
62. Watt Fuel Cell. 1kW portable, propane powered, hybrid SOFC system, which includes power management capabilities. https://www.wattfuelcell.com/about-us/.
63. Fuel Cells Works. 2019. Atrex energy successfully demonstrated 1.5 kW JP-8 Solid Oxide Fuel Cell (SOFC) power source to DOD. https://fuelcellsworks.com/news/atrex-energy-successfully-demonstrated-1-5kw-jp-8-solid-oxide-fuel-cell-sofc-power-source-to-dod/.
64. Horiuchi, K. 2013. Current status of national SOFC projects in Japan. *ECS Transactions* 57(1):3–10.
65. Kyocera. 2012. Osaka Gas, Aisin, Kyocera, Chofu and Toyota announce completion of world-class efficiency residential-use Solid Oxide Fuel Cell (SOFC) cogeneration system co-development and commercialization of "ENE-FARM Type S". https://global.kyocera.com/news-archive/2012/0305_woec.html.
66. IGU. SOFC cell-stacks. https://www.igu.org/research/fuel-cell-break-through.
67. NEDO. 4.2 kW SOFC cogeneration system. https://www.nedo.go.jp/english/news/AA5en_100255.html.
68. NEDO. SOFC system, designed for commercial and industrial use. https://www.nedo.go.jp/english/news/AA5en_100378.html.
69. IEA. 2018. China 13th renewable energy development five year plan (2016-2020). https://www.iea.org/policies/6277-china-13th-renewable-energy-development-five-year-plan-2016-2020?page=4&sector=Multi-sector.
70. Leah, R., Bone, A., Lankin, M., Selcuk, A., Pierce, R., Rees, L., Corcoran, D., Muhl, P., Dehaney-Steven, Z., Brackenbury, C., and Selby, M. 2013. Low-cost, REDOX-stable, low-temperature SOFC developed by Ceres Power for multiple applications: latest development update. *ECS Transactions* 57(1):461–470.
71. Leah, R.T., Bone, A., Lankin, M., Selcuk, A., Rahman, M., Clare, A., Rees, L., Phillip, S., Mukerjee, S., and Selby, M. 2015. Ceres power steel cell technology: rapid progress towards a truly commercially viable SOFC. *ECS Transactions* 68(1):95–107.

7 Research, Demonstration, and Commercialization Activities in the US, Europe, and Asia

7.1 REGIONAL FUEL CELL MARKET ANALYSIS

The Asia-Pacific region holds the largest share in the global solid oxide fuel cell (SOFC) market, and the market revenue from Asia Pacific was valued at 448.13 million USD in 2018, accounting for 58.3% of the total market share [1,2][1]. Under the Ene-Farm and NEDO sponsorship programs, the Japanese government is pushing the development to make the 2020 Tokyo Olympics a showpiece for its hydrogen and fuel cell strategy. Due to the COVID-19, the 2020 Tokyo Olympics have been pushed for the summer of 2021. In Japan, besides significant investment in SOFC development in more than 30 years, the Fukushima nuclear accident in 2011 led to a huge shift in primary energy consumption away from nuclear power, and thus fuel cell systems gained a prominent role to play in Japan's recent national energy plan. The public-funded programs (e.g., Callux in Germany and the Ene-Farm in Japan) played a significant role in the early adoption of SOFC. North America, with the US as the lead, is among the most important markets for SOFC with market size valuing at 186.35 million USD in 2018, and possible growth over 10% from 2019 to 2025. The SOFC market by the region is presented in Tables 7.1 and 7.2.

7.1.1 NORTH AMERICA

The power generation application is the largest application in the SOFC market and holds over 65% of the global market by value. The flexibility of the solid oxide platform to run with different fuels provides a huge advantage over other types of the fuel cell, and most units today use natural gas fuels for ultra-clean power generation. The system is also capable of supplying waste heat for onsite combined heat and power applications. These systems are also able to operate on renewable biogas, and most of these smaller size units of 1–2 kW are deployed in Japan. Another significant benefit is that the stacks can also be deployed in electrolysis systems, where they efficiently split water to produce hydrogen, and this presents the latest trend in fuel cell technology benefits. Solid oxide electrolysis is capable of

---

[1] Market research engine, solid oxide fuel cell (SOFC) market size, by Product (Planar SOFC, Tubular SOFC), by application (portable power generation, stationary power generation, transportation), Regional Analysis – Global Forecast 2019–2025.

TABLE 7.1

SOFC Market Revenue for Planar SOFC, by Region, 2018–2025 (USD Million)

Region	2018	2019	2020	2021	2022	2023	2024	2025	CAGR (19–25) (%)
North America	140.17	155.96	173.97	194.29	217.30	243.71	274.03	308.58	12.0
Europe	94.61	104.96	116.73	129.98	144.95	162.08	181.71	204.02	11.7
Asia Pacific	340.68	380.43	425.88	477.36	535.82	603.11	680.60	769.20	12.5
Row	6.33	7.07	7.91	8.85	9.91	11.13	12.52	14.10	12.2
Total	581.80	648.42	724.49	810.48	907.99	1020.03	1148.85	1295.90	12.2

Source: Market Research Engine.

TABLE 7.2

SOFC Market Revenue for Tubular SOFC, by Region, 2018–2025 (USD Million)

Region	2018	2019	2020	2021	2022	2023	2024	2025	CAGR (19–25) (%)
North America	46.18	51.53	57.65	64.57	72.43	81.47	91.87	103.75	12.4
Europe	30.66	34.12	38.06	42.50	47.53	53.31	59.94	67.49	12.0
Asia Pacific	107.45	120.34	135.13	151.91	171.03	193.08	218.53	247.72	12.8
Row	2.04	2.29	2.57	2.88	3.24	3.65	4.11	4.64	12.5
Total	186.34	208.28	233.40	261.86	294.22	331.49	374.45	423.60	12.6

Source: Market Research Engine.

operating much more efficiently than conventional electrolysis systems, requiring less power to produce the same amount of hydrogen. The huge benefit for reducing the cost of the system is a dual-mode operation where the cells are capable of operating reversibly – alternating between electrolysis and fuel cell mode. Hydrogen produced by electrolysis can be stored and later used to produce power, resulting in an energy storage system where storage duration can easily be prolonged by simply adding water and hydrogen storage capacity. With companies like Bloom Energy and Fuel Cell Energy (FCE) leading SOFCs development, North America is among the developed market places for early commercialization and adoption of a large-scale SOFC. As of 2018, Bloom Energy has installed about 300 MW of units. The company has continuously increased the size of its systems during these last years, currently producing the servers: ES-5000, ES-5400, ES-5700, and ES-5710, generating 100, 105, 200, and 250 kWe, respectively. The heart of these servers is a built-up box, composed of 40 cells of 25 We each, with 1 kWe stack, labeled as 'Bloom Boxes', fueled with natural gas or biogas. Bloom Energy's incredible shrinking power module now delivers 250 kWe. Bloom Energy's electrons-as-a-service program was launched already in 2011, and in this model, customers buy the power from the fuel cells instead of buying the fuel cells themselves. The term of the deal is normally 15 years. The North American market is dominated by the US, with 77.5% share in 2018. State incentives and resurrection of the U.S. Federal Investment Tax Credit for the stationary fuel cells are among key industry trends in the North American market. Significant government subsidies and regulatory compliance for fuel cells will further stimulate industry growth. The programs in the US are divided into two categories; one that focuses on research and the next generation of SOFC development under U.S. Department of Energy (DOE), the Office of Fossil Energy's Solid Oxide Fuel Cell and the other on tax incentives and implementation and deployment of SOFC technology. Since interception in 1995, the program is focused making progress toward achieving its technical goals: improve the efficiency of SOFCs to 60% without incorporating carbon capture and sequestration, achieve a proven lifetime of 40,000 h or more, achieve less than 0.2% per 1,000 h of degradation rate, decrease the stack costs of SOFCs to less than 225 USD kW^{-1}, and decrease system SOFC costs to less than 900 USD kW^{-1}. The degradation presents the largest challenge, and no company sponsored under DOE has achieved a proven lifetime of 40,000 h or more with less than 0.2% per 1,000 h of degradation rate [3]. DOE actively seeks a commercial entity to work cooperatively on the design, fabrication, testing, and commercialization of a 1–10 MW SOFC power system. Systems with this capacity are a considerable step up from current designs and require a significant financial investment. However, a 1 MW system could be developed by combining ten 100 kW modules, which is within current capabilities.

FCE core technology developed MCFC systems but acquired Canadian Versa Power and further collaborative development of the SOFC technology. FCE has integrated the SOFC components into fuel cell stacks as part of an FCE project under the U.S. DOE Solid State Energy Conversion Alliance (SECA) program. FCE has contracts with the U.S. Navy and also sub-contracts with the U.S. Defense Advanced Research Projects Agency program for the development of high efficiency and fuel flexibility portable power applications, an example of an unmanned undersea vehicle. The U.S. Navy is evaluating the use of SOFC power for propulsion and ship power of unmanned submarine applications as the high efficiency, virtual lack of emissions, as well as the quiet operating nature, which is well suited for stealthy operations. Some of the units developed for military applications are built by Mesoscopic Devices Ltd. (acquired by Protonex in 2007); 'MesoGen-75' and 'MesoGen-250' portable systems, at 75 and 250 W, respectively, with funding from the U.S. Department of Defense and the U.S. Navy. MesoGen-250 models were powered by propane or kerosene and designed to operate as a field battery charger, able to provide suitable power levels for radios, sensors, as well as auxiliary and emergency units on military vehicles. The next generation of the device that Protonex developed is the P200i power remote sensors, signaling, and communications systems. The unique feature of this system is that it can operate for years without human contact in a temperature range of arctic cold −30°C to 55°C (Figure 7.1).

Protonex claimed that when coupled with solar panels to minimize fuel consumption, the P200i withstands more cycles and operation hours than SOFC systems of

FIGURE 7.1   The Protonex P200i system. (Reprinted with permission from Ref. [4].)

their competitors. Later on, Army SBIR funded Protonex to develop a man-portable 2 kW SOFC system with 13 kW power spikes. The expectation was that it would power robotic vehicles, ground vehicle auxiliary systems, or exoskeletons. In 2016, Ballard Power Systems acquired Protonex. In 2017, it was determined that SOFC assets were not cored to Ballard's proton exchange membrane fuel cell business, and SOFC assets were transferred to a private, start-up company, Upstart Power, Inc., in 2017.

Other companies that are developing SOFCs for military applications are Acumentrics (US), and UltraCell (US), with the SOFC power range of 1 W to 100s of Ws, which is an attractive and potentially lucrative niche. Delphi is also focusing on powering vehicles, military applications, and stationary power. The Delphi SOFC can operate with natural gas, diesel, bio-diesel, propane, gasoline, and coal-derived fuel. Delphi is the only US company that has developed and demonstrated a practical and operational SOFC APU for heavy-duty commercial trucks. A single Delphi Gen 4 SOFC stack has a modular design and can provide 9 kW of electrical power, which could be integrated into larger power plants with electrical efficiencies ranging from 40% to 50% [5,6]. The Delphi truck with a SOFC unit is shown in Figure 7.2. The system net efficiency of a SOFC APU could reach above 30%, which will lead to a double saving in fuel consumption. Furthermore, the heat produced by the SOFC can be used for heating or cooling of the vehicles, and that would further increase the overall efficiency.

The major opportunities for SOFC market are the development of hybrid-SOFC technologies, and opportunity to use SOFC in a reverse mode to produce hydrogen as a solid oxide electrolysis cell (SOEC), rising demand for distributed power generation and growing acceptance by end-users for the energy storage in data centers and the military sector applications. The electrolysis systems would encompass solar and wind energy harnessed via photovoltaic arrays or wind turbines as the primary power source. The advantage of this system is that excess electrical energy and water are routed to the electrolysis system, converting the electricity to hydrogen, which is then compressed and stored. FCE is working

FIGURE 7.2 A demonstration model of the Delphi APU on-board of a commercial truck. (Reprinted with permission from Ref. [4].)

toward demonstrating the potential of SOEC systems to produce hydrogen at the cost of less than 2.0 USD kg^{-1} [7]. The focus is to develop an innovative SOEC and stack technology with ultra-high steam electrolysis current (>3 A cm^{-2}) and highly efficient hydrogen production from diverse renewable sources. A lot of progress has been made in the US under the DOE sponsorship.

The next market in the US is the CHP system, that holds a promise as well, featuring the local generation of electricity with relatively clean natural gas, and the subsequent use of waste heat from the generation process to satisfy local thermal loads, having the potential to substantially reduce residential primary energy usage and associated emissions. Although residential CHP systems are commercially available, widespread adoption has been hindered by their high cost and the lack of an attractive economic value proposition for homeowners.

The deployment of fuel cells will be strongly dependent on legislation because the costs prior to mass production are relatively high, compared to existing technologies. The presence of key SOFC manufacturer's needs for larger systems and energy is among key factors providing a positive outlook for the industry demand. Regulatory compliance and government subsidies for fuel cells will further stimulate industry growth. State incentives and the U.S. Federal Investment Tax Credit for the stationary fuel cells are among key industry trends in the North American market.

7.1.2 EUROPE

Fuel cell technology was invented in Europe in the mid-19th century, but this did not result in leading to large-scale commercialization of fuel cells. Although a strong involvement of the European industry as well as the European Union (EU) can be seen in the development of high-quality SOFCs, other parts of the world have higher numbers of commercial enterprises and more installed units. The lack of experience with field-units and commercial prototypes not only reflects the narrow commercial basis of the fuel cell industry in Europe but also results in a shortage of system operation data, field study, and exposure to everyday consumer conditions.

Even today, the European market holds smaller market shares as compared to the Asia Pacific and North America, with market revenue at USD 125.3 million in 2018. The absence of significant governmental incentives in certain European countries is among the key challenges. However, stringent emission restriction laws are anticipated to contribute to the European SOFC market. In order to reach the target of greenhouse gas emission reduction set by the EU, a reduction of the carbon footprint of the fuel has to be considered, and the use of renewable gases like biogas or hydrogen has to be considered. Moving further to promote the storage of intermittent renewable energies through the power-to-gas concept and electrolysis process, blending hydrogen with natural gas into the existing natural gas network is expected. There are numerous projects sponsored by EU, under EU Horizon 2020, that focus on SOFC development, cost of manufacturing reduction, and new potential applications [8]. The EU launched a new framework program (2014–2020) in which 2.8 billion EUR was allocated to the development of fuel cell and hydrogen technology [9]. The objective is to produce high-quality fuel cells with high reliability and long-term durability at a low cost. However, for mass production, one of the major barriers to decarbonizing the energy system is a large investment requirement, and there is less risk mentality in Europe comparing to the US. The European companies focus on smaller power generation between 1 and 10 kW. The German automotive supplier, Eberspächer, has developed an FC APU that runs on diesel, with the maximum output of 3 kW and an electrical efficiency up to 40% to onboard power components in a truck. The APU combines a diesel fuel reformer with a SOFC. The Eberspächer fuel cell system operates entirely independently from the truck's main engine, so the electrical loads can be taken off the generator when the vehicle is stationary. These loads include the air-conditioning system or the cab's refrigerator, which require continuous power [10].

In 2013, Bosch Thermotechnology placed in Germany, signed an agreement to use 700 W flat-tubular stack technology from the Japanese Aisin, that is currently tested in the frame of the European micro-CHP demonstration project [5]. Bosch Thermotechnology installed around 70 of these power-generating heating systems for demonstration purposes in Germany, the UK, the Netherlands, and France. Aisin supplied the power-generating module that had already been launched in Japan. Bosch Thermotechnology integrates it into an overall system meeting requirements of the individual European heating technology markets. The project is the largest European demonstration program for fuel cell-based solutions, facilitating the decentralized generation of power and heat for residential buildings. Bosh further developed technology and is currently developing SOFC with an electrical efficiency of 60%, generating electrical power between 5 and 11 kW. In addition, the system can operate using different energy sources – whether it is renewable hydrogen or methane or conventional natural gas [11].

Hexis/Viesmann from Switzerland develops SOFC-based CHP units for stationary applications with electrical power requirements below 10 kW. The company develops planar SOFC technology and its latest pre-commercial product is "Galileo 1000N", running on natural gas or bio-methane, as the system integrates a catalytic partial oxidation (CPOX) reactor. The nominal electrical power output is 1 kW (AC), and the thermal power output is 2 kW, with an electrical efficiency of up to 35% and maximum overall efficiency of 95% [lower heating value (LHV)] [12,13].

UK-based Ceres Power develops micro-CHP SOFC systems, based on metal-supported cells, allowing rapid start-up times and a great number of on/off cycles with little degradation for the residential sector. Ceres Power has built and developed relationships with key industry partners such as British Gas, Calor Gas, and Bord Gáis. In 2019, they signed a Joint Development Agreement with Doosan on SOFC power system development for commercial buildings [14]. Ceres Power, British Gas (UK), and Itho-Daalderop (Netherlands) are installing 174 micro-CHP units conducting a trial in the UK and Dutch homes from 2014. Select customers will have the opportunity to purchase a Ceres micro-CHP that has a strong backing of German giant Bosh, unit with full service and maintenance package provided by British Gas in the UK and by Itho-Daalderop in the Netherlands. Ceres power extended partnership in Asia by signing an agreement with Weichai Power, one of the leading automobile and equipment manufacturing companies in China, and they have completed the development of a first prototype range extender for Chinese electric buses [15]. In 2018, they signed another deal with Japanese giant Nissan to develop a range extender for electric vehicles, and they already have a deal in place with Honda [16]. The company reported a significant growth in revenue with total revenue and other operating income rising to 11 million EUR in six months to the end of December 2019 from 8.27 million EUR in the same period of 2018 [17].

Other companies in Europe that are significant players and worth of mentioning are: Convion Ltd. commercializing SOFC systems in the power range of 50–300 kW for distributed power generation fueled by natural gas or biogas; Elcogen is a privately owned company which focuses on commercializing anode-supported SOFC cells and a 1 kW stack to open markets; Power Systems Inc. is developing a complete 1, 2, 5, and 10 kW fuel cell power generators for stationary power applications; and Fraunhofer IKTS is designing HotBox solutions, and has very matured tested technology, which it can offer along with the stack modules for systems with the power range from 1 to 50 kW.

Another significant emerging player in Europe that is worth of mentioning is SOLIDpower, Italian-based company, specializes in development, manufacturing and commercialization of SOFC technology and systems for stationary applications, including micro-cogeneration and remote power. SOLIDpower acquired 100% of HTceramix SA, a spin-off of the Swiss Federal Institute of Technology in Lausanne (EPFL). In 2015, in Heinsberg, Germany, they acquired the business and employees of Ceramic Fuel Cells Ltd. (CFC). Over 750 SOLIDpower micro-CHP systems have already been sold globally, and contracts with utilities for further micro-CHP deployment are in place [6]. SOLIDpower has successfully installed four Bluegen BG-15 micro-CHP systems for field trials, in the frame of D2Service European project, to assess the improved serviceability of this generation of SOFC-based co-generators. The company is also looking to develop new markets in the US and Asia [18–20].

7.1.3 Asia

The Asia Pacific leads the global market with 58.3% share in 2018. Established markets, particularly in Japan and Korea, are among the key factors driving growth. In Japan, under the Ene-Farm program, about 300,000 SOFC-based CHP systems have been installed in homes to provide electricity and hot water [21]. In Ene-Farm units, the heat taken out of the fuel cell exhaust allows for total energy efficiency of 97%, claimed by Panasonic.

The Japan's Ene-Farm project is based on proton exchange membrane fuel cell technology with the largest players, such as Toshiba and Panasonic, while SOFC technology is mostly dominated by Aisin Seiki. The 2050 strategy for H_2 and fuel cell in Japan is to position H_2 as a new energy option [22]. The focus of Japan's Ministry of Economy, Trade, and Industry (METI) is to develop Japan's economy and industry by focusing on promoting economic vitality in private companies and to secure stable and efficient supply of energy and mineral resources. The METI's target is for the H_2 cost to be 3 USD kg^{-1} by 2030 and then further reduction to 2 USD kg^{-1} [23]. The Japan's strategy is to achieve a carbon-free society and present hydrogen to the rest of the world as a new energy choice that will lead global efforts for establishing a carbon-free society. Japan like the US had a focus on SOEC and is planning by 2050 to develop innovative technologies for highly efficient water electrolysis for hydrogen production as well as low-cost, highly efficient energy carriers, and highly reliable fuel cells. METI's strategic roadmap is presented in Figure 7.3.

In Japan, the research development activities are under New Energy and Development Organization (NEDO). In its role as an innovation accelerator, NEDO coordinates and integrates the technological capabilities and research abilities of industry, academia, and government and also promotes the development of innovative and high-risk technologies. Moreover, NEDO carries out research and development projects and sets targets based on changes in social conditions in order to realize maximum results. NEDO targets for research and development of fuel cells are presented in Figure 7.4. Some of the NEDO projects related to SOFCs are focused on the practical application of SOFCs and improvement of their commercial ability, a basic

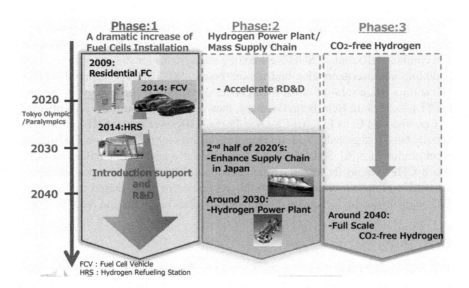

FIGURE 7.3   Strategic roadmap for hydrogen and fuel cell by METI [22].

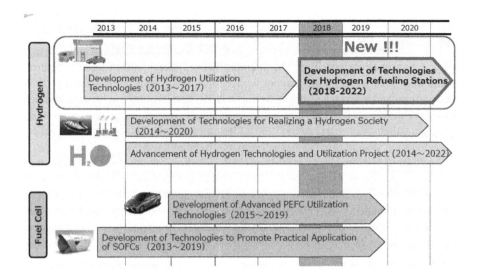

FIGURE 7.4 Current projects of NEDO program [22].

study on the rapid durability evaluation method for SOFCs, development of high-efficiency SOFC practical technology using the fuel recycling mechanism (Denso Corporation), etc. [24]. NEDO identified four remaining challenges for SOFC: (1) low cost and long life, (2) high efficiency, (3) create new customer value challenge (improve power supply value), and (4) expand applicable markets by responding to fuels (for future decarbonization of hydrogen society). Issues to be addressed are low cost and long life; short-term 2020, medium-term 2025, and long-term 2030–2040 with a stack cost >100,000, 100,000, 50,000, and 20,000–30,000 Yen kW^{-1} system, respectively [24]. In addition to cost reduction, the further increase in the power density is necessary, with demonstrated durability of 130,000 h. Some of the main SOFC players in Japan are Aisin Seiki, Kyocera, TOTO, Mitsubishi Heavy Industry, NGK/NTK Spark plugs, etc.

Aisin Seiki was established in 1965, and it comprises 181 consolidated subsidiaries, 66 of which in Japan and 115 overseas, and their focus is on micro-CHP systems. Their concept is based on flat sheet and tube cell, which operates at between 700°C and 750°C. Aisin supplied Ene-Farm since 2009. In 2012, Kyocera, Osaka Gas, Aisin, Chofu, and Toyota announced the completion of world-class efficiency residential-use SOFC co-generation system co-development and commercialization of "Ene-Farm Type S" [25]. Aisin's Ene-Farm Type S, for a residential fuel cell, using natural gas as a fuel, was launched in 2014. The system achieved a power generation efficiency of 46.5% (LHV), and an overall efficiency of 90% (LHV). The SOFC system includes a heating unit with a small storage tank of 90 liters with hot water, as well as a high-efficiency latent heat recovery-type unit for the back-up boiler. In 2017, Kyocera announced the launch of the industry's first 3 kW SOFC co-generation system [26]. The Kyocera system utilizes Kyocera's proprietary ceramic technologies to deliver 52% electrical efficiency, the highest of any comparable SOFC system currently on the market with exhaust heat recovery, overall efficiency of 90%. The Kyocera unit is shown in Figure 7.5. The new 3 kW SOFC system is built of modules of four of Kyocera's small and highly efficient 700 W cell stacks, which have a proven track record in Ene-Farm Type S residential SOFC systems sold in Japan. The product is only used in Japan and cannot be used in the U.S as it does not comply with US gas, water, or electrical regulations. In 2019, Kyocera partnered with Dainichi Kogyo to develop a fuel cell unit with a built-in hot water storage tank under a new program called "Ene-Farm MINI" [27]. The unit presents the smallest size fuel cell co-generation system that has been commercialized in the world.

Mitsubishi-Hitachi Heavy Industries (MHI) has been involved in the development of tubular-type SOFCs since the 1990s. They are the producers of the largest 200 kW in Japan, with a maximum power output of 229 kW and an electric efficiency of 52%. Mitsubishi Hitachi Power Systems, Ltd. (MHPS) is now a conglomeration of MHI and Hitachi, and is offering a company new business opportunities. The company's products besides of SOFCs include gas turbine-combined cycle and integrated coal

FIGURE 7.5   Kyocera 3 kW power generation unit. (Reprinted with permission from Ref. [26].)

gasification-combined cycle power plants, gas/coal/oil-fired (steam) power plants, boilers, generators, gas and steam turbines, geothermal power plants, air quality control systems, and power plant peripheral equipment [28]. In 2019, MHPS and NGK Spark Plug teamed up to set up a joint venture company dedicated to the manufacture and sale of anode-supported intermediate-temperature cylindrical cell stacks.

South Korea stands as the second large market shareholder after Japan, and the six generating companies have deployed almost 300 MW of fuel cell power to date, including the world's tallest, most energy-dense, and largest fuel cell parks [29]. Korea's strong economy and ubiquitous and well-maintained natural gas infrastructure have also supported the rapid proliferation of SOFC. KOEN (formally known as Korea Southeast Power Company) recently built a triple-decker fuel cell park in Bundang, with power tower generating 1.34 kW per square foot or approximately 1 MW for every 787 square feet. The site is the most energy-dense power plant in the world, with the largest SOFC systems being deployed in the country. This is a significant advancement comparing to the world's newest and most efficient combined cycle gas turbine plants. The park claims to be the world's most energy-dense power plant.

POSCO Energy is a comprehensive energy provider engaged in four key energy business areas: power generation, renewable energy, fuel cell, and resource development. POSCO Energy produces various fuel cell products ranging from 100 kW to 2.5 MW to provide customers with a wide range of fuel cell products to suit their needs. It currently supplies 100 kW, 300 kW, and 2.5 MW fuel cell products and is also developing other products applied to various areas, as well as the next-generation SOFC technology [30]. POSCO's system is presented in Figure 7.6.

SOLIDpower Group (Italy) has developed a Memorandum of Understanding with Korea Electric Power Corporation, to cooperate on building a greater share of SOFC,

Production
Electricity for 370 households

Installation area
95 m²

Fuel
LNG, Bio Gas, SNG

Applications
Small-scaled production facilities, mid- and large-sized buildings

FIGURE 7.6 POSCO's 300 kW SOFC product of 300 kW. (Reprinted with permission from Ref. [4].)

working to scale up system sizes and accelerate commercialization as part of the energy revolution in South Korea [31]. The project development will be located in the Bitgaram Energy Valley – a business park with more than 100 other companies and start-ups.

In South Korea, the research in SOFC has been going for many years, and the latest development is that the scientists at the Korea Institute of Science and Technology (KIST) have developed a high-performance, thin-film-based SOFC that can operate at mid- to low temperatures below 600°C using butane fuels. The flexibility of the fuels will allow for the new technology to expand the application range of SOFC to portable and mobile applications such as electric cars, robots, drones, and open new markets [32].

China has become the world's largest consumer of energy products; however, China's per capita energy consumption is far less than the world average. China is facing dual threats of energy deficiency and air quality and environmental degradation. Compared with Japan, South Korea, and other most developed countries, the development of fuel cell technology in China started rather late. In 2019, the Huaqing Jinkun Energy Company announced the first SOFC production line in China with the first batch of 200,000 single-cell production line [33]. There are several important types of research for a SOFC–biomass hybrid system that has been conducted in China as China is the biggest producer of biomass in the world [34]. China is rich in solar resources with an average daily radiation of 4 kWh m^{-2}. The SOFC–solar and SOFC–wind are hybrid systems, and integration with existing technologies can help China to move fast to the integration of sustainable technologies. Renewable energy usage took up around 8% of the total energy consumption in China in 2011. The Chinese government is planning to increase the proportion of renewable energy up to one-fifth of the country's electricity consumption by 2030, with solar energy playing a critical role [35].

Previously Chinese companies partnered with international companies to develop SOFC systems. Chaozhou Three-Circle Co., Ltd., established in 1970 in Chaozhou, China, developed anode-supported SOFCs, SOFC membranes, and stacks and partnered with Ceramic Fuel Cell Limited from Australia [36].

G-Cell Technology Co., Ltd. was founded in 2013 and is focusing on the development of 1 kW SOFC [37]. SOFCMAN is a manufacture of the SOFC raw materials and offers fine grade cathode/electrolyte/anode powders and high power density single cells [38]. India is a new and emerging market for SOFC deployment. The Bloom Energy along with real estate developer Atelier Global, natural gas companies GAIL (India) Limited and Indian Oil Corporation, and US–India Strategic Partnership Forum, recently announced a first-of-its-kind commercial real estate development in Bangalore that will be powered by clean, reliable electricity generated onsite using natural gas and Bloom Energy system [39].

7.2 CURRENT STATE OF TECHNOLOGY COMMERCIALIZATION AND LONG-TERM PERSPECTIVE

Despite many early applications of fuel cells, such as in the Apollo and Shuttle space missions, in 1966, General Motors (GM) developed the first fuel cell road vehicle. However, fuel cells have not achieved reliable operation or cost benefits compared to internal combustion engines or gas turbines [40,41]. A large number of government laboratories, universities, and private companies have conducted research and development on fuel cells. The fuel cell market in the last 30 years has been dominated by Japan and US companies. Both the US and Japan develop multiple fuel cell technology platforms, and there was a significant investment from the government to support research and development. The transportation sector is dominated by Japan's fuel cell vehicle, with a Toyota Mirai leading in putting cars on the road both in Japan and California. By the end of 2019, global sales of Mira cars totaled 10,250, the US with 6,200 units, Japan with 3,500 units, and Europe with 640 units [42]. Ene-Farm program has both proton exchange membrane fuel cell (leading companies Toshiba and Panasonic) and SOFC (leading company Aisin Seiki). Market entrants with proton exchange membrane fuel cells capture most of the market share at present. In the US, the mid-sized commercial market is represented by SOFC (Bloom Energy), phosphoric acid fuel cell (Doosan/UTC Power), and MCFC systems (FCE) at the MW sizes. There is also a significant difference in the size of the units and market in the two regions. In Japan, large, diverse, and presumably more stable companies such as Aisin, Kyocera, MHI, and TOTO are focusing on micro-CHP 0.7–2 kW, whereas US companies such as Bloom Energy, Doosan, and FCE are focusing on large systems over 200 kW. In 2019 under the Ene-Farm program, over 350,000 CHP units had been installed, in which Pansonic's proton exchange membrane technology represents about half and Toyota-affiliated Aisin Seiki is dominating with SOFC technology installation. It is important to emphasize that the Japanese government provides significant incentives for the installation of CHP units as a standard home unit costs about 16,700 USD, and most consumers have been hesitant to buy [43]. Prime Minister Shinzo Abe has set a goal of 5.3 million hydrogen-powered homes, roughly 10% of Japan's total, by 2030 with the objective of tackling environmental issues and increasing energy security. The Abe's cabinet endorsed legislation that would help with the purchase of CHP units with a price incentive of 2,970 USD.

In the US, the current technology entrants are dominated by SOFC technology, followed by MCFC and phosphoric acid fuel cell technology [44]. The U.S. DOE has invested in SOFC development and deployment since 1995, over 750 million USD [3]. Throughout this effort, DOE-funded R&D has identified gaps in state-of-the-art knowledge of SOFC power systems and also leverage on the state investment such as the Self-Generation Incentive Program in California which provides performance-based incentives for facilities that install qualifying distributed power and heating technologies such as fuel cells [45]. In early 2018, congress restored the Fuel Cell Investment Tax Credit and introduced the section 45Q Carbon Capture Credit that can further help fuel cell companies [46]. Legislation passed in 2019 in California should provide further lift and commits the state to 100% carbon-free electricity by 2045.

In 2018, Bloom Energy was marked the first fuel cell company to go public in a long time. We believe that Bloom Energy will not likely wind up in history alongside failed companies like Siemens-Westinghouse, General Electric (GE), Theranos, or Solyndra. The promises of hydrogen are convincing indeed as it is the most abundant element in the universe. The detractors of hydrogen would point out that fuel cell tanks can explode in an accident, that the fueling infrastructure is lacking, and the technology is expensive and unproven. They are correct, and all or some of these are roughly the same negatives as using gasoline, diesel, or battery-electric cars. Every fuel source has its own shortcomings, and hydrogens are certainly not outsized compared to other available technologies. Compressed hydrogen and fuel cells can provide electricity to a vehicle traction motor with weights that are 8–14 times less than current batteries and take up much less space (including the fuel cell system) than batteries for a given range. Fueling takes as little time as gasoline car fueling and fueling stations can be built in anticipation of future demand, much like what's being done right now with electric vehicle charging stations. The significant advantage of the fuel cell car is the ability to use them as a generator to supply household with electricity for a day or two in the event of an emergency [47]. The energy capacity of the fuel cell vehicle is 5 kg of hydrogen, compressed at 10,000 psi, is more than 150 kWh. A typical US household is using 32 kWh a day, but for a typical Japanese home that consumes 10 kWh it is a great opportunity to use a car with a power-out capability and might be a solution for the recreational cars/houses when power grids are completely offline or do not exist. The ability of a hydrogen-fueled vehicle to provide electricity would likely prove an interesting selling point in the US where many households have back-up generators to compensate for frequent power outages. At the moment, both fuel cell and battery electric vehicles are more expensive than the internal combustion cars, even taking local and federal government incentives into account. Significant cuts are needed to have this technology accepted by consumers. In 2019, Global Market Insights, Inc. reported that the fuel cell market is set to reach a valuation of 7 billion USD by the end of 2025. Growing demand for space heating across residential and commercial establishments along with fuel cell electric vehicles, portable charging devices, growing recreational activities, and increased development of hydrogen fueling stations will foster the fuel cell market. Favorable government policies pertaining to the development of sustainable energy technologies coupled with increasing government stimulus and funding from private and public organizations will stimulate industry growth.

Dr. Maric has been working in fuel cells since early 1990s in Japan, then in the US and Canada, and has been very engaged with European FR-06, FR-07, and Horizon 2020. She believes that many factors are going to play a role such as economy, government incentives, policy, environments, and nuclear disasters like one in Fukushima in 2011. On the other hand, the technology needs to demonstrate the high cost cut to be competitive. Execution remains key for industry players to land their perennial but long-postponed goals of profitability. Another important factor is that young generations and citizens have stepped forward to engage in community-based science, challenge the information and explanations given to them by government officials and other authorities regarding climate change, and protest existing policies. All these factors will play a critical role in the adoption of fuel cell technology by societies.

REFERENCES

1. businesswire. Global fuel cell market expected to grow with a CAGR of 22% during the forecast period, 2019-2024. https://www.businesswire.com/news/home/20200130005467/en/Global-Fuel-Cell-Market-Expected-Grow-CAGR.
2. E4tech. 2014. The fuel cell industry review 2014. http://www.fuelcellindustryreview.com/archive/TheFuelCellIndustryReview2014.pdf [accessed on 10 December, 2015].
3. DOE. Report on the status of the solid oxide fuel cell program. Report to congress August 2019. https://www.energy.gov/sites/prod/files/2019/09/f66/EXEC-2019-002655_Signed%20Report%201.pdf.

4. McPhail, J., Kiviaho, J., and Conti, B. 2017. The yellow pages of SOFC technology: International status of SOFC deployment 2017. VTT Technical Research Centre of Finland Ltd., Finland.

5. Andersson, M., and Sunden, B. 2017. Technology review-solid oxide fuel cell report 2017. https://energiforskmedia.blob.core.windows.net/media/22411/technology-review-solid-oxide-fuel-cell-energiforskrapport-2017-359.pdf.

6. Hennessy, D. 2010. US DOE–Hydrogen, fuel cells and infrastructure technologies, *DOE Peer Review 2010*, Washington DC.

7. Ghezel, H. 2019. Modular solid oxide electrolysis cell system for efficient hydrogen production at high current density, *DOE Peer Review 2019*, Washington DC.

8. Fuel Cells and Hydrogen (FCH). EU Horizon 2020 projects. https://www.fch.europa.eu/page/all-h2020-projects.

9. Wang, J. 2015. Barriers of scaling-up fuel cells: cost, durability and reliability. *Energy* 80:509–521.

10. Barrett, S. 2014. Eberspächer unveils diesel reformer/fuel cell for truck APU. *Fuel Cells Bulletin* 2014(10):2–3.

11. BOSCH. High-temperature fuel cell systems. https://www.bosch.com/research/know-how/success-stories/high-temperature-fuel-cell-systems.

12. Mai, A., Grolig, J.G., Dold, M., Vandercruysse, F., Denzler, R., Schindler, B., and Schuler, A. 2019. Progress in HEXIS' SOFC development. *ECS Transactions* 91(1):63–70.

13. HEXIS. Galileo 1000 N system. http://www.hexis.com/en/galileo-1000-n.

14. ReportLinker. 2020. Global fuel cells industry. https://www.reportlinker.com/p05379572.

15. proactive. 2019. Ceres Power holdings - first range extender bus prototype. https://www.proactiveinvestors.co.uk/LON:CWR/Ceres-Power-Holdings-PLC/rns/38578.

16. proactive. 2016. Ceres Power signs second deal with Japanese car maker. https://www.proactiveinvestors.co.uk/companies/news/121426/ceres-power-s-nissan-deal-a-huge-endorsement--121426.html&c=4909987209651187960&mkt=en-us.

17. proactive. 2020. Ceres Power expects full-year revenues to grow by around a third. https://www.proactiveinvestors.co.uk/companies/news/914985/ceres-power-expects-full-year-revenues-to-grow-by-around-a-third-914985.html.

18. Solid Power. SOFC market developments. https://www.solidpower.com/en/news/all-news/details/news/solidpower-announce-changes-to-the-board-of-directors.

19. SOLIDpower. 2018. SOLIDpower agrees German distribution with Bosch's Buderus. *Fuel Cells Bulletin* 2018(7):12.

20. Energy Newspaper. SOFC market developments. http://www.energy-news.co.kr/news/articleView.html?idxno=42977.

21. Ene-Farm program. SOFC-based CHP systems. https://fuelcellsworks.com/news/

22. Yokomoto, K. 2018. Country overviews (Japan). In *6th International Workshop on Hydrogen Infrastructure and Transportation*, U.S. Department of Energy, Boston, MA. https://www.energy.gov/sites/prod/files/2018/10/f56/fcto-infrastructure-workshop-2018-3-yokomoto.pdf

23. METI. The basic hydrogen strategy. http://www.meti.go.jp/english/press/2017/1226_003.html.

24. NEDO. Research and development of fuel cells. https://www.nedo.go.jp/english/archives2019_index.html.

25. businesswire. 2012. Kyocera, Osaka gas, Aisin, Chofu and Toyota announce completion of world-class efficiency residential-use Solid Oxide Fuel Cell (SOFC) cogeneration system co-development and commercialization of "ENE-FARM Type S". https://www.businesswire.com/news/home/20120315005562/en/Kyocera-Osaka-Gas-Aisin-Chofu-Toyota-Announce.

26. Kyocera. 2017. First commercially available SOFC cogeneration system in the 3–5 kW class for institutional applications, based on research by Kyocera (as of June 1, 2017). https://global.kyocera.com/news-archive/2017/0702_bnfo.html.

27. Fuel Cell Works. 2019. World's smallest high efficiency household fuel cell cogeneration system "Ene-Farm" developed. https://fuelcellsworks.com/news/worlds-smallest-high-efficiency-household-fuel-cell-cogeneration-system-ene-farm-mini-developed.

28. Mitsubishi Hitachi Power Systems (MHPS). Mitsubishi Power, Ltd. Products. https://www.mhps.com.

29. Green Car Congress. 2020. KIST team develops low-temperature high-performance solid-oxide fuel cell that operates on butane gas; possible EV applications. https://www.greencarcongress.com/2020/04/20200428-kist.html.

30. POSCO. Fuel cell products and next-generation SOFC technology. https://www.poscoenergy.com.

31. SOLIDpower. 2017. SOLIDpower, KEPCO to push for more fuel cells in South Korea. *Fuel Cells Bulletin* 2017(1):5–6.

32. Power Grid International. South Korea fuel cell research and development. https://www.power-grid.com/2018/08/15/feature-south-korea-flies-flag-for-fuel-cells/#gref.

33. Fuel Cell Works. 2019. China: first 200,000 solid oxide fuel cell production line put into service in Xuzhou. https://fuelcellsworks.com/news/china-first-200000-solid-oxide-fuel-cell-production-line-put-into-service-in-xuzhou.

34. Lu, Y., Cai, Y., Souamy, L., Song, X., Zhang, L., and Wang, J. 2018. Solid oxide fuel cell technology for sustainable development in China: an over-view. *International Journal of Hydrogen Energy* 43(28):12870–12891.

35. Chiu, D. The east is green: China's global leadership in renewable energy. https://www.csis.org/east-green-chinas-global-leadership-renewable-energy

36. CCTC. SOFC developments and partnerships. http://www.cctc.cc.

37. GCell. G-Cell Technology Co., Ltd. 1 kW SOFC development. http://www.Gcell.com.

38. SOFCMAN. SOFC research and development. http://www.sofc.com.cn.

39. MERCOM. 2019. Bloom energy to develop the first natural gas-powered solid oxide fuel cell project in India. https://mercomindia.com/bloom-energy-natural-gas-solid-oxide-fuel-cell.

40. Dumoulin, J. 2000. "Gemini-V information". NASA - Kennedy Space Center [accessed on 2 August, 2011]. https://science.ksc.nasa.gov/history/gemini/gemini-v/gemini-v.html

41. Eberle, U., Müller, B., and Von Helmolt, R. 2012. Fuel cell electric vehicles and hydrogen infrastructure: status 2012. *Energy & Environmental Science* 5(10):8780–8798.

42. Wikipedia. Toyota Mirai. https://en.wikipedia.org/wiki/Toyota_Mirai.

43. Watanabe, C. 2015. Ene-Farms use hydrogen to power homes but don't come cheap. https://www.bloomberg.com/news/articles/2015-01-15/fuel-cells-for-homes-japanese-companies-pitch-clean-energy.

44. Jiao, K., Park, J., and Li, X. 2010. Experimental investigations on liquid water removal from the gas diffusion layer by reactant flow in a PEM fuel cell. *Applied Energy* 87(9):2770–2777.

45. US Department of Energy (DOE) Fuel Cell Technologies Office Multi-Year Research, Development, and Demonstration Plan. 2012. Technical plan – fuel cells, Section 3.4.4. Available at http://energy.gov/sites/prod/files/2014/12/f19/fcto_myrdd_fuel_cells.pdf [accessed 1 June, 2015].

46. Green Tech Media. 2018. Fuel cell power plays from California to Kenya. https://www.greentechmedia.com/articles/read/fuel-cell-power-plays-from-california-to-kenya.

47. Green Car Reports. 2014. 2016 Toyota mirai power-out jack could run your home in emergencies. https://www.greencarreports.com/news/1095522_2016-toyota-mirai-power-out-jack-could-run-your-home-in-emergencies.

Index

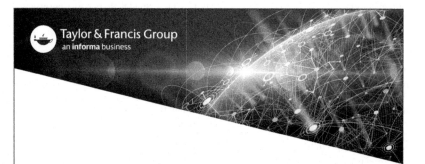

Printed in the United States
By Bookmasters